*Otherlands*

# Otherlands

## A Journey Through Earth's Extinct Worlds

# THOMAS HALLIDAY

RANDOM HOUSE

*New York*

Published in the United States by Random House, an imprint and
division of Penguin Random House LLC, New York. Simultaneously
published in hardcover in Great Britain by Allen Lane, an imprint of
Penguin Press, a division of Penguin Random House Ltd., London.

RANDOM HOUSE and the HOUSE colophon are registered trademarks
of Penguin Random House LLC.

Hardback ISBN 978-0-593-13288-3
Ebook ISBN 978-0-593-13289-0

Printed in the United States of America on acid-free paper

randomhousebooks.com

2 4 6 8 9 7 5 3 1

FIRST EDITION

# Contents

# List of Maps

| EON | ERA | PERIOD | EPOCH | DATES |
|---|---|---|---|---|
| PHANEROZOIC | CENOZOIC | Quaternary | Pleistocene | 2.58 million – 12,000 ya |
| | | Neogene | Pliocene | 5.333 – 2.58 mya |
| | | | Miocene | 23.03 – 5.333 mya |
| | | Palaeogene | Oligocene | 33.9 – 23.03 mya |
| | | | Eocene | 56 – 33.9 mya |
| | | | Paleocene | 66 – 56 mya |
| | MESOZOIC | Cretaceous | | 145 – 66 mya |
| | | Jurassic | | 201.3 – 145 mya |
| | | Triassic | | 251.9 – 201.3 mya |
| | PALAEOZOIC | Permian | | 298.9 – 251.9 mya |
| | | Carboniferous | | 358.9 – 298.9 mya |
| | | Devonian | | 419.2 – 358.9 mya |
| | | Silurian | | 443.8 – 419.2 mya |
| | | Ordovician | | 485.4 – 443.8 mya |
| | | Cambrian | | 541 – 485.4 mya |
| PROTEROZOIC | NEOPROTEROZOIC | Ediacaran | | 635 – 541 mya |

# Introduction
## The House of Millions of Years

'Let no one say the past is dead.
The past is all about us and within'

– Oodgeroo Noonuccal, *The Past*

'What tempest blows me into that deep
ocean of ages past, I do not know'

– Ole Worm

I am looking out of the window, across farmland, houses, and parks, towards a place that for hundreds of years has been known as World's End. It has this name because of its past remoteness from London, a city that has now grown to absorb it. But not too long ago this really was the end of the world. The soil here was laid down in the last ice age, a gravelly mixture deposited by rivers that once flowed into the Thames. As the glaciers advanced, they diverted its course, and the Thames now enters the sea more than 100 miles south of where it used to flow. From the ridged hills, clay crumpled by the weight of ice, it is possible, just about, to mentally strip away the hedgerows, the gardens, the streetlamps, and imagine another land, a cold world on the edge of an ice sheet extending hundreds of miles away. Below the icy gravel lies the London Clay, in which even older residents of this land are preserved – crocodiles, sea turtles, and early relatives of horses. The landscape in which they lived was filled with forests of mangrove palm and pawpaw, and waters rich in seagrass and giant lily pads, a warm, tropical paradise.

The worlds of the past can sometimes seem unimaginably distant. The geological history of the Earth stretches back about 4.5 billion years. Life has existed on this planet for about four billion years, and life larger than single-celled organisms for perhaps two billion years. The landscapes that have existed over geological time, revealed by the palaeontological record, are varied and, at times, quite other to the world of today. The Scottish geologist and writer Hugh Miller, musing on the length of geological time, said that all the years of human history 'do not extend into the yesterday of the globe, far less touch the myriads of ages spread out beyond'. That yesterday is certainly long. If all 4.5 billion years of Earth's history were to be condensed into a single day and played out, more than three million years of footage would go by every second. We would see ecosystems rapidly rise and fall as the species that constitute their living parts appear and become extinct. We would see continents drift, climatic conditions change in a blink, and sudden, dramatic events overturn long-lived communities with devastating consequences. The mass extinction event that extinguished pterosaurs, plesiosaurs and all non-bird dinosaurs would occur 21 seconds before the end. Written human history would begin in the last two thousandths of a second.[1]

At the beginning of the last thousandth of a second of that condensed past, a mortuary temple complex was built in Egypt, near the modern-day city of Luxor, the burial place of the pharaoh Ramesses II. Looking back to the building of the Ramesseum is a mere glance over the dizzying precipice of deep geological time, and yet that building is well known as a proverbial reminder of impermanence. The Ramesseum is the site that inspired Percy Bysshe Shelley's poem 'Ozymandias', which contrasts the bombastic words of an all-powerful pharaoh with a landscape of what was, when the poem was written, nothing but sand.[2]

When I first read that poem, I had no knowledge of what it was about, and mistakenly assumed Ozymandias to be the name of some dinosaur. The name was long and unusual, and it was hard to figure out a pronunciation. The descriptive language used in the poem was that of tyranny and power, of stone, and of kings. The pattern, in short, fitted that of my childhood illustrated books about prehistoric life. At '*I met a traveller from an antique land who said: two vast and trunkless legs of stone stand in the desert*', I thought of a plaster jacket being applied to the remains of some terrible beast from prehistory. A true tyrant lizard king, perhaps, now broken into bones and fragments of bones in the badlands of North America.

Not all that is broken is lost. The lines '*on the pedestal these words appear: "My name is Ozymandias, king of kings; Look on my works, ye Mighty, and despair!" Nothing beside remains.*' might be seen as time having the last laugh over a self-important ruler, but the world of that pharaoh *has* been remembered. The statue is evidence of its existence; the content of the words, the details of its style, clues to its context. Read like this, 'Ozymandias' gives us a way to think about fossilized organisms and the environments in which they lived. Take out the hubris, and the poem can be read as being about finding the reality of the past from the remnants that survive to the present. Even a fragment can tell a story in itself, a piece of evidence for something beyond the lone and level sands, for something else that used to be here. For a world that no longer exists but is still discernible, hinted at by what lies among the rocks.

The Ramesseum itself was originally known by a name that translates as 'The House of Millions of Years', an epithet that could easily be appropriated for the Earth. Our planet's past also lies hidden under the dirt. It wears the scars of its formation and change in its crust, and it, too, is a mortuary, memorializing its inhabitants in stone, fossils acting as grave marker, mask and body.[3]

Those worlds, those otherlands, cannot be visited – at least, not in a physical sense. You can never visit the environments through which titanic dinosaurs strode, never walk on their soil nor swim in their water. The only way to experience them is rockwise, to read the imprints in the frozen sand and to imagine a disappeared Earth.

This book is an exploration of the Earth as it used to exist, the changes that have occurred during its history, and the ways that life has found to adapt, or not. In each chapter, guided by the fossil record, we will visit a site from the geological past to observe the plants and animals, immerse ourselves in the landscape, and learn what we can about our own world from these extinct ecosystems. By visiting extinct sites with the mindset of a traveller, a safari-goer, I hope to bridge the distance from the past to the present. When a landscape is made visible, made present, it is easier to get a sense of the often-familiar ways that organisms live, compete, mate, eat and die there.

One site has been chosen for each geological epoch up to the last of the 'big five' mass extinctions 66 million years ago, which together constitute the Cenozoic, our own era. Preceding that mass extinction, one site has been chosen for each geological period (which comprises several epochs) back to the beginnings of multicellular life in the Ediacaran more than 500 million years ago. Some sites have been chosen for their remarkable biology, some for their unusual environment, some because they have been preserved so precisely and give us an unusually clear glimpse of how life existed and interacted in the past.

Journeys have to begin at home, and the course of this journey will pass from the present day backwards in time. We will begin in the relatively homely surrounds of the Pleistocene ice ages, when glaciers locked up much of the world's water in ice, lowering sea levels worldwide, before travelling progressively further back in time. Life and geography will become less and less familiar. The geological epochs of the Cenozoic will take us back through the early days of humankind, past the largest waterfall ever to have existed on Earth and a temperate, forested Antarctica, then right up to the end-Cretaceous mass extinction.

Beyond that, we will meet the inhabitants of the Mesozoic and Palaeozoic, visiting dinosaur-dominated forests, a glass reef thousands of kilometres long, and a monsoon-soaked desert. We will explore how organisms adapt to entirely new ecologies, moving onto land and into the air, and how life, in creating new ecosystems, opens the possibility for even more diversity.

After a brief visit to the Proterozoic, some 550 million years ago, in the geological eon before our own, we will return to our own Earth, that of the present day. The landscapes of the modern world are changing rapidly, thanks to human-caused disturbances. Compared with the radical environmental upheavals of the geological past, what might we expect to happen in the near and more distant future?

We cannot experiment easily on the planet to discover what changes happen at a continental scale in a high-carbon atmosphere, nor do we have enough time to see for ourselves the long-term effects of global ecosystem collapse before mitigating it. Our predictions must be based on accurate models of how the world works. Here, Earth's dynamism throughout geological history provides a natural laboratory. Answers to long-term questions can only be found by looking at the times in which the past Earth mirrors what we expect of future Earth. There have been five major mass extinctions, isolation and rejoining of continental landmasses, changes in ocean and atmospheric chemistry and circulation, all of which add data to our understanding of how life on Earth functions over geological timescales.

Of our planet, we can ask questions. The biology of the past is not just a curiosity to be glanced at with a bemused eye, or something foreign and otherworldly. Ecological principles that apply to modern tropical rainforests and the lichen-world of the tundra equally applied to the ecosystems of the past. Although the cast is different, the play is the same.

Taken in isolation, a fossil can be a fantastic lesson on anatomical variation, on shape and function, and on what an organism can do with simple tweaks to a general developmental toolkit. But just as the statues of antiquity stood within the context of a culture, no fossil, whether animal, plant, fungus or microbe, ever existed in isolation. Each lived within an ecosystem, an interaction among myriad species and the environment, a complex mishmash of life, weather and chemistry also dependent on the spin of the Earth, the position of the continents, the minerals in the soil or the water, and the constraints imposed by an area's past inhabitants. Recreating the worlds in which fossils were laid down, and in which their makers lived, is a challenge that palaeontologists have been attempting to address since the eighteenth century; attempts which have gathered in both pace and detail over the last few decades.

Recent palaeontological advances have revealed details of past life that would have been thought impossible not long ago. By delving deep into the structure of fossils, we can now reconstruct the colours of feathers, of beetle shells, of lizard scales, and discover the diseases these animals and plants suffered. By comparing them with living creatures we can establish their interactions in food webs, the power of their bite or strength of their skull, their social structure and mating habits, and even, in rare cases, the sound of their calls. The landscapes of the fossil record are no longer merely collections of impressions on rocks and taxonomic lists of names. The latest research has revealed vibrant and thriving communities, the remnants of real, living organisms that courted and fell sick, showed off bright feathers or flowers, called and buzzed, inhabiting worlds that obeyed the same biological principles as those of the present day.[4]

This is, perhaps, not what people have in mind when they think of palaeontology. The image of the Victorian gentleman collector, travelling off to other lands and other cultures, hammer in hand, ready to break open the earth, is pervasive. When the physicist Ernest Rutherford supposedly declared, rather dismissively, that all science was 'physics or stamp collecting', he surely had in mind the rank and file of taxidermized beasts, drawers of butterflies with immaculately open wings, and looming skeletons, bolted together with industrial iron. Today, though, a palaeobiologist is as likely to be spending their day in front of a computer or using circular particle accelerators to fire X-rays deep into fossils in a lab as out in the heat of the desert. My own scientific work has mostly happened in basement museum collections and within computer algorithms, using shared anatomical features to try and work out the relationships among the mammals that lived in the aftermath of the last mass extinction.[5]

It is far from impossible to obtain insights about the history of life only from the life that exists in the present day, but it is like trying to understand the plot of a novel by reading only the last few pages. You will be able to infer some of what has gone before, and find out the present situation of the characters that make it to the end, but the richness of the plot, numerous characters and major beats of the story might be missing. Even including fossils, most of the history of life remains obscure to non-specialists. Dinosaurs and the ice age animals of Europe and North America are widely known, and those with a little more familiarity with the subject will have heard of trilobites and ammonites or perhaps of the Cambrian explosion. But these are fragments of the whole story. In this book, I want to fill in some of the gaps.

This book is, necessarily, a personal interpretation of the past. The long-gone past, true 'deep time', means different things to different people. To some, it is exhilarating, a dizzying vertigo to think of the time taken for the trillions of plankton to settle, compact and rise into the chalklands of Kent and Normandy – countrysides made of skeletons. To others, it is an escape, a chance to think of ways of life outwith those we experience now, of a time before the worries of human-caused extinction when the dodo was merely a future possibility. Everything that we will see is nonetheless grounded in fact, either directly observable from the fossil record, strongly inferred, or, where our knowledge is incomplete, plausible based on what we can say for sure. Where there is disagreement, I have selected one of the competing hypotheses, and run with it. Even so, a flurry of wings in a thicket, a half-seen hide or the sensation of something moving in the dark, is an integral part of experiencing nature. A little ambiguity can generate as much wonder as a fixed truth.

The reconstructions here are the result of the work of thousands of scientists over more than 200 years. Their interpretations of fossil remains are ultimately what has led to the factual elements of this book. To a palaeobiologist, the bumps, ridges and holes in bone, exoskeleton or wood give clues necessary to build a picture of an individual organism in life, whether that organism is alive today or not. To look at the skull of an extant freshwater crocodile is to read a character description. The buttressed processes and arches evoke Gothic architecture, here resisting not the weight of a cathedral roof but the powerful force of the jaw muscles. The high-set eyes and nostrils speak of low swimming, peering and breathing just above the water surface; the long series of teeth, pointed but round, and set in a long, sweeping snout, suggest a feeding style of swiping, grabbing and holding prey, suitable for catching slippery fish. The scars of life are there, with fractures sustained knitted together. Lives leave their marks in detailed, reproducible ways.

Going beyond the individual specimen, and deciphering the characteristics of past ecosystems, the interactions, niches, food webs and flow of minerals and nutrients, is now commonplace in palaeobiology. Fossilized burrows and footprints can reveal details of movement and lifestyle that anatomy is silent on. Relationships between species help to tell us what factors were important to their biology and distribution, and what drove their evolution. The patterns and chemistry of sand grains in sedimentary rock record the environment – was this cliff face once a meandering river delta with the ever-changing courses of rivers snaking through a mudflat, or a shallow sea? Was that sea a sheltered lagoon, where fine silt drifted slowly to the floor in still water, or a place of crashing waves? What was the atmospheric temperature at the time? What was the global sea level? In which direction was the prevailing wind? All these, with the necessary knowledge, can be easily answerable questions.[6]

Not all of these types of information are available in any given location, but sometimes, many of the strands come together such that a palaeoecologist can build a rich picture of a landscape, from climate and geography to the creatures that inhabit it. These pictures of past environments, as vibrant as any today, often hold important lessons for how we approach our contemporary world.

Many parts of the natural world we take for granted today are relatively recent arrivals. Grasses, the main component of the largest ecosystems of the planet today, only arose at the very end of the Cretaceous, less than 70 million years ago, as rare parts of the forests of India and South America. Grass-dominated ecosystems did not emerge until about 40 million years before the present. There were never dinosaur grasslands, and, in the northern hemisphere, grass simply did not exist. We must drop preconceived ideas of what a landscape looks like, whether they arise because we have imprinted modern species on the past, or lumped together creatures that, although extinct, lived millions of years apart. More time passed between the lives of the last *Diplodocus* and the first *Tyrannosaurus* than passed between that of the last *Tyrannosaurus* and your birth. Jurassic creatures like *Diplodocus* not only did not see grasses, but never saw a flower either; the flowering plants only diversified in the middle Cretaceous.[7]

Today, with the biodiversity crisis brought on by habitat destruction and fragmentation, combined with the ongoing effects of climate change, we are very familiar with the idea that more and more organisms are going extinct. It is frequently said that we are in the midst of a sixth mass extinction. We are now used to hearing about widespread bleaching of coral reefs, melting of Arctic ice sheets, or deforestation in Indonesia and the Amazon basin. Less commonly discussed, though also extremely important, are the effects of land drainage on wetlands or the warming of tundra. The world that we inhabit is changing at the level of the landscape. The scale and ramifications of this are often difficult to comprehend. The thought that something as vast as the Great Barrier Reef, with all its vibrant diversity, might one day soon be gone, sounds inherently improbable. Yet the fossil record shows us that this sort of wholesale change is not only possible, but has repeatedly happened throughout Earth history.[8]

Today's reefs may be coral, but in the past clam-like molluscs, shelly brachiopods and even sponges have been reef-builders. Corals only took over as the dominant reef-building organisms when the mollusc reefs succumbed to the last mass extinction. Those reef-building clams originated in the Late Jurassic, taking over from the extensive sponge reefs, which had in turn filled the reef-building niche after brachiopod reefs were entirely wiped out by the end-Permian mass extinction. From the long-term perspective, continent-scale coral reefs might just end up being one of those ecosystems that never returns, a distinctively Cenozoic phenomenon, brought to a close by the human-driven mass extinction. Now, the future of coral reefs and other threatened ecosystems is in the balance, but the fossil record, in showing us how fast dominance can become obsolescence and loss, acts as a memorial and as a warning.[9]

Fossils may not seem like an obvious place to obtain insights into future life. The strangeness of fossil imprints, biological hieroglyphs, lends a distance to the past, a kind of uncrossable boundary, across which is an enticing other that can never be reached. The poet and academic Alice Tarbuck, in her poem 'nature is taxonomy which all small bones resist' captures this distance, saying 'give me leviathan trace, give me roiling sea-beast'. She yearns for 'footprints that lead down centuries, into the basement of what might be', and rejects the museum-label naming of classification with 'Let nobody sing taxonomy'.

Even as one of those who spend parts of their working life placing organisms in the series of phylum-class-order boxes, I too feel more of an affinity with the living thing than with the classification. A name can be evocative or meaningful but, for the most part, cannot evoke the sense of an organism. Latin names are mere markers, the Dewey decimal system of biology. A number would suffice, and indeed, this is essentially how the system works. For every single species and sub-species, there is somewhere in the world an individual specimen that marks what it means to be, for example, an Italian red fox. The definitive individual of *Vulpes vulpes toschii* is ZFMK 66-487, housed in the Alexander König Museum in Bonn. To be considered part of this sub-species you must be close enough, in anatomy and genetic make-up, to this particular Platonic fox, an adult female collected from Monte Gargano in 1961. Practical this may be, but it tells you nothing about the high-wire artistry of a city fox on a rickety garden fence, the purposeful hurry of padding adults, the mythological cunning of Reynard, or the carefree outdoor sleeping of cubs. And this is a creature we see around us today. What hope from name alone for those that are gone? My challenge in presenting them is to bridge the gap between the name and the reality, between the guinea's stamp and the gold. To see ancient life forms as if they were commonplace visitors to our world, as quivering, steaming beasts of flesh and instinct, as creaking beams and falling leaves.[10]

Today, where an extinct creature is portrayed as alive, it is frequently as a monster, something villainous, with an insatiable appetite. This dates back to the early nineteenth century sensationalizers of geology. Some were so keen to promote their vision of a dramatic and vicious past that woolly mammoths and ground sloths, known even then to be herbivorous, were presented by some as voracious meat-eaters. The mammoth, for instance, was introduced to the public as a powerful predator, lurking ominously in lakes to ambush their turtle prey, while the docile herbivorous ground sloth became 'huge as the frowning precipice; cruel as the bloody panther, swift as the descending eagle, and terrible as the angel of night!' Even today, the depiction of mindless, barbaric aggression of prehistoric animals continues in countless films, books and television programmes. But the predators of the Cretaceous were filled with no more bloodlust than a lion is today. Dangerous, certainly, but animals, not monsters.[11]

What the sedate collection of fossils as curios and the portrayal of extinct organisms as monsters have in common is a lack of real ecological context. Plants and fungi are typically absent, and invertebrates get only the most cursory look. And yet, the rock record of the Earth contains that context, revealing the settings in which extinct creatures lived, settings that shaped them into the forms that now seem so unusual. It is an encyclopaedia of the possible, of landscapes that have disappeared, and this book is an attempt to bring those landscapes to life once more, to break from the dusty, ironbound image of extinct organisms or the sensationalized, snarling, theme-park *Tyrannosaurus*, and to experience the reality of nature as one might today.

To consider the landscapes that once existed is to feel the draw of a temporal wanderlust. My hope is that you will read this in the vein of a naturalist's travel book, albeit one of lands distant in time rather than space, and begin to see the last 500 million years not as an endless expanse of unfathomable time, but as a series of worlds, simultaneously fabulous yet familiar.

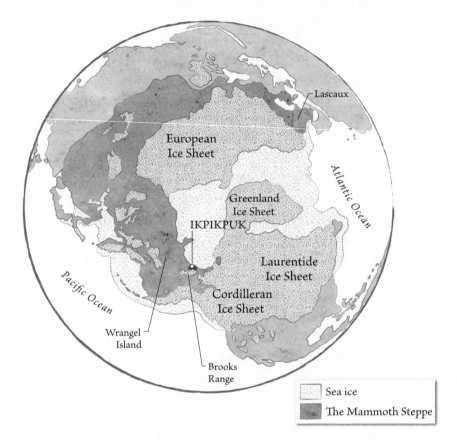

Lascaux

European
Ice Sheet

Greenland
Ice Sheet

IKPIKPUK

Laurentide
Ice Sheet

Cordilleran
Ice Sheet

Atlantic Ocean

Pacific Ocean

Wrangel
Island

Brooks
Range

Sea ice

The Mammoth Steppe

# 1.  Thaw

*Northern Plain, Alaska, USA*
Pleistocene – *20,000 years ago*

'Day and night, summer and winter, in foul weather or fine weather,
it speaks of freedom. If someone has lost his freedom, the steppe
will remind him of it'

– Vasily Grossman, *Life and Fate*

'Telipinu too went into the moor and blended with the moor.
Over him the *halenzu*-plant grew'

– Hittite myth (tr. H.A. Hoffner)

Dawn is near breaking in the Alaskan night, where a small herd of horses, four adults and three foals, huddle against the frigid north-easterly wind. By this time, the sun has been gone for well over ten hours, and the air is skin-tighteningly cold. Two of the mares are taking their turn on sentinel duty, keeping their vigil against the dark while their family rests or forages. They stand together, flank on flank, nose to tail, a good way to reduce stress while staying close and warm while still watching in all directions. It is spring, but even through winter the ground has not been snow-covered, and it is carpeted instead with a profusion of dead grass and blown sand. The flatlands between the Brooks Range of northern Alaska and the coast of the perpetually frozen Arctic Ocean are exceptionally dry. Rain and snow alike have mostly passed this land by. A fickle stream making its way across pebbles, scarcely dribbling from the higher land to the south, is almost inaudible above the gale. Even this stream gives up before it reaches the sea, disappearing entirely as it is absorbed by encroaching dunes. The river flow is variable

*Arctodus simus* and *Mammuthus primigenius*

from day to day but will peak in the next few months, being dependent on the thaw from the hills. In winter, there is little to eat; the ground is four fifths bare earth, one fifth dry brown stalks, and what meagre food there is is coated with abrasive dust. Even so, the desiccated remains of the summer's plenty are enough to support several small herds of these short-legged horses. In such anaesthetic temperatures as are found on the North Slope during the height of the last glaciation, overly long limbs would risk hypothermia. The Alaskan horses are closer to the size of ponies, resembling modern Przewalski's horses, but slenderer of limb. Their coats are shaggy and dun, their manes short, black and bristly. Those that are sleeping still move, absent-minded tail flicks twitching in the faint light of the aurora overhead. These are the truest inhabitants of the arid north, those that remain no matter the conditions. The summer visitors to the North Slope – large congregations of bison and caribou, and rare, scattered groups of muskoxen, moose and saiga – have left, less able than horses to survive on such poor fodder. Even for the horses, sustaining a living through the northern winter is difficult, made more so by the pregnancy of one of the mares. Each small herd contains a single male and several females, and the birth of foals is timed to coincide with late spring. Mortality is high, and life expectancy is half that of modern-day wild horses. Fifteen years is a typical lifespan for these Alaskan horses, living near the edge of their limits, in the face of a yowling wind.[1]

That wind blows from a sand sea of 7,000 square kilometres in the eastern half of what will become Alaska, bordered in the west by the Ikpikpuk River, a river that still exists in the modern day. Across this frigid desert crawl ridged dunes, 30 metres high, in rows 20 kilometres long. They blast their sand westwards across the steppe, coating the foothills of the Brooks Range in an icing-sugar dust of the loose, windblown sand-silt mixture known as loess. In the cold parts of the Pleistocene world, there is so little food during the cold months that every herbivore, from caribou to mammoth, ceases to grow. Like trees, their bones and teeth lay down growth marks, a physical scar of seasonality, a count of winters endured.

They subsist on what they can find, using little energy and relying on their bulk to hold out until better times return. Where there are herbivores predators will be lurking. At any moment, a pair of grasping paws could spring from the brush, a bite to the neck could crush the life from them. Across these scrubby landscapes, a small number of cave lion prides control large territories. They prowl silently across the steppe, shoulders sloping up and down with each footfall, and for the horses, there is little way of knowing whether they are near. Lion hunts rely on stalking and stealth, so the dark draws them closer. The mares are vigilant, any noise sending ears flickering over domed, pale foreheads.[2]

Three lions roam the Earth in the Pleistocene, and, of them, the African lion – the only survivor into modern times – is the daintiest. Across the other side of the Laurentide Ice Sheet and throughout North America, as far south as Mexico and even reaching South America, lives the American lion, the largest of the three. Slightly spotted, dusty-red beasts, up to 2.5 metres long, they are recent immigrants, descended from ancestors that moved across from Eurasia about 340,000 years before the present. Throughout the steppes of Europe and Asia, though, and here in Alaska, the major risk to these horses and caribou is from the Eurasian cave lion, *Panthera leo spelaea*, which diverged from lions of the modern day some 500,000 years before the present. Much of what we know about their appearance we have gathered from art – there are hundreds of detailed paintings and sculptures by northern Eurasian humans who documented many of the species of the mammoth steppe. About 10 per cent bigger than an African lion, the Eurasian cave lions are paler and shaggier, with a rough, coarse fur covering a dense, wavy, almost white undercoat, two layers of insulation against the cold. Neither male nor female is maned, although both have short beards; and males are substantially larger. Because animal remains tend to accumulate and lie undisturbed in caves, we know them as cave lions, but it is in the open that they are at home, wandering the steppe in small social groups, hunting caribou and horses.[3]

All cats are ambush predators, their anatomy adapted for stalking

4

and surprising prey, with at most a short sprint. This kind of stalking requires stealth, but in the openness of the steppe, stealth is difficult, and, compared with other cats, cave lions are relatively good at chasing down prey. Drawings of cave lions often show their markings – dark lines emerging from their eyes like cheetahs, helping them to avoid being dazzled by sunlight, and a clear division between their darker backs and pale underbellies.[4]

Today, lions, elephants and wild horses are not associated with the northern parts of North America. Then again, neither are snowless grounds, rainless skies or sand seas. When imagining parts of the natural world, we tend to think of them as a whole, every part of the ecosystem defining a sense of place. What would the Sonoran Desert of south-western North America be without the giant saguaro cactuses, tarantulas and rattlesnakes? If you are familiar with a place, there is a sense of intrinsic correctness in its elements. While this sense is very strong, ecosystems are built piecemeal. The aggregations of species that produce a feeling of place also provide a sense of time. A community – the census of organisms from microbes to trees to giant herbivores – is a temporary association of living things that depends on evolutionary history, climate, geography and chance.

I grew up on the edge of the Black Wood of Rannoch in the Scottish Highlands: steep, quartzite-studded slopes covered with musky bracken cloisters and cushions of blaeberry, woods with stained-glass ceilings of birch leaves or cracked pine pillars; a fragment of temperate rainforest among moor and open hill. I have a strong nostalgia for the inhabitants of that place – marten and diver, siskin and deer. To me, they are avatars of childhood, and separating the place from the wildlife is near impossible. But these are only those creatures that shared the wood and the world in my time, and, taking the long view, nature forswears such nostalgia. Thousands of years into the Pleistocene, while herds of wild horses roam the wild expanse of Alaska, Rannoch is a dead place, a glacial scour under 400 metres of ice. Before the ice advanced, and while the ice remains, it is not the place I know; my sense of the Black Wood is as tied to our current geological epoch, the Holocene, as it is to the bedrock on which it grows.[5]

Fossil communities do not map neatly onto modern preconceptions. A species' range today might reflect where its ancestors lived, but equally it might not. Camels and llamas, for example, are each other's closest relatives, having separated about 8.5 million years before the present. Llamas are descendants of the tribe (in the Linnaean sense) that remained in the camelids' ancestral homeland of the Americas, while camels crossed over the Bering Strait to Asia and beyond. Even up until 11,000 years before the present, during the warmer parts of the cyclical glaciations of the ice age, herds of camels were wandering what would become Canada. At this moment in the Pleistocene, near the largest extent of ice, camels inhabit as far south as California – we know this from those unlucky enough to become trapped in the natural asphalt seeps at La Brea, where tar has bubbled from the ground for thousands of years.[6]

The first people have already arrived in the Americas; the footprints of a gleeful group of children, running through tufts of ditchgrass into the mud of a chalky lakeshore, 22,500 years before the present, are still visible in the white sands of New Mexico. As the populations of these first Americans grow in numbers, they will hunt both native camels and horses. As a result, like so many large mammals of the Pleistocene, they will have become extinct only a few short thousand years after human arrival. For now, those human populations are still small, and there is little direct evidence of exactly where they lived. At the time of the most recent glacial period, which reached its maximum extent about 25,000 years before the present day, humans thrive in the low plains of Beringia, moving along the southern Alaskan coast where the ice is scarcer into this new continent full of resources. To the north of the ice sheet, in the dry eastern edge of Beringia, hundreds of kilometres east of Ikpikpuk, there may be campfires lit by small communities of eastern Beringian humans – lakes there preserve chemicals characteristic of human faeces and of charcoal – but these are few and far between. As the climate changes, and humans gain an ever-deeper foothold in the continent, many of the native species will not survive for long, battered by the warming world and these versatile new predators.[7]

6

Traces of historical associations can long outlast actual contact. In the dense, subtropical forests from India across to the South China Sea, venomous snakes are common, and there is always an advantage in pretending to be something dangerous. The slow loris, a weird, nocturnal primate, has a number of unusual features that, taken together, seem to be mimicking spectacled cobras. They move in a sinuous, serpentine way through the branches, always smooth and slow. When threatened, they raise their arms up behind their head, shiver and hiss, their wide, round eyes closely resembling the markings on the inside of the spectacled cobra's hood. Even more remarkably, when in this position, the loris has access to glands in its armpit which, when combined with saliva, can produce a venom capable of causing anaphylactic shock in humans. In behaviour, colour and even bite, the primate has come to resemble the snake, a sheep in wolf's clothing. Today, the ranges of the loris and cobras do not overlap, but climate reconstructions reaching back tens of thousands of years suggest that once they would have been similar. It is possible that the loris is an outdated imitation artist, stuck in an evolutionary rut, compelled by instinct to act out an impression of something neither it nor its audience has ever seen.[8]

In the case of the lorises and cobras, and that of the Arctic camels, it is climate, along with geography, that has defined their evolutionary history and their interactions with other animals. An ecosystem is not a solid entity – it is made up of hundreds and thousands of individual parts, each species with their own tolerance to heat, salt, water availability, acidity, and each with their own role. In its broadest sense, an ecosystem is the network of interactions between all living members of the community and the land or water that forms its environment. Alone, a species has properties of its own, but the interactions of an ecosystem bring complexity. We call the possible survivable conditions for any given species its 'fundamental niche'. When interactions with other organisms limit that niche, we call the reality of a species distribution its 'realized niche'. No matter how wide the fundamental niche is, if the environment changes, and passes outwith the limits of that niche, or if the realized niche slips to a size of zero, that species has gone extinct.[9]

The Pleistocene North Slope in winter is one time and place where the environment passes out of the fundamental niche of many creatures. Horses survive here thanks to their ability to subsist on poor fodder, as long as there is enough of it. Sleeping and rousing in fits and bursts, they spend up to about sixteen hours a day feeding to ensure they get enough nutrition. Mammoths also thrive on poor-quality food, though their digestion is less efficient, and they require more bulk than the sparse winter grazing can provide. In times of scarcity, they are known to turn to eating their own dung, to access any remaining nutrition. Bison, living in herds thousands strong elsewhere, have to let their food ferment in their four-stomach digestive system, so they cannot eat as much as quickly. That means their food needs to be of a higher quality, and in winter these arid northern plains cannot provide.[10]

It is the physical geography of this corner of the world that has resulted in the dry, windy weather. The constant ankle-stinging wind that hisses through the Ikpikpuk dunes is part of a vast, anti-clockwise gyre of wind centred far to the south-west of here. By the time it has whipped up the water of the Pacific and driven clouds over central Alaska and the Yukon, what moisture it once held has been lost. Most of the rain has dropped over the moister bison plains that run close to the great wall of ice that separates this land from the rest of North America. That ice sheet covers almost all of modern-day Canada and reaches southwards, forming a frozen barrier from the Pacific to the Atlantic. It is in places up to 2 miles deep, and the carving and gouging forces it exerts on the landscape are even now excavating what will become the Great Lakes. As the ice melts, the water pooling against the southern border of the Laurentide Ice Sheet will be released, cutting out new riverbeds, eroding the deposited moraines of glaciers, and forming spectacles such as the Niagara Falls.[11]

The water locked up in this continental ice sheet, and its near neighbour in northern Europe, has been drawn from oceanic reserves. Sea levels worldwide are lower than those in the modern day by about 120 metres, and so growth of the ice has exposed shallow seabeds, building so-called 'land bridges' between continents.

Alaska may be isolated from North America, but just such a bridge connects the Alaskan wildlife with Asian communities to the west, bringing it into a continuum that covers half the circumference of the world. The Bering Strait, that stretch of water that in the present separates Alaska from Chukotka in Russia's Far East, is dry and hospitable, and gives its name to the biological province of Beringia. Beringia may be a cold land in winter, but it grows bright and warm in the hotter months. Meadows of wildflowers bloom throughout the spring and summer. Most of the trees are shrubby: short willows write wordless calligraphy on the wind with flourished ink-brush catkins, while dwarf birch shrubs hide ptarmigans. Above, skeins of snow geese wing and cry their way to the sea. In autumn, the more sheltered parts of Beringia shine with a pouring of molten gold as the cottonwoods and aspens turn yellow, set off by the blue-green of tall spruce. These lowlands are refugia for many species of plants and animals, a part of the world with a fairer, milder climate where those that cannot tolerate the extended ice age cold can survive. In places, bog-dwelling sphagnum moss oozes, while elsewhere silver-haired prairie sage releases its warming scent under the hoof-step of bison.[12]

The total area of the Beringian land bridge that will be sunken by the sea – including the land north of what will become Russia – is vast, about the size of California, Oregon, Nevada and Utah put together. This province is itself merely one part of an extensive biome – a landscape made up of consistent communities of plants and animals, and with a relatively consistent climate – that begins in eastern Beringia and ends on the Atlantic coast of Ireland. From the depths of the exposed Beringian plain, up into the hills of Alaska, the air cools and dries, the plants grow shorter and hardier, but the grassland continues. On its eastern margins, the edge of the sea of dunes at Ikpikpuk marks one end of the largest contiguous ecosystem the world has ever seen – the mammoth steppe.[13]

The steppe continues to exist because of this very connectivity. Ice age weather patterns are volatile, with conditions often wildly different from year to year. If you were to drive tent pegs into the loose ground and set up camp for years in one place, the populations would seem to go through extreme cycles of boom and bust, with the weather and plant life one year favouring horses, then bison, then mammoths, and so on. Because the mammoth steppe is contiguous, species can move to follow their ideal climates, and stay within the bounds of their niches. In a wildly variable environment, mobility is crucial to long-term survival. Somewhere on the continent, there will always be refuge. Throughout the high Arctic, there is a constantly repeated pattern of local extinction followed by re-establishment from just such refugia. Even in the modern day, the largest Arctic herbivores, reindeer and saiga, take part in the biggest terrestrial migrations on the planet. Elsewhere, in the Mongolian steppe, a Beringia-like environment where humans herd goats and other livestock, the climate is still volatile, its winter temperatures unpredictable year on year. As climate change makes the Mongolian steppe warmer and drier, the grasslands are becoming less productive, restricting the areas in which herds can be grazed. Because migration distances are increasingly limited, people are increasingly vulnerable to various kinds of harsh winter or *zud* – enough snow to prevent grazing, not enough snow for drinking water, frozen ground, cold winds – that can devastate herd populations and herder livelihoods. In a variable environment, the ability to up sticks and move elsewhere is critical, for wild animals and humans alike. As climates change in the modern day, that way of life is under threat, in a way that directly mirrors the demise of the mammoth steppe.[14]

The continuity of Beringia will be broken. Ultimately, the seas will rise; about 11,000 years before the present day, Beringia will drown. The steppe that ringed the world will be severed into smaller, less connected chunks as the vast taiga forests of spruce and larch grow northwards, the tundra shifts south, the weather warms, and long-distance migration between those favourable fragments of land suited to cold-adapted species is no longer possible. Migration cannot save a population if there is nowhere to go. If wiped out, there is no surviving group from which to replenish the lost creatures, and so they become locally, and eventually globally, extinct. Others may persist but must reduce the area over which they roam. In

Alaska, of all the species that once roamed the mammoth steppe, only the caribou, brown bear and muskox, this last solely through reintroduction, have survived.[15]

As day breaks, the expanse of the mammoth steppe emerges. The weak sun rises, topping the dunes one by one. Soon, every grain on the leeward side is casting a shadow, and the dunes glitter. The recumbent horses snort and stand, shaking themselves quickly awake; they never sleep deeply or for long. Broad, dark hooves shuffle impatiently, their edges flared; with less walking over winter, they have not been worn down, and are rather overgrown.[16]

Under a clear, crisp sky, summer begins to unfurl. Foals and thaw lakes appear, and thundering squadrons of caribou and bison return to the north, bound for the new vegetation. The vast mammoth herds return, too – the populations of mammoth make up nearly half of the herbivore mass on the North Slope. The sun rapidly warms the air, and the horses head for a low cloud swirling beyond a hillock. The hanging mist signals the presence of a rare pool, formed from the melt that has gathered in a warmer, sheltered hollow. Kept in shadow, the groundwater has until recently been frozen, but standing water in the river floodplain is a magnet for those that need to drink, and home to a diverse insect community – diving beetles, pill beetles and arid-adapted ground beetles are all common around the Ikpikpuk River.[17]

In the sunshine, the weather is fine, not only drier and more fertile, but also warmer than modern Alaska. This may be the ice age, but Beringia is a relatively warm spot, with a continental climate – similar to modern-day Mongolia. There is a real distinction between coastal and continental places. Throughout the year, seawater temperatures do not vary a great deal, so they act as sinks or sources for heat on the nearby land, producing winds and cloud cover that limits the variability of the weather. Inland, the summer heat is stored more easily by the earth, and so continental climates maintain high temperatures in summer. By the same token, the land cools rapidly, which makes for frigid winters. This is why, for example, coastal St Petersburg today is on average 19°C in July and −5°C in January while continental Yakutsk, at only a slightly more northerly latitude, has an average of 20°C in July but −39°C in January. The North Slope of Pleistocene Alaska is more like Yakutsk than St Petersburg – it is warm in summer, cold in winter, and always dry. There is no nearby unfrozen sea, so the continually cloudy, drizzly world of modern Alaska cannot form. Without snow and rain, glaciers cannot form either, which is why it is an ice-free corridor into the rest of the world.[18]

Fresh shoots replenish the dry grass, and the herd of horses push westwards. With the caution of prey, they never stray far from one another; while some eat, others stare, but after a stationary winter, their horizons are expanding once again to hundreds of square kilometres. As the group tops a crest, there is a skip of panic among them, and they instinctively cluster around the youngest, a schiltrom of hooves and teeth. Across the horizontal stripe of green between the shadowed slope and the sky, an *Arctodus* is moving.

Compared with brown bears, even at their grizzliest, the short-faced bear *Arctodus simus* is big. The biggest of the Alaskan short-faced bears weighed over a ton, three times the weight of the biggest modern-day predator on land, the Siberian tiger, and four times that of an adult male grizzly bear. The eponymous short face and long-limbed stride are in part an optical illusion created by scale. Bears have short, sloping backs and deep jaws, and when a brown bear is scaled to the size of a short-faced bear, these features are accentuated. Certainly, the biggest modern bear, the polar bear, has a long snout, but this seems to be an adaptation to an exclusively meaty diet. *Arctodus* are not common on the North Slope, and their behaviour is poorly understood. Until recently, it was thought that their long limbs might be an adaptation to running, suggesting that *Arctodus* was a giant pursuit predator, a wolf pack combined into one terrifying individual. Others, citing the short-faced bear's close relationship with the tree-dwelling, almost exclusively vegetarian Spectacled Bear, have painted *Arctodus* as a gentle herbivore, a rootling, footling giant. Still others consider them to be scavengers, living the lifestyle of a bully, a kleptoparasite, stealing the carcasses from other carnivores who have made the kill. The reality is probably far closer to that of a larger brown bear, eating a mixture of small and large prey as well as plants.[19]

Nevertheless, of all the American populations of *Arctodus* from Alaska to Florida, the Beringian community is most likely to be found eating meat. Where the winter has removed much of the ground vegetation, the bear's flexible diet bends towards predation and scavenging. With its sheer size, an adult *Arctodus* is capable of dominating a kill site, preventing other predators from coming too close. Shoulders rolling, it plods towards the pool, where the giant carcass of an old woolly mammoth, dead from the cold, gives off a sickly, molten odour. It is a welcome prize. Prodding with broad, powerful paws, the bear tugs at the fur of the dead mammoth, stripping it to expose the sinewy meat. It is slow and laborious work; mammoth hide is thick and covered with two layers of dense fur. In death, even the icon of the Pleistocene megafauna looks diminutive against its consumer. Mammoths may have been 3 metres tall at the shoulder, but, rearing on their hind limbs, the largest *Arctodus* can stretch a metre further.[20]

Bears are fearsomely powerful beasts. Wherever humans have lived alongside the brown bear, mythologies have grown up around them. The founding myth of Korea is dependent on the patience of a bear who would be content to eat only wild garlic and ssuk, a type of mugwort, *Artemisia*, for 100 days. Both these plants are found in the Eurasian mammoth steppe. Even the names given to bears are shrouded in euphemism wherever humans and bears coexist, a linguistic theory called taboo deformation, avoiding the 'true' name to prevent manifesting the animal. To the Russians, who venerated the bear and took it as a national symbol of power and cunning, it is *medvědi*, the 'honey-eater'. Germanic languages, including English, use varieties of *bruin*, the 'brown one'. Across the world, the euphemism 'grandfather' is used. The bears to which these names refer are brown bears, the ancestors of the North American grizzlies. They, like their fellow Eurasian migrants, humans, are only now arriving, venturing throughout this land, and encountering *Arctodus*.[21]

Across the mammoth steppe, the large populations of herbivore herds, joined together, paint a picture of a thriving community. There are certain fundamental rules that all ecosystems must follow. Energy, usually harnessed from sunlight or, rarely, from the breakdown of minerals, must flow into the ecosystem to replace what is lost through activity and decay. The organisms that can access this energy are the producers, and those that cannot are consumers, feeding on other living things in order to survive. The more energy the producers produce, the more consumers can be supported. The Beringian steppe is remarkably productive. In the inhospitable far north of Siberia, about 10 tons of animals – equivalent to about 100 caribou – are supported in every square kilometre, far more than can survive in equivalent cold places in the modern day. The number of predators in an ecosystem is always lower than that of producers – in the summer on the North Slope, this reaches extremes; only 2 per cent of the animals here are carnivorous.[22]

For the short-faced bear, the mammoth carcass is particularly welcome, as prey have been declining in recent years. The number of bison that have made their way into the North Slope has begun to decrease, and the population of horses is declining, too. Underfoot, the world is beginning to soften, and the hegemony of grass is nearly at an end. Around the thaw pool are the beginnings of formation of peat. This is a worrying sign for all the creatures that live in this dusty, windswept world. Most of the mammoth steppe is like an enclosed courtyard, surrounded on all sides by dry, solid walls. Across its northern extent, the Arctic Ocean is frozen, with glaciers covering North America, Scandinavia and Britain. On the steppe's western flank, the Atlantic is frozen, and in the south, the many ranges of mountains from the Pyrenees, through the Alps, the Taurus and Zagros mountains, into the Himalayas and the Tibetan plateau form a near-continuous wall. This mountainous barrier shelters an entire continent from the monsoons to the south, with their harsh winter droughts and summer downpours, a high-pressure air system over Siberia maintaining year-long aridity. Beringia is the weak point, the place where the Pacific can throw moisture into the shallow, exposed strait. In the past, this has not been a problem; ice advances and retreats cyclically, and the steppe has grown and shrunk with it, existing in a stable equilibrium. But after 100,000 years of existence, this time it is different. This is the beginning of a transformation, the beginning of the end of the mammoth steppe.*

As the ice sheets melt and the sea levels rise, there is more water available for evaporation, more water that can be added to the landscape. Now, the variable climate sometimes produces warmer, wetter summers than usual, bringing humidity into Beringia, and with it summer clouds, and autumn rot. The mammoth steppe's existence has relied on aridity, on the clear and endless blue skies. When the summers are warm and wet, there is more chance that the water will fail to drain away, forming local bogs, decomposing the plant material and producing peat. The growth of peat begins a destructive cascade for a steppe. Sand sticks together, and blown dunes become wetter, stable hillsides. Soils dampen, acidify and lose their fertility. Wet ground stays cooler, and frost rises from beneath, pushing the water table nearer the surface and out above as clouds, which drop snow, insulate the ground from what sunlight remains and make it cooler still. Cold engenders cold, and as fungi slow in their decomposition of plant life, more and more turns into more peat, and the circle continues.[23]

Emerging bogs also become barriers to migration, mires in which unsuspecting large herbivores can easily become trapped and drown. For the migratory herds of horses and caribou, the spread of peat means a nightmare for navigation as well as a loss of food, a runaway transformation of hard ground covered with grasses into soft, unforgiving wetland. The plants that thrive in peatlands jealously guard what little nutrition they can absorb, and grow defensive

---

* The loss of the mammoth steppe began about 19,000 years ago, but became particularly rapid 14,500 years before the present, during a sudden and humid warming known as the Bølling-Allerød Interval. This is associated with the point at which Antarctica began to deglaciate.

prickles, spines and hairs. In places, trees spread – moisture-tolerant plants like birches, alders and willows. As Beringia is submerged, this is the fate of the mammoth steppe.

On the North Slope of Alaska in modern-day conditions, the change from bare sand to a stable, long-term peat soil takes only a few hundred years. From Ireland to Russia to Canada, the ancient mammoth steppe is almost entirely gone, replaced by permafrost and peat bog. Steppe-tundra ecosystems still remain in isolated parts of Siberia, where relics of smaller creatures, from small mammals to snails, live in a patchwork of habitats defined by the level of moisture. Today, the North Slope of Alaska is a mix of sedges, mosses and woody dwarf-shrubs, a semi-arid but water-saturated plain. Rain- and snowfall amount to only about 250 millimetres a year, roughly the same as San Diego, California, but the moisture stays in the soil, a high water table above the solid permafrost beneath. In summer, the soil thaws down to 50 centimetres depth, producing transient lakes and soft peat, with unpromising forage for the likes of horses or mammoths. Modern Alaska, with its sparser and more heavily defended vegetation and its waterlogged ground that sinks under the print of hooves, is just not survivable for wild horses any more. For the first time since they appeared in North America 55 million years before, horses will become regionally extinct, not returning until they arrive on European ships, only a few hundred years before the present day. The climate has shifted beyond their niche-space, just as it has for mammoths and mastodons, and even, in Alaska, bison. Caribou and muskoxen, those who inhabited the wetter parts of the mammoth steppe, are among the few large species still to live wild in Alaska in the present.[24]

Woolly mammoths survived on a small Beringian island called Wrangel, now part of Russia, until about 4,500 years before the present. That island is and was, however, too small to support a viable population for a long period of time, and, by the end, the Wrangel mammoths, the last surviving family anywhere in the world, were in serious genetic trouble. After 6,000 years of total isolation in a small community that numbered somewhere between 270 and 820 individuals, they were hugely inbred. From the DNA that is preserved in the Russian frost, we can read a catalogue of their genetic disorders. Their sense of smell was severely impaired, and their fur translucent, shining like satin but less able to protect them against the cold. They had problems with development and in their urinary systems, perhaps also their digestive systems. In all, we know of 133 genes for which no individual in the population had a functioning copy. Wrangel, too, was by this time a sedge-dominated peatland; the mammoths could not outlive their steppic landscape for long.[25]

The mammoth steppe is an entrancing vision of life gone by, attracting attention as a romantic vista filled with beasts we feel we can almost understand. Lonely, and buffeted by the Arctic wind, the mammoth is a universal symbol of a lost past. Somehow, because we as humans saw them, we as humans painted them, hunted them, perhaps revered them, they are a tangible link to Earth's history, even when gone for ever. Indeed, there are trees still living that emerged from their seeds while mammoths walked the Earth. The extinct past is closer than we often care to think, and alongside the decline of the Pleistocene came the rise of human civilizations. Humans may not yet have reached the Americas, but elsewhere they are capturing the detail of life of the Pleistocene world. Even as the horses of the North Slope grit their teeth in the wind, daubs of paint are being applied to a cave wall in France, scrubbed clean for the purpose, to represent the wild horses of Lascaux. A few thousand years later, a human will pick up a piece of antler to make a spear-thrower, an atlatl, decorating it with the features of a maned and bearded steppe bison, turning its head and licking, with stretching, curved tongue, the bite of some irritating insect on its back. The cultures of the Pleistocene humans in the north have largely faded, but there are parts of the globe where shadows of the previous epoch are still remembered, still passed on. On the underside of a rock-shelter in northern Australia called Nawarla Gabarnmang, the 'cleft in the rock', stylized wallabies, crocodiles and snakes are painted. The oldest was created, at a minimum, 13,000 years before the present, and painting continued into the twentieth century, a site preserving the cultural memories of the Jawoyn people over scarcely imaginable timescales. By the time the mammoth steppe finally came to an end, when Wrangel's mammoths glinted on cliffs overlooking the flooded plains of Beringia, the Great Pyramid of Giza and the Norte Chico in Peru had already existed for generations, and the civilizations of the Indus valley were centuries old.[26]

At about the time the last Wrangel mammoths died, the Mesopotamian city of Uruk was ruled by Gilgamesh, the Sumerian king and protagonist of the oldest written story, one of the oldest works of literature in any form. The story of Gilgamesh is one in which humankind attempts to escape nature. In it, the arrogant and powerful Gilgamesh, alongside his friend, the wild man Enkidu, trap and kill Humbaba, the guardian of the gods' cedar forest, in order to fell its trees and strengthen the walls of Uruk. Enkidu, the wild, untamed counterpart to Gilgamesh's seemly, royal urbanity, falls ill and dies, and Gilgamesh spends the rest of his tale searching futilely for immortality, before realizing the impossibility of his desire.

Nothing in nature is for ever, and the largest biome of the Pleistocene world will sink into mire. Gatherings of species in time and space may give the illusion of stability, but these communities can only last as long as the conditions that help to create them persist. When conditions of a biome change, whether its temperature, acidity, seasonality or rainfall, any number of its constituent species can lose a foothold there. For some, this means migration, following the environment across the landscape, as many plants did at the end of the last glaciation. Some environments, though, are not moved, but lost. When changes happen too rapidly, or pass a critical tipping point, runaway alterations can destroy even the most widespread landscape on the planet, and with it the communities it supports. This does not necessarily mean total disaster or an ecological blight, but can sometimes mean new combinations of creatures and landscapes, new worlds. Moss-dominated tundra, still occupied by caribou and saiga, peatlands inhabited by willows, alders and voles, and the atmospheric coniferous taiga forests of

Siberia will fill the vacuum. To the roaming horses of the North Slope, and to the cave lions that pursue them, the steppe must seem immovably wide, but when seen at the scale of deep time, permanence is an illusion. As the ice retreats, all it takes is a drop of rain, and the hard land beneath the stamping hooves will soon give way. All it takes is a flicker, and the aurora dies.[27]

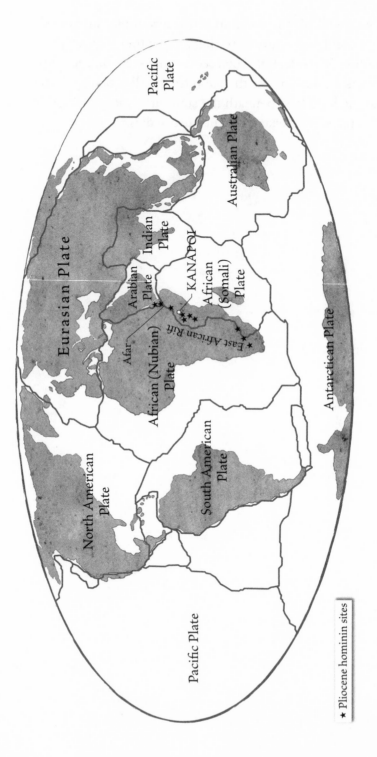

Pacific
Plate

Eurasian Plate

Indian
Plate

Arabian
Plate

KANAPOI

African
(Somali)
Plate

Australian Plate

Afar

African (Nubian)
Plate

East African Rift

North American
Plate

South American
Plate

Antarctican Plate

Pacific Plate

★ Pliocene hominin sites

## 2.  Origins

*Kanapoi, Kenya*
Pliocene – *4 million years ago*

| | |
|---|---|
| 'Merewo tiombotim | 'Turaco, beast of the forest |
| Merewo mito ket | Turaco in the tree |
| Mito mosop cherumbei | Which is the highland waterfalls |
| Merewoni mi mosop | Turaco in the highlands |
| Kayekei katanyon' | Dawn is coming to our home' |

– traditional Marakwet song (tr. J. K. Kassagam)[1]

| | |
|---|---|
| 'Hû ojeka syryrýva tape | 'Black and streaming the road |
| Che akỹ. Ajeity. Syryry.' | fragments before me. |
| | Wet I plunge on, |
| | all before me flowing.' |

– Miguelángel Meza, 'Ko'ê' / 'Dawn' (tr. Tracy K. Lewis)

The swifts arrive with thunder at their backs. Winter birds, they appear, loudly, and in great numbers, pursuing the emerging insect swarms just as the wet season begins after more than four months without rain. The arrival of the migratory birds marks the return of fertility and life, the continuation of a seasonal pattern that will last for millions of years to come. The relieving rhythm and endless cycle of rain and dry, rain and dry. By the present day, people from places as far apart as South Africa and Wales will have come to link the flight of swifts with the coming of rain. Now, the birds barrel through the mountain air of the East African highlands that will one day be part of Kenya and Ethiopia. The rise of these highlands, along with that of the Tibetan plateau thousands of miles away, has diverted the winds that once watered north-western Africa,

*Sivatherium hendeyi*

changing the patterns of rainfall across the region and beginning the slow decline of the Sahara and Sahel into desert.[2]

The great lake, Lonyumun, is evidence enough that rain can be plentiful here. Viewed from its stony shores, it may as well be an ocean; distant, blue-hazed mountain tops can be seen on a cloudless day, dipping their feet into the horizon. Only from the air are the limits of this lake, the shape of this drowned valley apparent. As the swifts descend with scything screams, the blue-green lozenge provides them with a first glimpse of their destination. Lake Lonyumun is spread wide and shallow, well over 300 kilometres from north to south and some 100 kilometres wide. It fills a vast crack in the continent, the East African Rift. Rising plumes of particularly hot magma in the Earth's mantle below hit the crust and spread out like steam hitting a ceiling. The drag of these magmatic currents is, slowly but incessantly, splitting Africa apart. The Somali plate, which holds the entirety of the eastern African coast, is separating from the Nubian plate, holding most of the rest of Africa. Further north, at Afar in Ethiopia, the Arabian plate is also splitting, a three-way junction that leaves a deep depression. The jagged line that runs down from Afar will one day split entirely open, heralding the birth of a new ocean where the rift now runs.[3]

Now, the fractured land fills with rainwater, forming a series of rift lakes that change with climatic fluctuations. In the modern day, the area of Lake Lonyumun will house another lake, Turkana, from which no water ever flows. Lake Turkana is an alkaline, salty body of water, surrounded by volcanoes, as this region has been for millions of years. Its jade green, algal surface is often blown into a rage by powerful desert winds. Pliocene Kenya is wetter, and Lake Lonyumun is wider, spilling over the high ground towards the Indian Ocean. It is fed by rivers that carve through a base of laminated claystones, dense assemblages of mollusc shells, and thick, solidified sandbars, precursors to rivers that will still exist in the modern day – the Omo, the Turkwel, and the wide and gentle Kerio. The volcanic hills of the Pliocene are now being eroded, buried beneath this oxygen-rich river system.[4]

It is in this dynamic world of diverging continents and seasonal thunderstorms that the earliest humans will emerge. Far in the future, there will be species of *Homo* here: individuals like the *Homo ergaster* juvenile known as Turkana Boy, and '*Homo rudolfensis*'\*, although it is possible that these are just variations within *Homo erectus*. But in the Pliocene, in Kanapoi among the acacias, where the Kerio flows into Lake Lonyumun, lives *Australopithecus anamensis*, the 'southern ape from the lake', perhaps the oldest hominin of all.[5]

Between the thorn-guarded galleries of acacias, the river's flow is ponderous and muddied. Swooping low over the lake's surface, the swifts scoop gnats and flies, snatching at the water to drink, and daring anything to fly so fast and free. They circle carelessly in the treeless air above the wide water as it crawls towards Lake Lonyumun. This is the closest that the migratory species will ever come to landing here. So at home are swifts in the sky that they can remain airborne for ten months at a time, feeding, mating and, by resting only half of their brain at a time, even sleeping on the wing. They fly at over 100 kilometres per hour, among the fastest at level flight, outpaced only by free-tailed bats. Their legs and feet are minute – reduced to claws that cling onto walls, trees and cliffs, but cannot function on flat ground. The only time that many swift species will land is to rear chicks, and only because laying eggs in mid-air is hardly a recipe for evolutionary success. Even then, their nests are conjured from thin air, made from what debris they can catch in flight. When not breeding, they whirl in rings above the earth, dive for flies with gaping, froglike mouths, and perform whifferdales and rolls, glancing in and out of sight. The rigours of parenthood are a summer occupation, one for their European sojourn, not here at Kanapoi, where they live only as gaping shrieks on the wind.[6]

The rain brings other creatures out from hiding. In a fiery flash, a

---

\* Lake Turkana was known during Kenya's colonial era as Lake Rudolf, a name given to it by the first Europeans to reach the lake. The name 'Turkana' refers to one of the predominant cultures of the region, while the Turkana people themselves call the lake Anam Ka'alakol.

kingfisher breaks the river surface, feathers silvered with sunken air. Rebounding from the splash it has made with a fish in its beak, it flaps downstream to find a place to perch. Shovelnose frogs, fat little things, lumpy-backed and moss-coloured, come together to mate, the male climbing on the back of the female as she digs down into the ground away from the river. Once the eggs are laid and ferti-lized, the males will leave and the females will continue to burrow, down towards the water table, carrying their tadpoles with them. As the river rises with the rain, the water will fill the hole from the bottom up, giving the tadpoles a safe, private pool in which to grow. Mice scamper through the greening grass, wary of the ambush of small carnivores – dwarf mongooses, dark-striped genets, and the earliest *Felis*, the wild ancestor of the domestic cat.[7]

The low-running sleek of otters slide through the water, and the rain intensifies, as if it will never end. Its splashes form a low-lying fog of spray across Lake Lonyumun. The bull otter *Torolutra*, as big as a sea otter, and a hunter of fish – catfish, lokel and the young of the idji – is very much at home in the rippling current. Wherever *Torolutra* are found, there are also their bigger cousins – bear otters. With its muscular, flattened tail, an *Enhydriodon* swimming in the river would seem to be a floating, mossy log until it coils into a gleaming arch and dives. Searching for hard-shelled prey – molluscs, crabs and the like – there are two species of bear otter in Kanapoi. Both have rounded pestles for teeth, and use them to crush the same type of prey. It is believed that the two species can coexist only by dividing up their prey by size – the smaller of the bear otters going after younger individuals and smaller species of shellfish. The larger, *E. dikikae*, has the dimensions of a modern lion – 2 metres from whisker to tail and some 200 kilograms in weight. Beneath the water, half buried in sediment, are round freshwater mussels, *Coe-latura*, and the giant otter is searching for these. The juveniles are too small to concern the otter, but full-grown *Coelatura* are up to 6 centimetres long – a nutritious snack even if packaged with a crunch. *Enhydriodon* are less exclusively aquatic than most of their otter relatives, spending time relaxing on the bank, but still rely on the presence of large bodies of water in which to find their food. They are equally at home in the river and the more open waters of Lake Lonyumun itself.[8]

The river, delta and lake are all teeming with fish, large numbers of which are eaters of shellfish. Interspersed among the layers of clay that are gradually turning to stone beneath the current river-bed are dense tracts of mollusc shells, their hard parts slowly petrifying as their descendants grow above. In the river delta, one in every three fish is a mollusc-eating characin, *Sindacharax*, and in the lake, almost half of the fish are *Clarotes* catfish. With all the nutrients drawn down by the seasonal rain, the carpet of molluscs in Lake Lonyumun and the Kerio is the major pillar on which the ecosystem rests. The lake is shallow; there are no deep-water fish here, and the inflow from the river makes it well-mixed and aerated. Separation from the Nile has caused Lake Lonyumun to develop its own endemic species, although this separation has recently begun to break down.[9]

This lake has become a haven for waterbirds. Competing with *Torolutra* for fish, the sinuous neck of a darter snakes through the water, swimming awkwardly back to the shore, the rest of its body submerged. Its feathers are unoiled to reduce buoyancy and to help it hunt underwater more effectively, but this means that its feathers are not waterproof. It hauls itself onto the bank, a waterlogged waterbird, its feathers in need of drying before it can fly up to its roost. As the rain subsides, and the ground is pungent with relief, its fellow darters are already standing along the river's edge; their widespread wings are banners, slowly steaming in the sun.[10]

Wizen-headed storks, larger versions of the hunched, cloak-winged marabous, emerge along the bank or soar overhead, scanning for food. Even in these early days, marabou storks appear wherever humans live, from the Pliocene of East Africa to the Pleistocene of Flores in Indonesia, to modern-day cities across the world. They are unfussy in their diet, known for their habit of inhabiting landfill sites and dumps, and are scavengers on carcasses, which earns them the nickname of 'undertaker', but helps to remove diseases from the environment. Storks, with their large size and lazy flight, have regularly inspired the attention of human folklore. In the medieval Slavic religion, migratory winter birds were believed to depart for a paradise land called Vyraj. Among them, the white storks in particular were thought to carry human souls to the afterlife and back for a form of reincarnation.[11]

A second plunge from the kingfisher hardly causes a splash in the current. It emerges once more, this time unsuccessful, and lands on the back of a giant, silhouetted beast. The metallic blue bird stares at the water, taking advantage of this new fishing platform, which appears to be entirely unconcerned about its new associate. Two and a half metres at the shoulder, it stands cautiously in the muddy shallows, wary of the possible presence of gigantic horned crocodiles. Its short, whorled hair is matted down by the rain, with dark, long-lashed eyes shadowed by two bulbous projections. From the top of its head emerge two more, curving outwards and backwards to give the impression of an upturned sliver of a crescent moon. Not all giraffids are slender and long-necked; *Sivatherium* has the stockiness of an ox. Though they are an extremely rare part of the Kanapoi community, their kin are found from here in East Africa to the Himalayan foothills of India. They are the heavy-set relatives of giraffes and okapis, with full-grown males weighing well over a ton. But where sivatheres lack the charismatically gangly un-grace of a giraffe, they make up for it in the ostentation of their heads.[12]

All members of the giraffe family, including okapis and sivatheres, have bony lumps on their skulls called ossicones. They function, like keratinous horns or bare antlers, for display and as weapons, but unlike horns or antlers, these are permanently covered by hair and skin. Male okapis have two, short and thin, almost antenna-like, one above each eye. Giraffes all have two, also rather short and straight, sticking up between their ears, and some individuals, particularly in East Africa, also have a single thick boss in the middle of their forehead, between their eyes. The Kanapoi *Sivatherium* has two pairs of ossicones, above the eye and between the ears, and neither is diminutive.[13]

The sivathere gingerly lifts one leg from the river, disturbing a dozing otter who hauls herself into the water, suddenly weightless. Shins coated in mud, the sivathere strides out onto more solid ground, towards the shade to browse. Around the banks of the Kerio, the rain has corralled the dust into a sheen of clay, but the slopes of rising hillocks are kept drier by well-drained sand. Where there is clay, the soils are fairly impervious to water, and this allows hollows to transform into muddy basins, and in the rain, the clay minerals expand, making the slopes less stable. The land undulates, the higher ground sparsely specked with tree-shrubs and meadows of grass, while damper gullies are filled with forbs – non-grass fodder plants. In a long strip, flanking the river, water persists all year round as deep groundwater, even at the height of the dry season, so trees can flourish by sending long, vertical tap roots down to the secret subterranean aquifers. Growing tall, they are a snaking avenue disclosing the course of the river for miles around. Where the Kerio slows and drains into Lake Lonyumun, the water table is closer to the surface, and the canopy descends; shrubs compete with the trees, dissolving into thickets interspersed with wet, sandy ground, robed in sedges.[14]

The local variations in soil chemistry, aspect and drainage have created a quilted landscape, with patches of taller trees and shrubs broken by swards of grass. A varied environment supports greater species richness, and in Kanapoi, there is a far higher proportion of generalist herbivores than will be seen again in East Africa. Plants are in the middle of an industrial revolution, and the herbivores are only just catching up.[15]

Plants feed themselves using photosynthesis – using solar power to convert carbon dioxide and water into carbohydrates. The water is drawn from the ground, but the carbon dioxide has to come from the air, and so leaves have holes in them – stomata – through which the gas enters. As long as the stomata are open, carbon dioxide can enter the leaf, the process can continue, and energy can be captured, but this comes at a cost; open stomata leak valuable water through evaporation, and the plant wilts. The hotter the environment and the scarcer the water supply, the more this is a problem. But it is one that several plant groups have, by the Pliocene, solved.[16]

To turn light into food takes several steps, but the key is an extremely inefficient enzyme called RuBisCO. Where photosynthesis needs to be as efficient as possible, as in hot and arid places, many plant species worldwide concentrate the various required chemicals around RuBisCO in special cells deeper in the plant, away from the leaky stomata. This takes energy to do, but can make the whole process six times faster, which saves water.[17]

Ten million years before the present, the proportion of such plants worldwide was less than 1 per cent. In the modern day, nearly 50 per cent of the world's primary productivity – the amount of new energy harnessed by photosynthesis – is performed by the roughly sixty groups of plants that independently discovered this spatial sugar assembly line, known in scientific parlance as $C_4$ photosynthesis. These plants, including many crop plants from grasses like maize, sorghum and sugar cane to amaranths like quinoa, spread as a result of the changing atmospheric conditions – the lower atmospheric carbon dioxide concentrations that have existed during our time as a world with polar ice make concentrating the reaction more attractive. Since $C_4$ plants have become more prevalent, and because they are poorer sources of nutrition, herbivores have had to adapt their feeding behaviour to the changing flora.[18]

The mosaic of dry and open bush, shrubland and tree-lined river channels at Kanapoi allows different species to specialize to different types of plant. Most herbivorous species have yet to fully adapt to the recent innovation that is $C_4$. Several species – far more than in any modern ecosystem – are mixed browsers and grazers, such as *Simatherium*, perhaps the ancestor of the African buffalo, the muskox relative *Makapania* or the early relative of wildebeest and hartebeest *Damalacra*. *Sivatherium hendeyi* is a browser, eating only the shrubs and trees that grow close to the lake and river, although their descendants will one day move to grazing too. The Kanapoi giraffes, a full-sized and pygmy species, stay true to their long-necked monopoly on tree-tops and are both browsers. Impala and three-toed horses graze in the open spaces between the trees, along-side the lowered heads of wandering, warty, half-ton pigs, restless gaggles of ostrich chicks, and huge herds of trunked proboscideans – the kin of elephants.[19]

Proboscideans are certainly diverse in Kanapoi. There is not just *Loxodonta adaurora*, closely related to and barely distinguishable from the African elephant, but also *Elephas ekorensis*, a cousin of Indian elephants and mammoths. Among the trees strut stately, short-legged *Anancus* with their long, straight, forklift-truck tusks that almost touch the ground, and unlikely *Deinotherium*, whose short tusks curve backwards and are used to scrape bark from trees. Most are browsers, as elephants are today, but *Loxodonta adaurora* is a grazer. Quite why *Loxodonta* species switched to a $C_4$ diet and then back again is not clear, but it might be down to the sheer number of competing proboscidean species. As the trees thin and the savannah opens to its full extent, the only surviving elephant in Africa will be the descendant of the species that learnt to graze. Today, African elephants are ecosystem engineers, true foresters that control the density and cover of trees throughout their range, defining the niche-space in which their neighbours must live.[20]

With giant giraffids, 10-ton deinotheres, huge otters and outsized pigs, Kanapoi is filled with large herbivores. This level of diversity can only be supported because the area is so rich in food resources; the area around Lake Lonyumun produces new plant material faster than any other African fossil site from the last 10 million years.[21]

A stone's throw from the eastern bank of the Kerio is a low patch of tree-shrubs. Acacia shade dapples the ground, while above, paths of light run like snail trails where the crowns fail to touch. The ground is dry and covered in a natural hay meadow: tussocks of buffel-grass with their drooping, fluffy seed-heads, covered in sharp burs; thin and wispy dropseeds rising weakly over their vibrant leaves; and the rough vertical foxtail brushes of *Tetrapogon*. The skeleton of a dead tree is enticingly scarred, a low hollow revealing where the interior has rotted away, leaving only the hard outer tis-sue and a faintly fungal reek. Within, a family of nocturnal mastiff bats sleeps. When night falls and the swifts rise to sleep at higher altitudes, these bats will take over the relentless pursuit of the flying insects over Lake Lonyumun.[22]

A scream of alarm from a turaco sets a commotion among a troop of *Australopithecus*. Disturbed from chewing on leaves, the hominins scramble to their feet and run to climb lianas to safety in a spreading, broad-trunked edurukoit acacia – *Australopithecus* is the first hominin to walk and run exclusively on two legs. Hostile smiles display their enormous canines as they hold on to the branches and lean over at the source of their fear and anger, while the swifts circle overhead in their endless ringing dance. There is something threatening them from low in the grass – a python, for which an australopithecine would be a good-sized meal.

Although upright, *Australopithecus* are rather different from modern humans. Their body hair is still long, with loss of hair thought to be associated with humans' later adaptation to long-distance running. Their faces are still very ape-like, with jaws that project forwards, and a sloping forehead tapering the head behind strong brows, to thick necks. The tallest are only about 150 centimetres tall, the same size as a chimpanzee, though they are less muscular, and there is a much bigger size difference between males and females than in modern humans. Their feet are not yet perfect for running; australopithecines have a slightly in-turned stance, helping them to climb the trees in which they sleep.[23]

Having lost the initiative, the python retreats towards the river once more, sliding back towards the cracked trunk and buttressed roots of an echoke fig tree as the rain starts up again. The australopithecines in the tree calm now, but remain in the branches, too spooked to return to the ground any time soon. Their food is mostly soft to tough plant material, with nothing hard, nothing really brittle, and no grassy $C_4$ plants.[24]

*Australopithecus anamensis* is the earliest species to be unarguably more closely related to humans than to chimpanzees and bonobos. A few other candidates are older, but there are disputes about whether they are closer to chimpanzees or humans, or whether they split off from our lineage earlier than chimpanzees. The life of *Au. anamensis* in Kanapoi is the beginning of the evolution of an initially diverse group of which we are the last survivors. The species to which the famous fossil 'Lucy' belongs, *Au. afarensis*, are direct descendants of the Kanapoi hominins that lived about 3.2 million years before the present.[25]

In ancient Athens, a thought experiment was proposed concerning the Ship of Theseus, preserved for posterity as a museum piece. As part of its preservation, decaying timbers are replaced from time to time, until eventually none of its original material remains. Plato asks whether the identity of the original ship is retained by the preservation piece, or whether, having been entirely replaced, it can be thought of as the same ship at all. An extension to this experiment supposes that the removed timbers are treated, the rot removed, and the ship is later rebuilt using the original material. Which of the ships is the same as the original, or do they both inherit the identity of the original Ship of Theseus?[26]

Ever since the earliest attempts at classifying the natural world, humans have been labelled separately from the rest of life, as something apart, something special. The trouble with taxonomic labels is that they, like communities of organisms, are not constant through time. In the modern day the distinction between humankind and our nearest relatives, the genus *Pan*, comprising chimpanzees and bonobos, is clear. But every species shares a common ancestor, and each lineage is its own Ship of Theseus.

Were we to look at the population of apes that existed before the ancestors of chimpanzees and humans went their separate ways, we would see a single species, and we might give that species a name. Frequently, when a new species is born, it results from 'budding' – an isolated population of a species changing relatively rapidly, with the ancestral species persisting in every meaningful way elsewhere. In such a case, we might continue to use the original name for the relatively unchanged population, but from the perspective of the 'new' species, the number of generations from the pool of shared ancestors is hardly distinguishable. It is only with the benefit of geological hindsight that we can determine that a population in a slice of past time should be considered to be different. In real time, a species is a dynamic plurality, the sum of its component populations and individuals within and between which genes flow.[27]

For humans, defining the point at which we are confident of claiming 'humanity' is difficult. What, after all, distinguishes us from other animals? There was no moment at which humanity suddenly arose – the populations that led to *Pan* and the populations that led to *Homo* did not undergo a sudden shift – the mixing of two populations simply reduced to the point where no genes flowed. We, like every other species, are the culmination of a series of partial replacements, the deaths and births of individuals in an ever-shifting population, with a continuity back in time and forward, connecting all living things together.

To talk of the first humans is to hammer a signpost into an ancient river saying 'no humans beyond this point', no matter the ever-flowing stream around its base. There is nothing essential to humanity, no single feature that intrinsically caused one creature to be a human where its parents were not. If we were to fast-forward, accelerate through time and follow these *Au. anamensis* as the shared average features of the community shift into those of *Au. afarensis*, the unimportance – or at least ambiguity – of this notion of species along the axis of time would be laid bare. In the time dimension, distinctions between Linnean ranks become meaningless. However hard you try to define every point before the signpost as non-human, and every point after the post as human, the river flows continually onwards.[28]

Instead, we might use natural marker points – the places where river systems divide. Along the continental divides of the world, creeks and streams fork, never to meet again. In the highlands of what will become Ethiopia and Kenya, a stream splits around a rock that blocks its path. By chance, the water that passes to the left will tumble down the eastern side of the hill, join Lake Lonyumun, and eventually flow out into the Indian Ocean. The water that passes to the right will flow west, becoming a tributary of the Nile, crashing northwards to the Mediterranean Sea. Before the rock, every drop is intermingled; afterwards, the two streams will be separated for ever. Just past that rock, there is nothing intrinsically Mediterranean about that water, just as there is nothing intrinsically chimpanzee about the first species on the route to modern *Pan*, and nothing intrinsically human about the first species on the route to modern *Homo*. The earliest chimp relatives and earliest human relatives were necessarily more similar to one another than either is to chimps or humans. But if we are to place an identifying point for the beginning of humanity, a marker to say 'these were the first', the divide between *Pan* and *Homo* makes as much sense as any other, and this is the approach that palaeontologists use.

*Australopithecus anamensis* are among the first creatures we find within that river of humanity more closely related to us than to anything else that exists in the modern day. Although upright, *Australopithecus* are smaller than modern humans, about 130 to 150 centimetres tall, but still spend plenty of time in the trees, and retain the projecting jaws of non-human apes. No doubt as capable as chimpanzees with simple tools like stone hammers and anvils, they nevertheless predate the earliest human knapped flint tools by half a million years. There are large body-size differences between males and females within their mixed social groups. As they shift into *Au. afarensis*, their canines will reduce both in root size and in the extremity of their shape, the enamel will thicken, and the jaws shift to be wider-set. Precisely how the australopithecines and later hominins grew and evolved in such a way as produced us is not well known; the course of the river has yet to be fully charted, and several paths will dry and evanesce to nothing, but *Homo sapiens* will eventually appear not too far from its headwaters in the East African Rift.[29]

The same is true of so many creatures in the cradle of Kanapoi. Those African elephants that can be found on the plains of Kanapoi, *Loxodonta adaurora*, are close relatives of modern African elephants, *Loxodonta africana*, but are a lineage that never made it to the present. The impala that graze on the grasslands are similar enough to modern impala to be assigned the same genus, *Aepyceros*, and are probably their direct ancestors. The giraffes here are almost identical to modern giraffes, a little smaller and with smoother foreheads, but with the unmistakeable loping stride and long, ungainly necks.[30]

Much, of course, will change in the interim, and many species will be lost. As creatures adapt and evolve, their niche-space shifts, and some will come to overlap and compete. Circumstantial evidence has led some to suggest that the bear otters of East Africa will ultimately meet their extinction because of hominins.* The idea is that as the genus *Homo* arises, as tools become an ever-more important part of the ecology of hominins, the diet of humans will change from the pure herbivory of the australopithecines. This ever-more carnivorous niche will bring hominins into conflict with the other carnivores of East Africa, including the bear otters. The rocks record a decline in large carnivore number and diversity that peaks in intensity 2 million years before the present, just as the first species of *Homo* emerge from the Rift. The large carnivores that will survive to the present are the specialist meat-eaters – big cats, hyenas and wild dogs – that prey on large, dangerous herbivores. Those that will be lost – otters, a bear, giant civets – are mixed feeders on plants, molluscs, fish, fruit – precisely the niche that we will ultimately make our own. If this is true, the bear otters of Kanapoi are destined to become perhaps the earliest species extinction caused by hominins.[31]

Nature lovers often see the world as a dichotomy between some primordial, natural Eden and the cityscapes of the modern day. Humanity is seen as an external force, something separate to the ideal of 'nature', which must be escaped to experience the wild, and

---

* The idea that this was the cause of bear otter extinction is, it must be said, controversial because of the small size of the data set. As a general ecological principle, though, competitive exclusion is a real phenomenon, and has been linked to the rise and fall of other groups, such as the loss of borophagines (hypercarnivorous relatives of dogs) following the arrival of big cats in North America from about 20 million years ago.

something which can only wreak a destructive force on the world. To take this view is to deny the naturality of humanity. Ever since our emergence, we have been fighting our corner, exploiting our own ecological niche, part of which is as a modifier of habitats, an engineer of ecosystems, altering the worlds in which we find ourselves to suit our biological requirements.

In Kanapoi, we see one of the earliest worlds largely recognizable as our own. The continents are almost at their modern positions, and the world is cool and ice-capped. Pliocene Earth resembles the recent interglacial periods, including the modern day. Kanapoi is a cradle, but not just of humanity. We are but one of the families that benefited from the ecological variability of East Africa, part of the earliest endemic African mammal community, with carnivores like hyenas, genets, mongooses and wild cats. Among hoofed mammals, zebras, wildebeest, elephants, antelopes and giraffes all trace their heritage to the shores of Lake Lonyumun at Kanapoi. Even among primates, Kanapoi is not just home to hominins. Early baboons, slight, long-limbed and resembling mangabeys are here, too. The lake makes Kanapoi unique even among contemporaneous African fossil sites; no other site has such a diversity of aquatic and aerial birds.

From the western hills, the rivers bring minerals that settle around the mussel floor of Lake Lonyumun, and fertilize a highly productive landscape. Though beneficial, this influx will ultimately destroy Kanapoi. As more and more silt pours in and settles on the floor, the lakebed rises, the water choked and dried. In all, the lake will last only 100,000 years. But the lake and its inhabitants have an echo. Half a million years after it dries, the rifting of Africa will open space again for a new lake, Lokochot, where *Kenyanthropus*, perhaps the first tool-using hominin, will make its home. It, too, will fill with silt, but from the mudflats that follow, the lake Lorenyang will grow. Along its shores, *Homo habilis*, the earliest species to share our genus, will live. The lifespans of the lakes are generally short, but Lorenyang will last for nearly half a million years. Eventually, 1.5 million years after Lorenyang has turned to floodplain, the present-day Lake Turkana will develop from about 9,000 years before the present, where communities of us, of *Homo sapiens*, still live, diverting the course of the modern Kerio to irrigate fields of $C_4$ grasses; sorghum and maize.[32]

The swifts still wheel over the Kerio valley, and stilts and storks still stride along the edge of the great rift lake. East Africa still has some of the most suitable landscape for dense concentrations of large herbivores in the world, and indeed the herbivores that live there are still very diverse. This regional diversity masks a bigger issue. There are equally suitable hotspots in India, in eastern Australia and around the Great Lakes of North America, where large herbivores should exist, but do not. Even in the rich landscapes of Kenya, the presence of large herbivore communities is seriously under threat. Many of the great beasts of the Pliocene past are already long gone – sivathere and bear otter, giant pig and scimitar-toothed cat – and no intrinsic trait guarantees the continued survival of those still alive in our present. But even now, in the Rift Valley, there are glimpses of familiarity with what has gone so recently before, of the conditions in which we as humans slowly emerged. The planet may have lived long before us, but Kanapoi is the first world that humanity can claim as home.[33]

Possible salt lake fragments

Floor of the Mediterranean basin

# 3.   Deluge

*Gargano, Italy*
Miocene – *5.33 million years ago*

'Our description commences where the sun sets and at the
Straits of Gades, where the Atlantic Ocean, bursting in, is
poured forth into the inland seas'

– Pliny the Elder, *Natural History* (tr. J. Bostock and H. T. Riley)

'I will love thee still, my dear, till a' the seas gang dry'

– Robert Burns, 'A Red, Red Rose'

The air shimmers with the rising thermal wind, blowing the sweet
smell of juniper around the cliff edge. Cedar branches flick gently
like cattle tails. The sound of cricket song and the scent of salt on
the breeze permeate the evening. Ahead, there is nothing but sky.
The plain below, difficult to resolve through a kilometre of refract-
ing heat, is laid out brown and white, the arid landscape coarse-cut
by rivers slowly carving their way towards the next drop into the
abyss. Beyond that, the bare expanse stretches yawningly to the
horizon. In the other direction, the fading sun is dropping towards
a blurred ridge of mountains, scarcely visible at the limits of sight.
The canyoned flats are nothing compared with the great river-cuts
beyond these Apennine forebears. There, the Rhône's deep, steep
valleys are many times deeper and wider than will be reached by
the Grand Canyon, which is only now beginning to be excavated
by the Colorado River a continent and an ocean away.[1]

Looking out from Gargano in the latest Miocene, more than 5
million years before the present day, it is hard to countenance the
idea that, in a little over a year, swirling brine will be washing these

*Hoplitomeryx matthei*

stones. Even harder is to visualize this towering mountain, alone and proud, sending boats out into the intangible air, this sky becoming the centre of trade and warfare, filled with people, goods, armies and ideas for thousands of years. This clifftop will hold communities of fisherfolk, as a limestone promontory surrounded by the Mediterranean Sea. For now, the basin is drained, a salty, dry, inhospitable land reaching down kilometres into the depths of the Earth. From the Levant to Gibraltar, from the North African coast to the Alps, the Mediterranean has run dry.[2]

Neither is this for the first time. As the tectonic plate beneath Africa and Arabia has pushed northwards, the once mighty Tethys Ocean has grown narrower and narrower, reduced to a small, enclosed sea between Afro-Arabia, Asia and Europe – the Mediterranean. The only connection between this sea and the rest of the world's oceans is a narrow gap between what will be Spain and Morocco – the Straits of Gibraltar. Through the last million years, the push of Earth's plates has periodically closed the gap, with drastic impacts on its environment.[3]

To the south and east, high temperatures and little standing water mean that rain is sparse, and what little does fall is as likely to evaporate quickly as it is to reach a river. The picture to the north is more promising, but only slightly. The position of Europe's mountain ranges – the Sierra Nevada, Alps and Dinaric Alps – means that there is much more land to their north. The strip between the sea and the mountains is narrow, a catchment that leaves little rainfall able to flow into the Mediterranean. Some of the great rivers of Africa and Europe may pour into the Mediterranean, but there are few of substantial size. Of all the rivers that pour into the Mediterranean, only the Nile, the Po and the Rhône are of note, discharging between them about 600,000 cubic metres of water – approximately seven times the volume of London's Royal Albert Hall – every minute. The total amount of fresh water added to the Mediterranean in one form or another is about 600 cubic kilometres per year, or eighty Loch Nesses. This may seem large, but the hot climate evaporates seawater at a faster rate, dwarfing the influx with the

evaporation of 4,700 cubic kilometres every year. The Bosporus –
that narrow passage linking the Mediterranean and Black Seas – does
not yet exist; a spit of high ground separates the Mediterranean
from the smaller Paratethys Sea, which extends from Romania to
Central Asia. The imbalance in water flow can only be corrected by
a constant, compulsive current from the Atlantic through the narrow Straits of Gibraltar. When this is closed, as it was, on and off,
for the last 700,000 years of the Miocene, the sea evanesces into
almost nothing in a mere 1,000 years. All that remains is a small lake
in the eastern Mediterranean, fed by a river system that flows out of
Turkey and Syria.[4]

The sheer volume of water moved from the Mediterranean
causes sea levels around the world to rise. Islands become mountains, as rivers flow futilely into ever-evaporating salty lakes in the
depths of a valley in places 4 kilometres below sea level. This is the
deepest land in the world. Descending into the abyss, the ever-larger
weight of the atmosphere pushes down, as winds fall over the cliffs.
When a pocket of air moves downwards, the air pressure rises. Like
the air within a combustion engine, the increase in pressure causes
the air mass to shrink and to heat. For every kilometre a wind falls,
its temperature increases by about 10°C. This is a cool period in
Earth's history, but even so, on a hot day, maximum summer air
temperatures 4 kilometres down at the base of the plain could reach
a hellish 80°C – some 25°C hotter than the hottest temperature
ever recorded in modern times in Death Valley, California. The floor
of the Mediterranean basin itself is now made of salt, laid down in
places more than 3 kilometres deep – a total volume of over 1 million cubic kilometres of glittering gypsum and sodium chloride.
Nothing beyond extremophiles – those microscopic organisms that
thrive where none else can – survive on the floor of this Mediterranean valley.[5]

For humans, the waters of the Mediterranean have been connective, bringing together cultures from Europe, Asia and Africa,
linking cities and civilizations with transport far quicker than land.
For land-dwelling animals, though, the sea acts as a barrier. Not an

entirely impassable obstacle, but water slows migration and isolates communities far more than even extensive terrestrial habitat barriers like deserts. As the sea has receded, fragile island ecosystems have become exposed to one another across relative high grounds that exist between the peaks of these uplands. The Baleares – Mallorca, Menorca, Ibiza and Formentera – are connected by a plain only about 1 kilometre deep that stretches to the mainland of Spain, and northwards up to France and the Rhône canyon network. Sardinia and Corsica are similarly connected to northern Italy, while Sicily and Malta ride a high saddle of land that rises into the Apennines, linking Africa and Europe. The Hellenic island arcs from Crete to Rhodes have not yet risen to their modern heights, while Cyprus is a volcanic table-top plateau in glorious isolation.[6]

Here, at Gargano, the ancient limestone massif stands apart from the Italian ridge, a solitary carbonate watchtower guarding the Adriatic sky. Until recently, this too was an island, separated from the rest of Europe. Having evolved in isolation, it is now a land of dwarves and giants, a unique landscape at risk of disappearing for ever. It has been an island for most of the last few million years, linked only ephemerally to nearby Scontrone, until the Mediterranean receded and passage opened up to the mainland. Initially lush and fertile, the lack of large bodies of water means that there is little evaporated water to fall as rain, and this region has become ever more arid. There are no lakes, no standing water to speak of, although streams still flow in the right seasons. The cedar trees spread wings of shade over the rocky slopes, while unhappy, green-tipped clumps of mountain hemlock still hold on in the deeper valleys. The reduced rainfall has not been good for some of the conifers, and drier outcrops host only shrubland of pistachio, box, stooping carobs and gnarled olives, the fruits of drought.[7]

Among the vegetation, grey-white heads rise and fall with an undulating rhythm, a flock of about a dozen enormous geese, pale adults and dark juveniles not long out of their infant down. The largest are twice the weight of a mute swan, and each forages with the single-minded purpose of autumn. Geese and ducks are notoriously voracious feeders, with an instinctive tendency towards gorging until their crops are full, a trait that will be exploited by French farmers in millennia to come but which serves a purpose in the wild. The urge to migrate vast distances requires plenty of energy, and so geese bulk up, readying themselves for their long journey. These geese, though, are not going anywhere. Their enormous size and proportionally small wings means that they are, like many island birds, flightless, although their ancestral instinct to gorge remains. Winter is coming; the goose is getting fat.[8]

Gargano seems to have been colonized over the water, with ancestrally small animals – mice and dormice, for example – blown across on bits of floating plant, and birds flying over. This type of colonization usually occurs by chance, sampling the mainland with all the randomness of a sweepstake. Other islands, like the Baleares, were colonized over land, during short-lived periods of drying, with larger animals more likely to have walked over before the community became isolated again once the water rose. Island communities are subsamples of mainland communities and are often imbalanced and distinct. Pressures of the food chain can unbalance them still further. Small populations of prey mean that carnivores cannot support themselves easily, so there are virtually no carnivorous mammals in Gargano. No cats, no weasels, no dogs, or bears, or hyenas. A small population of otters lived here only a few hundred years ago, when tides still washed the chalk shore. Perhaps they are still here somewhere, improbably surviving in a dark, wet corner of the mountain.[9]

In the absence of these mammals, birds fill the roles of the large predators and herbivores – a latter-day dinosaur-dominated landscape. Many of these birds are visitors from elsewhere, ephemeral components of the landscape, like swifts and pigeons, but the island has created some natives of its own. The largest herbivores on the island are the short-winged geese – *Garganornis*, the 'Gargano bird'. Two birds take exception to one another – too close, perhaps, while feeding, and rear up. Wings held out like a judoka, a baritone honking, each tries to bite the other's wings to prevent blows being struck.

It is over almost as soon as it has begun – the smaller goose recognizes a lost cause. Another goose, skulking near the edge of the group, has a limp wing. *Garganornis*, like all ducks, geese and swans, needs little reason for a fight, and a well-placed blow from their wings can shatter bone. They may be useless for flying, but the wrists of *Garganornis* have prominent bony knobs hidden underneath their plumage, a mace decorated with feathers. The geese use them to compete with one another if neither side will stand down.[10]

A distant whistling mewl signifies the presence of a raptorial predator. Against the cloudless white sky – it hardly rains any more in summer since the sea retreated – the stubby, curved, buzzard-like wings are silhouetted, soaring above empty space. The lazy flaps of the thermal glider, the distinctively plaintive call. That shadow belongs to the largest of all the predatory dinosaurs here – *Garganoaetus freudenthali*, or Freudenthal's Gargano eagle. Properly a relative of buzzards and their kin, there are two species of 'eagle' endemic to this island, and Freudenthal's is a giant – larger than a golden eagle. It is among the largest raptors ever to live, though not quite approaching the size of New Zealand's pouākai, that Pleistocene hunter of moa with a 3-metre wingspan, an eagle so terrifying as to persist in Māori folklore long after its extinction. The geese are not concerned by the raptor. They are too large, too strong. The eagle eye is fixed on something else.[11]

A crackling of parting twigs, and a jagged face appears through the scrub. A small deer-like creature, no more than half the height of an adult human, bows to the river to drink among the rushes. His head fits poorly with an image of abasement, adorned as it is with a crown-like array of horns. There are five in all, two long ones between the ears, one short and projecting sideways above each brow, and, imposingly, a single long horn between his eyes. He lifts his head regally, and against the rust brown of his jaw, dripping in the evening sun, are shining white daggers of canines. *Hoplitomeryx*, the 'armed deer' is, like many of its contemporaries, sabre-toothed. Those teeth are not for hunting. Like modern-day musk deer or the Chinese water deer, which also have sabre teeth, they are largely for fighting one another. With the rut approaching, this male will need both sabre teeth and horns in prime condition to find a mate.[12]

Those are horns, too, not antlers, although the origin of horns, antlers and the ossicones of giraffids was probably a single evolutionary event. In the late Miocene, antlers – specialized external bones that are shed and regrow every year – are a relatively new innovation. Having arrived from Asia across the verdant grasslands of eastern Europe, the antlered deer are experimenting, shifting from a simple, functional, single tine or fork to ever more ornamental designs. To grow new bones every year requires enormous amounts of calcium. The demand is so intense that modern-day red deer on the Hebrides are known to wait outside shearwater burrows in spring, crunching down on the chicks as they emerge above ground for the first time and obtaining calcium from their bones, while white-tailed deer in North America are notorious nestling predators of a variety of small songbirds.[13] Antlers are expensive.*

The horns of *Hoplitomeryx* are more like those of sheep or cattle, a bony core covered with a permanent sheath of keratin. The five-horned arrangement of *Hoplitomeryx* is unique, but because horns are sexually selected traits as much as they are weapons, other bizarre horny structures are found across the world. Males of one group of Miocene North American deer-like animals, the synthetocerines, have a single, very long horn, balanced circus-like almost at the tip of their snouts, split at the tip like a barbecue fork.[14]

The sabre-toothed deer of Gargano generally forage alone for their diet of soft leaves and stems under the cover of the conifers, as the Gargano eagle circling above is their main predator. Those horns on their heads are not just for show; they cover the places that predatory raptors like eagles tend to strike at; two above the eyes, two over the neck, and one on the nasal bone. Most *Garganoaetus*

---

* They do have benefits, though; antlers grow through a mechanism that is very similar to cancer, but because deer can keep that growth in check, they are extremely resilient to cancers, with 20 per cent the rate of cancer in other wild mammals.

kills of *Hoplitomeryx* will occur after the upcoming rut, when the deer come together and the small herd ventures into open spaces, which are becoming more common as the island dries. It is only the fawns who fall victim; an adult of even the smallest *Hoplitomeryx* species is still 10 kilograms, a heavy prize to carry off even for a bird with a 2-metre wingspan.[15]

Even without the threat of the eagle, to be under cover is appealing in the remaining heat of the day. The uplift of Mesozoic limestone that forms the Gargano promontory has, over millions of years, been rained upon, slowly dissolving the rock, eroding it into cave systems, water absorbing the earth, seeping through, coating caves in layers of soft, silvery calcium carbonate, and forming the dripstone columns of stalagmites, ruffled curtains, and fissures in the earth. Humid, cool and damp to the touch, it is a refuge for all, and will become their tomb. Landscapes like this – karstic systems – are formed patiently, with small surface cracks growing, fracturing, opening caves and swallowing water courses. Those subterranean rivers and streams wash in all manner of animal remains along with pebbles and fragments of the environment, which commonly become trapped in fissures, covered in a fine powder of limestone, infused, transformed and preserved.[16]

The promontory of Gargano, despite having been above the water-line for millions of years, is itself made of the sea. The dazzling lime was once part of the continental shelf of an entirely lost landmass, Greater Adria. Geologically part of Africa until separating some 200 million years before the present, Greater Adria crossed the narrow ocean and embedded itself in southern Europe. What is now Gargano, and more broadly Puglia, Calabria, Sicily and beyond, was once the deep-water margin of this Greenland-sized continent. Caught and spun in between the same long-term collision that is bringing Africa and Europe together and which has dried out the Mediterranean, the plate that was once Greater Adria is now mostly buried, in places driven down into the crust more than a thousand kilometres below the Alps. In the present day, only fragments remain, scattered from Spain to Iran, the offshore shelf of a lost continent. Embedded in the walls of this cave, and the thousands of caves like it along the edges of Europe, are the last remnants of the Adrians, fossilized calcitic shells of sea snails and clams, held in an alabaster matrix of microscopic shells of planktonic organisms.[17]

A black-masked songbird begins to sound its mellifluous twilight tune outside, in the waxy shade of a laurel. The cave is cool and the air filled with the damp musk of stalactites. On the ground lies a pellet from the giant barn owl *Tyto gigantea*. Contained within it are the bones of the local giant dormice, enormous mice and the imperial pika, a gigantic version of the modern, diminutive mountain-dwelling relatives of rabbits and hares. Everything on Gargano is mis-sized. The giant barn owl, whose slender mainland relatives are some 30 centimetres tall from beak to feet, is a full metre long, as big as an eagle-owl. The pikas are much bigger than their close mainland relatives, while the mice, the dominant species here in terms of number, weigh between 1 and 2 kilograms. At the larger end there are also the giant geese and buzzards. Some of those animals that are not giant are dwarfed, like the sabre-toothed deer, and a now rather stranded population of small crocodiles, recently arrived by swimming from Africa and trapped in an unsuitably water-free environment.[18]

Island dwarfism, one half of the general rule that island animals tend towards a medium size, was first noticed in a Cretaceous fossil site in Hațeg, Romania. At the time the limestone of the Gargano caves was being laid down beneath the European seas, Hațeg was a largish island, and housed dwarfed dinosaurs. Their small size was thought to come from the lower resources of islands, with enormous creatures unable to survive on the limited nutrients available. This is not limited to creatures as big as dinosaurs. Over time, in the absence of ordinary predators, many large animals whose size would otherwise offer protection against predation – such as deer, and, on other islands, hippopotamuses and elephants – become smaller as food is scarcer. Small animals, which cannot store energy or water as easily, become larger, aiding the survival of the population through periods of scarce resources. The pattern has been repeated across the islands of the world, across the Miocene islands of the Mediterranean, and throughout evolutionary history, although, as with all biological rules, there are exceptions. In the Miocene, giants inhabit islands all over the world; metre-tall flightless parrots live in New Zealand, and 3-metre-tall elephant birds, whose closest living relatives are the diminutive kiwi, roam Madagascar.[19]

On different Mediterranean mountains, the niche of small mammalian herbivore has been filled by supersized or shrunken versions of whatever happens to have colonized the island upon isolation. On Gargano, there are the herds of *Hoplitomeryx*. On Mallorca, a minute goat with a disconcerting forward-facing stare prunes the box shrubs. Box is notoriously toxic, containing large amounts of alkaloid compounds that normally deter predators. *Myotragus* has a behavioural solution to this toxicity, however: it eats small quantities of clay from the riverbed, which neutralizes the toxic alkaloids in the leaves. This abrasive mud antidote wears down their teeth, so they have evolved rodent-like ever-growing incisors, and molars with very high crowns, which explains the meaning of their name, 'mouse antelope'. The pressures of island life often produce such unusual responses. Physiologically, *Myotragus* is even rather unlike most mammals. To avoid problems with fluctuating nutrient supplies, the dwarf goat can vary its metabolic rate. They grow slowly, speeding up only when times are good, exactly as ectotherms, or 'cold-blooded' creatures do. On Menorca, the role of medium-sized herbivore is filled by a giant rabbit, *Nuralagus*, wombat-like in its hapless, hopeless, rollicking gait and tumbleweed figure.[20]

An ecstasy of loose feathers, and the songbird outside the cave is gone, clasped in the long snout of a pale predator. Round-bottomed, bare-tailed, and with an outsized, whisker-bristled head, it has ambushed the warbler at the peak of its song. The skin of its jaws folds around the limp corpse, and the eerie hunter darts off. Gargano's lack of cat-like animals has left the way open to a unique small mammalian predator – the so-called Terrible Moon-Rat, or

*Deinogalerix*. Moon-rats, or gymnures, are a group that in the modern day is restricted to Asia. They are awake from dusk to dawn and, among other living mammals, their closest relatives are hedgehogs, though none of them is spiny. For the most part, they are similarly sized to hedgehogs, and have the same diet of slugs, worms, insects and other invertebrates. Unlike hedgehogs, all gymnures produce a strong ammoniac smell reminiscent of rotten garlic, the better to establish territories and to deter enemies when frightened. This is no hindrance when your prey are invertebrates and birds, often with poor senses of smell. On the island of Gargano, the two species of *Deinogalerix* are the closest mammals get to being top predators, feeding on smaller mammals and birds as well as invertebrates.[21]

In the west, the dam has broken. Pliny the Elder reported the Roman legend that the Straits of Gibraltar were a channel carved from the rock by the sword of Hercules. At the dusk of the Miocene, that channel is being carved, hundreds of metres deep and hundreds of kilometres long, but it is being carved by the ocean. Two plates, gripping one another for years, have built up so much tectonic tension that they have slipped past one another in parallel. This strike-slip has jolted down the level of the wide, flat isthmus at Gibraltar, and opened a sluice, 9 miles wide, to the full volume of the Atlantic. Water flows at 40 miles per hour down, down the natural weir into the western Mediterranean. Once the dam has broken, there is no going back, as the water erodes a path ever deeper. But the Mediterranean basin is not evenly deep throughout, and natural barriers block the water from filling the sea evenly like a bath. The high ground on which Malta and Sicily sit, as well as the ranging peaks of the Apennines, prevents any water from getting into the eastern Mediterranean for the time being. On Mallorca, the dwarf goat ceases browsing on its meal of toxic box to look out over the tumultuous mist cloud below. The arch-backed giant hares of Menorca startle to the noise. The rate of flow slows as the sea refills, carving out channels in the new seabed, refreshing the desiccated evaporite deposits of the floor. The major islands, one by one, begin to take their modern shape. Plants and bacteria that tolerate the cliff-faces and valley floor are drowned. The Mediterranean, though, must clear one last hurdle before it can entirely isolate Cyprus and fill the Aegean and Adriatic.[22]

South from Gargano, along the eastern Italian ridge of the Apennines, the weather begins to change. As the Tyrrhenian Sea fills, the dry sky sucks its moisture, forming rain-heavy clouds. Despite the turn in the weather, the deep chasm to the south and east remains undisturbed. Over the saddle that separates the Italian mountains from the Sicilian massif, there are dark lakes on the flatland to the north, and in the distant west, the shimmer of a coast. The Mediterranean to the west is almost full, but the east is as dry as it ever was.

Four months after the Straits of Gibraltar first opened, this begins to change, as to the south a standing plume of mist, hundreds of metres tall rises from the eastern edge of Sicily, visible from many kilometres away. The roar comes further south still, near the modern-day site of Siracusa. The Maltese-Sicilian sill is a vast natural dam, a barrier between the two deepest basins in the Mediterranean Sea. Across its wide expanse are now scattered sea-lakes. As seawater begins to spill over the dam, the eastern basin will be filled by the greatest waterfall ever to have graced the Earth. It is 1,500 metres – nearly a mile – high, one and a half times the height of the modern-day Angel Falls in Venezuela. The water pours over the escarpment at speeds of 100 miles per hour, and much of it turns to mist before it even reaches the ground. Unlike the Straits of Gibraltar, where the descent into the western Mediterranean basin is gradual, weir-like, this is a true, sheer drop, where the force of an entire ocean is channelled into a single, 5-kilometre-wide site. Even with this constant deluge raising the eastern Mediterranean by a metre every two and a half hours, it will take over a year before the eastern Mediterranean is filled, until Malta, Gozo and Sicily are finally cut off from Africa and Italy, and Gargano becomes an island once more.[23]

The return of the sea has formed new islands, which in time will attract new colonizers and evolve into more mis-sized communities. Well into the Pleistocene, there will be isolated Mediterranean islands with unusually sized organisms. In their own sweepstake fashion, hippopotamuses will reach Malta, Sicily and Crete over the water, and become dwarfed to tiny forms. In many islands, dwarf elephants will roam. With a single, large nasal opening to support the trunk, and eye sockets not entirely surrounded by bone, their skulls will provide a mystery to early civilizations, who will imagine giant, one-eyed cyclops living in the caves of the Mediterranean. Towering above these dwarf elephants, 2 metres from beak to tail, will be the Sicilian giant swan *Cygnus falconeri*.[24]

In the modern day, the Mediterranean is still an almost enclosed sea, dependent on the constant attention of the Atlantic for refilling. If the Straits close again for the duration of a millennium, the Mediterranean will once again dry up. Strangely, a century ago, the idea was mooted as a deliberate engineering project – Atlantropa. The aim was to build dams across the Straits of Gibraltar, at Sicily, and at the Bosporus, lowering the Mediterranean by 200 metres and using the resultant hydroelectricity to power the entirety of Europe. The project was one steeped in colonialist aims, and which entirely failed to take into account the damage this would cause to fragile Mediterranean ecosystems. As Africa continues to push northwards into Europe, it is highly likely that the Straits' full closure will happen naturally over the next few million years. The entire area of North Africa, southern Europe and the Middle East is a relative low ground, bounded by mountains that prevent their rivers from reaching the oceans. This makes it the site of numerous seas, termed 'endorheic', into which water flows but only leaves through evaporation. As well as the Mediterranean, these include perhaps the most famously endorheic body of water – the Dead Sea. There, the River Jordan spills its water into a desert valley, from where it rises into the air, leaving salts and minerals behind in famously dense waters. A better modern analogue for the Mediterranean at the end of the Miocene is the former Aral Sea, one of the last remnants, along with the Black and Caspian Seas, of the ancient Paratethys, a sea that once overflowed most of Europe. Once fed by the Amu Darya and Syr Darya, and with no outlet, it has slowly dried up as the waters from those rivers were diverted for agriculture. Split into two by the disappearing water, the South Aral no longer has any surface water flowing in, and is a stagnant, dwindling pool of groundwater. The ecological communities of the South Aral have collapsed, as have the communities of humans that depended on them. Once diverse fisheries are now empty, replaced by intolerable waters and toxic, wind-blown deserts of salt.[25]

The refilling of the Mediterranean, called the Zanclean flood, 5.33 million years ago, marked the end of the Miocene and the start of a new epoch, the Pliocene. The contours of Gargano had ensured the survival of the community throughout the drought, an isolated haven above inhospitable plains, but its separation was also its downfall. After the Mediterranean refilled, the Apulian plate kept pushing northwards, and the changing height of the land through tectonic movement meant that, by the middle of the Pliocene, Gargano found itself below the waves, its unique creatures wiped out. Tectonic movement has continued to raise and lower the land in fits and jolts. When Gargano was later lifted up once again, it joined the Italian mainland, and the creatures of mainland Europe moved in.[26]

The loss of the dwarves and giants on Mediterranean islands is a common tale throughout the 5 million years between the Zanclean flood and the modern day. The Sardinian pika – the last of the larger Mediterranean species of *Prolagus* – was almost wiped out through competition and predation from invasive species introduced by the Romans, only holding on in isolated communities until becoming extinct perhaps within the last 200 years. Sardinia's own dwarf deer, a relative of Ireland's giant elk, was eradicated within a hundred years of human colonization, some 9,000 years ago. The youngest known dwarf goat died about 4,000 years ago, only 150 years before the first evidence of humans on the islands. No dwarf hippopotamuses or elephants younger than the Pleistocene have been found, and invasive animals, brought to the islands through swimming or, more frequently in association with humans, have robbed the islands of the Mediterranean of much of their diverse endemic fauna. Yet, whenever species arrive on isolated landmasses, the island rule of dwarfism and gigantism still applies. The Corsican deer is an endangered subspecies of red deer introduced only 8,000 years ago, yet it is half the height of typical mainland red deer. The St Kilda field mice that live on that isolated Hebridean island, introduced by hitchhiking on Viking longboats only about a thousand years ago, are already much heavier than those on the mainland.[27]

In future, as the ice at the poles melts, as the sea rises, and Gargano is, perhaps, once again cut off from the rest of Italy, the exiles from the mainland will, once more, turn these ancient limestone crags into a land of dwarves and giants.

# 4. *Homeland*

*Tinguiririca, Chile*
*Oligocene – 32 million years ago*

'I have had a dream – and the dream I had was deeply disturbing. In the
mountain gorges, the mountain fell down on me, wet, like flies'

– Sîn-lēqi-unninni, *Epic of Gilgamesh* (tr. Maureen Kovacs)

'You can't cross the sea merely by standing and staring at the water'

– Rabindranath Tagore, *The King of the Dark Chamber*

Across the dusty vista, ripples dance through the grass, stalks bowed
as if by an unseen, brushing hand. A cool wind blows through the
world, with the promise of new horizons. For life on land, true horizons have been hard to come by until very recently. One family of
plants has changed all that. In the Oligocene of South America, the
first grasslands on the planet have recently appeared. Although
grasses have been around in South America, Africa and India since
about 70 million years before the present, they were minor parts of
tree-dominated landscapes, relatively unimportant aspects of a
tropical and jungly flora, restricted to the southern parts of the
world. As Antarctica has finally separated from its neighbouring
continents, the routes of ocean currents have changed, formerly
strong winds have fallen and others risen where none before existed.
Throughout its history, the world has flipped between two stable
states, an 'icehouse', when there is permanent ice at the poles, and
a 'greenhouse', where that ice is absent. The world of the modern
day is an icehouse, and that shift towards cold began in the Oligocene. Although this is a global pattern, South America in particular
has become colder and more arid. Grasses, with traits that make

57

*Santiagorothia chilensis*

them already well-adapted to the new climate, have come into their own, and in these low-altitude, semi-arid floodplains on the foothills of the nascent Andes, they make up a major part of the view for the first time.[1]

The oceanic crust at the bottom of the Pacific has been moving eastwards and subducting, slipping under South America, causing the continent to buckle into new peaks. The Andes began to climb in the Cretaceous. As the coastal lowlands of western South America folded, the rock strata tilted and bent as if it were cardboard. In the modern day, by when this place, Tinguiririca, will have grown into a huge volcano, the beaches of the Cretaceous, risen high into the Altiplano, will have twisted and turned by a full 90 degrees, so that their petrified layers of settled sand now plunge directly into the earth. At the modern-day site of Cal Orcko in nearby Bolivia is a rock wall where footprints, laid by dinosaurs in a Cretaceous river, appear to climb gecko-like up a sheer vertical cliff. But all this is yet to come. The Andes are for now fairly small, not yet even 1,000 metres high, and as they grow, so too does the influence of grass, a way of life that will soon colonize the world. Where here once was forest is now only a scattering of copses, a few patches of open woodland, dwarfed by wide expanse and pure space, marked only by the undulating line where the sky meets the ground. Tinguiririca is not an easy analogue of a modern ecosystem. Grasses are common, but so are palms. Among the environments of the modern day, it most closely resembles a savannah, with sparse trees and wide spaces, but the creatures that live here divide up the ecological space differently, with three times as many leaf-eaters than in any other modern fauna, and with very few tree-climbing mammals.[2]

High peaks are rain generators. As air is forced up over a mountain range, it cools and condenses, compelling the water it holds to drop out as rain on the windward side. By the time the air has passed over the ridge of those peaks, it has become dry, and casts a rain shadow on the leeward side, where very little falls from the sky. In the modern day, the Andes cast a very strong rain shadow, resulting in the extreme aridity of the Atacama Desert. Though the

Oligocene Andes are half the height of the modern range, the lee-ward sides of even much smaller mountains can have half the rainfall of the windward side. Combined with a naturally occurring pressure high over the early central Andes, this means that rainfall in Oligocene Tinguiririca is very seasonal. Through the dark-soiled, peak-rimmed plain of Tinguiririca, a winding river passes but once a year, descending from gullies and cascading stream beds higher in the volcanic uplands. Now, the riverbed is dry, the ashen mud mosaic-cracked, natural tiles plastered across a flat, 20-metre-wide channel. The heat of the dry season has cooked the little squares of mud and curled their corners; once the river returns a terracotta flotilla will rise on the waters, and the little earthen boats will rush, helter-skelter, towards the Atlantic. Stalks pierce the mud in older channels abandoned by the meandering flow; from above, the for-mer curves of the riverbank are traced by rows of riverine plants, a lagged series of historic images, a flick book of the past routes that water has taken through this floodplain, year after year.[3]

Along the edge of the current riverbank, where an underground reserve of water keeps the earth moister, fluffy-headed grasses rise amid rushes and amaranths, growing in places into spiny gallery forests of palms and mesquite. Away from the channel, crackling shrubs mingle with the grasses, with some hardy succulents green-ing rocks across the scrubby, brown ground. It is here and now, in the Chilean Andes of the Oligocene, that the earliest cactuses are believed to be diverging from their purslane sisters, a family born of this ever-drying world. What grasses remain are brittle and short, but they are just about surviving, helped by a recent drifting of mineral-rich volcanic ash. They will not have to wait long for the rain to come. The wet season is just beginning – the sky to the north is dark, and a striated sheen of heavy rainfall obscures the peaks to the north, the abating smoke covered by the relief of fresh water. The air has cooled with the arrival of the clouds, and the mountain lakes are already refilling.[4]

For now, the grassland is occupied by herds of herbivores, a pano-ramic scene reminiscent of the most diverse parts of the Serengeti, gathering under the trees for shade as they await the rising of the river. These, though, are not the loping masses of zebra, of wildebeest, of rhinoceroses, giraffes or hippopotamuses, but smaller, daintier creatures. South America is an island continent, and its animals are unique. In the earliest grasslands, they are particularly distinct. Scratching through the dense and dry stems, a cluster of dun, fox-sized herbivores, long-faced and long-tailed, are grazing together. They have surrounded a larger, solitary, shaggy-coated beast at the edge of the narrow strip of forest. If the grasses were in full growth with the onset of the wet season, the shorter grazers would disappear within them completely, but as a massed herd, they keep swathes of the landscape cropped close, maintaining the natural lawn throughout the year. In semi-arid grasslands such as this, grazing favours plants that can grow back quickly, and weeds out those that are eaten faster than they can return. Abandoned by herbivores entirely, the forests would spread further from the river, but the constant pruning of young plants means that few saplings can survive for long, with only pockets of trees in the scrubby terrain. Grasses are rapid exploiters of fallen rain, able to get to it faster than trees. This means that grasslands tend to occupy middling climates; where rain is high or evaporation low, there is enough water for all, so forests still abound; where rain is low or evaporation high, the environment becomes a desert shrubland, an outback. In Tinguiririca, a full forest would die of thirst long before the rains returned, but grasses persist, and when the rains come, they are intense enough to green the valleys and plains and produce, almost instantaneously, a carnival of flowers. Variability in rainfall, times of plenty and times of drought, are the hallmark of most grass-dominated worlds. The new patterns of wind and water in the atmosphere have created a perfect cradle for meadows here in South America, in the aftermath of the warmer, more humid, and lusher Eocene.[5]

Slowly, in fits and starts, the herd drifts among the tussocks, no individual moving for long, but, through a combined and continual search for new shoots, a communal movement. Not so the shaggy beast. It turns, and falls back contentedly onto its hind legs, sitting, cross-legged and bear-like, in the sun. Its arms are long and muscular, and end in contorted hands with long, curved claws. It tugs an unfortunate sapling towards it with its hands, and chews on it as if thoughtful. *Pseudoglyptodon* is a sloth, but one rather unlike its modern-day cousins. The two genera of sloth that exist in the modern day are remnants of a once great and diverse order, the majority of which dwelt on the ground. The tree-hanging sloths, the so-called two-toed and three-toed sloths, are not closely related within sloths, but have separately adapted themselves to a lifestyle in the canopy. An arboreal sloth eats only leaves, and so spends 90 per cent of the time either feeding or resting and digesting. Life is passive, avoiding expending too much energy in movement or, indeed, in holding on – the curved claws are perfect for passively hanging under branches, or holding onto trees while sitting in them. Some ground sloths use those same claws for digging, as well as foraging and defence. By the Miocene, the epoch from 23 to 5 million years before the present, sloths will reach their zenith, with some ground sloths even slowly adapting to a marine lifestyle off the coast of Peru, using high nostrils, dense bones and a beaver-like tail to live somewhat like hippopotamuses, walking along the sea floor to find seaweed.[6]

Sloths are one of South America's oddities – alongside armadillos and anteaters, they form an endemic group of mammals called xenarthrans or 'strange joints' in allusion to the distinctively complex joints within their backbone. Besides these, there is the multispecies herd of herbivores that surrounds *Pseudoglyptodon*, all of them part of an order of mammals grouped loosely by their enigmatic nature – the South American Native Ungulates, or SANUs. Referring to them by an acronym makes them sound like some kind of quango, but this reflects our uncertainty as to exactly what they are. As unified as any committee, the evidence of how they relate not only to mammals in the rest of the world but also to one another is limited – they aren't clearly closely related, but are frequently occupying similar niches.[7]

The animals grazing around this *Pseudoglyptodon*, for example, are very similar in many ways to the hyraxes of Africa and the Middle East, but clearly different in so many others. True hyraxes are squat, square-jawed beasts, like a rugged, short-eared rabbit, with eyebrow marks that give them a perpetually cynical expression. These South American animals, dubbed *Pseudhyrax* – 'false hyrax' – have the same square jaws, but are longer of limb, more graceful, with something of a deer in their faces. A few *Santiagorothia* are among the mixed herd, too; they are lithe, hare-like creatures, with long bodies and limbs. They eat the low vegetation cautiously and with eyes alert, constantly on the lookout for borhyaenids: pouched predators and relatives of marsupials with hyena-like crushing jaws and grooved, ever-growing canines.[8]

That *Santiagorothia* is hare-like and *Pseudhyrax* is hyrax-like is down to convergence, the parallel evolution of unrelated and isolated groups to the same general anatomy. Whenever a new environment like the one in Tinguiririca originates, there are only so many ways of making a living in that world, and so the same solution is frequently happened upon. One problem of the open plain is cover. Unlike in the forest, there are fewer places to hide, and speed becomes an asset. So it is that small-bodied animals like hares and *Santiagorothia* become flexible and long limbed, or that large-bodied animals reduce their digits into long, thin, hoofed limbs for efficient running. There are SANU equivalents for almost every group of hoofed mammal from the rest of the world – elephant- and hippopotamus-like astrapotheres and pyrotheres wallow in the remaining humid jungles of the north, while the long-limbed litopterns are developing into analogues of antelopes, horses and camels, just as antelopes, horses, and camels are independently evolving their modern forms elsewhere in the world.[9]

Such strikingly similar forms are usually produced across distant continents, when different lineages have no contact with one another. The distance means there is no competition in which one species is excluded, and convergent species can be very distantly related indeed. Armadillos from South America, for example, were

for a long time argued by many to be closely related to the pangolins of Africa and Asia, until it was confirmed that their armoured
bodies, big claws and reduced teeth are just adaptations to a similar
lifestyle. Today, we know that pangolins are more closely related to
dolphins, bats or humans than they are to armadillos. But no matter
how isolated a community is, there are always those that have
arrived from elsewhere. Among the native South Americans of Tinguiririca are recent immigrants, travellers from across the Atlantic
Ocean, half a world away.[10]

A prolonged climatic cooling has engulfed the world, and life is
adjusting. The extinction of species in one part of the world typically opens up an opportunity for other species to spread, their
ranges increasing along the path of least resistance to fill areas now
abandoned. In Europe, beavers, hamsters, hedgehogs and rhinoceroses are migrating in from Asia, wiping out native European groups
like omomyids, the nocturnal relatives of tarsiers and monkeys, and
several families of hoofed mammal. South America has no contact
with other landmasses across which migration can easily occur, and
so it has developed a unique fauna and flora, just as Australia has
today. The grasses themselves are a South American innovation in
plant biology. Still, its isolation is not complete, and there are newcomers arriving on the continent that have spread from far afield,
through an unlikely route. In Tinguiririca, we find the traces of
these new arrivals, African creatures walking among the grass of
America.[11]

The great rivers of Africa pour into the Atlantic, and in high
weather trees and other plants are washed from their eroding banks.
These trees are frequently filled with creatures; insects, birds and
mammals. Sometimes, whole banks of vegetation are washed away
in their entirety, or aquatic plants naturally aggregate, held together
in a natural raft, which is washed out to sea. To see a large island
raft pass downstream to the ocean on a rain-swollen flow is a natural marvel, a slow-motion drama. Trees still standing, held up by
the intertwining of roots that have knitted together the soil, an

undergrowth filled with creatures oblivious to their imminent voyage. Around it, smaller, unattached patches swarm like tugboats around a ferry. Against the backdrop of the undetached forests, the movement is an incessant and unyielding parallax. Only a rough patch of rapids or collision with the bank at a bend in the flow will halt the procession, and if none are encountered, the island raft will eventually emerge from an estuary into the open ocean, and will be carried from shore with the momentum of the river current. The odds of anything good then happening to the inhabitants of these floating islands are minute, but they are good enough that several of these rafts, blown by lucky winds, and each carrying a small population or a pregnant female, arrived in South America.[12]

By the time of Tinguiririca, one of the South American groups descended from African rafters has begun to diversify – monkeys. Every monkey in the Amazonian rainforest, from spider monkeys to howler monkeys, tamarins to marmosets, owes its existence to a few lucky survivors from their own presumably difficult and traumatic ocean voyage. The distance to cross from Africa to South America at the time was considerably lower, about two thirds of the width of the modern Atlantic, but this is still a huge distance when relying on rain and pooled water in leaves for a supply of drinking water. Even assuming continuous movement in exactly the right direction, the communities of rafting monkeys must have survived at sea for over six weeks. Having arrived, they had spread to the west coast well within the Eocene. Nor are monkeys the only mammals to have rafted across from Africa. Caviomorph rodents are diversifying in their new home too, and two species can be found in Tinguiririca. All native South American rodents, from capybaras to agoutis to guinea pigs, are descended from a population that crossed a thousand miles of ocean and survived, also arriving at least by the late Eocene.[13]

This route is surprisingly common; several utterly bizarre species with distributions across Africa and South America are too young to have simply been present in both when the continents themselves separated from about 140 million years before the present day.

Caecilians, burrowing amphibians hardly famous for their ability to survive for even short periods without fresh water, are among the transatlantic travellers. Oceanic dispersal has even been shown to have happened in freshwater fish; two closely related species of goby, understood to share a fairly recent common ancestor, are only known in Madagascar and Australia, respectively. To add an extra layer of mystery, they are blind cave-dwellers that cannot live anywhere else. In modern North America, too, seemingly uncrossable barriers have been traversed – the Devil's Hole pupfish, found only in a single cave in Nevada, is closely related to species in Death Valley and the Gulf of Mexico. Those lineages diverged only 25,000 years before the present, and there is no direct freshwater route between the eastern and western sites, so it has been argued that eggs might have been transferred by migrating waterbirds. Long-distance dispersal may be rare, but in a world where enough attempts occur, only one attempt needs to succeed. What is remarkable is how many appear to have been successful.[14]

Some of the spiny shrubs that scatter the grassland have gaps near their bases, little doors formed by the movements of small creatures. Inside, well-used tunnels descend into colonies of *Eoviscaccia*, descendants of Eocene rodent arrivals. With its soft fur, coiled tail and wire-whiskered face, it is specifically a relative of chinchillas and viscachas, living under the ground here in large family aggregations. Chinchillids like *Eoviscaccia* are not yet strongly adapted to the cooler climates found in either high altitudes or southerly latitudes. By the late Oligocene, they will reach Patagonia, and from the Miocene, the rise of the Andes will generate the high-altitude homes that modern viscachas are known for. Nonetheless, they are common enough in these parts, if elusive. More common are *Andemys*, a close relative of agoutis, but less specialized than those free-running, deer-like creatures. Compared with *Eoviscaccia*, they prefer a browsing diet, feeding among the softer tree leaves rather than the harsh grass. They have not been on this continent long, but the South American rodents are already ecologically diverse, impressive given their presumably tiny initial population.[15]

Later, of course, grasses will leave South America and settle around the world. Their characteristics make them exceptional dispersers. Their seeds are small and easily spread by wind and by hitching a ride on or within animals. They grow to reproductive age quickly, the seeds are starchy, holding plenty of energy for the developing embryo, and they are able to survive burning, freezing and near-continuous grazing. Grasses spread long distances easily, are hard to kill once established, and able to modify environments to their advantage, making them some of the most effective colonizers and most successful groups of species on the planet.[16]

When we hear radical stories of long-distance dispersal, it is all too easy to place a human mindset onto the events, and it is worth spending a moment to address this. There is a temptation to describe these rodents and monkeys as hopeful adventurers, with a narrative of pioneering spirit and survival against the odds in an unknown and inhospitable land, an inappropriate framing that owes much to the era of colonialism. Where an animal or plant from one part of the world appears in another, some might use the language of invasion, of a native ecosystem despoiled and rendered lesser by newcomers. Frequently, this is an appeal to nostalgia, to the landscape known in childhood, contrasted with the altered, often depleted world of today. It brings with it an implication that what *was* was right and what *is* is wrong.

What is important in conserving an ecosystem is conserving the functions, the connections between organisms that form a complete, interacting whole. In reality, species do move, and the notion of 'native' species is inevitably arbitrary, often tied into national identity. In Britain, 'native' plants and animals are categorized as those that have inhabited Britain since the last ice age. In the United States, however, 'native' plants and animals are those that have existed there only since before Columbus landed in the Caribbean. These plants and animals have legal protection over and above 'aliens', but there is no easy distinction between native and non-native ranges for species, and non-native plants are not necessarily damaging to native diversity. Dwarf nettles, for instance, are not considered a 'native' British plant, but they are near universally present and have been recorded in Britain well into the Pleistocene. The milk thistle *Lactuca serriola*, which grows wild across Eurasia and North Africa and is the ancestor of cultivated lettuce, is considered a native plant in Germany but is explicitly an 'ancient introduction' in Poland and the Czech Republic, and has been described as 'invasive' in the Netherlands.[17]

So it is that even neutral biological terminology, that of dispersal and migration, carries with it an uncomfortable ring of political language. Looking back through time, the folly of sharing metaphors between those who are against immigration of individual humans and those that seek to conserve an ecosystem is laid bare. There is no such thing as a fixed ideal for an environment, no reef onto which nostalgia can anchor. The human imposition of borders on the world inevitably changes our perception of what 'belongs' where, but to look into deep time is to see only an ever-changing list of inhabitants of one ecosystem or another. That is not to say that native species do not exist, only that the concept of native that we so easily tie to a sense of place also applies to time.

This has not stopped some current geographic entities from extending their identity into the past, and the interaction between national politics and palaeontology do have real effects. Argentinean palaeontologists of the early twentieth century went against the scientific consensus of the day to suggest, incorrectly, that humans originated in South America. Wrong this may have been, but it was part of an attempt to reject the northern-centric belief (also incorrect) among the palaeontologists of Europe and North America that the continents of the south were places where evolutionary progress lagged behind. Even today, our conception of evolution is dominated by the story in the global north, where a longer history of study and a greater concentration of rich institutions have produced a far more complete picture of the fossil record.[18]

Hominid fossils in particular have been used to influence national identities even into the twenty-first century, such as the early humans found in the Sierra de Atapuerca in Spain. Today, most of the states of the USA have an official state fossil, from Illinois' *Tullimonstrum*, of which more later, to Alaska's woolly mammoth. West Virginia has chosen Jefferson's ground sloth, *Megalonyx jeffersonii*, which, as a sloth, is part of an order native to South America, not North America. However, its status as an American fossil icon comes from it being used as a counter-example to an even earlier prevalent idea deliberately and inherently infused with racist assumptions; that the animals of the Americas as a whole, not just South America, were somehow degenerate with respect to those in Europe. What is native to an area and what is not is a function of the scale at which you choose to look, and tying long-extinct species or ecological concepts to present-day artifices like borders and flags is a game in which one must tread carefully.[19]

This is perhaps particularly true in the case of the transatlantic Oligocene dispersals because they include our close primate kin. It is all too easy to read human motives, however subconsciously, into past events, and we must avoid putting our own ahistorical spin on what was, although certainly dangerous and unlikely, a journey guided entirely by chance.

The wind is bringing rain down from the heights, a swirling demon of cloud darkening the skies. With the first drops, the sloth looks up, shuffles, and resumes feeding. The typothere herd in the open begins to drift towards a patch of trees by a dogleg in the riverbank to shelter. The smell of the relieved earth blooms as the combined percussive beat of the raindrops brings a sighing to the air. But under the breath of that sigh is another sound, a shushing sound, a sound of pouring water and pounding hooves, increasing in volume to a roar. From a prominent perch on a mesquite, a bird calls sharply and takes flight, followed by several others, and alertness spreads instantly through the mixed herd on the ground. Shrubs shiver as *Eoviscaccia*, taking no risks, disappear into the safety of their burrows.

Down the river echoes the whiplash sound of cracking wood, and then arrives a dancing wave, 3 metres high. Alertness turns to flight, as the sloth moans and rolls onto all fours, and the typotheres startle and scatter. The spate surges forward and, striking the curve by the tree, rises into the air in a black bulge over the bank, collapsing to the ground, followed by another, as if dense, damp, velvet cloths are being thrown over the grass. The rhythmic slopping of the mudslide, folding in on itself like boiling porridge, turns into pouring, the smooth force of the water spreading across the landscape and filling in the valleys at tens of metres per second. The delicate squares of riverbed clay are shattered into pieces, boulders bob as if weightless, and tree trunks are carried as if no heavier than twigs, catching, submerging or breaking everything in its path, carving new paths in the underlying soil, and turning the valley floor into a rough, grey turbulence.

Heavy rains in the volcanic peaks around Tinguiririca have caused a flash flood, and as the flood descends, it picks up fine ash to make a lahar, a slurry of volcanic concrete, eroding the banks of the river to become larger, faster, heavier, until it is a destructive force of terrific speed.* Where the sloth sat is now a 2-metre-deep bubbling flood; the typotheres and rodents are nowhere to be seen. Those trees growing closest to the riverbank have toppled as their supporting soil washes away. A few of the sturdier ones still stand above the water, shaking uncontrollably with the continued roar of the lahar. In the now heavy rain, the grass is gone, the landscape is nothing but flood; the only indication of the river's course is a standing wave as the flood pours over the dogleg, and a curving stripe of faster flow that marks the river channel.[20]

After an hour, the drama is at an end. It is hard to comprehend that so much mud could have come so far from the mountains. Across the grassland are splayed fingers of ooze. As the river emptied its contents onto the flatter land, the flow slowed and solidified. Now still, it is hard, bare earth, the once freely bounding boulders immovably set in place. None of the animals caught in the path of

---

* The lahars that followed the Mount St Helen eruption in 1980, for example, were measured at speeds of up to 100 kilometres per hour.

the lahar has survived; they too are forever set in stone. The slate has been wiped clean, coated in ash, sand and soil, ready for the grasses to recolonize the valley.

Yet, for all that the animals here will be preserved as fossils, none of the grasses of Tinguiririca will ever make it directly into the fossil record – no body fossils, no pollen, nothing at all. A lahar is too coarse, the preservation too scrappy to reveal the soft tissues of plants or the lacy wings of insects. Those creatures that could take flight are gone, escaping the oncoming spate. Only fragments of land-bound mammal bones and broken teeth will survive the ravages of the rock record. But despite their absence from the physical cast list of the Tinguirirican ecosystem, grasses have left their marks. Environments shape their inhabitants as much as their inhabitants shape them. Remove all life from a place but its herbivorous animals, and line them up in a row, smallest to largest. The distribution of body sizes can be turned into a graph called a cenogram, and the shape of that graph is remarkably accurate at predicting the relative openness and aridity of an environment. Tinguiririca, seen purely as mathematical space, still loudly proclaims its openness. Even with only one or two specimens, the presence of grass would be clear. Look deep into the mouths of the mammals of Tinguiririca, and you will see that they have been doing something new, something driven by the way in which grasses and herbivores interact with one another and the world around them.[21]

Plants are very keen to get certain parts of their bodies eaten. Fruit is sugary and garish so that it can be found and eaten and the seeds dispersed. Flowers are bright, strong-smelling, and contain nectar to attract pollinators; some plants will even listen for their pollinators and quickly produce extra sugar to boost their nectar's sweetness when the sound from a passing insect's wings gently vibrates their petals, a botanical equivalent to a market hawker's shout. Grass has no such desire to cooperate; grasses are wind-pollinated, and their seeds are wind- or water-dispersed. They produce no enticing flowers and only nutrient-poor fruit – grains. That humans have turned grasses, from wheat to rice, corn to rye, into a staple component of our diet is testament to hundreds of generations, tens of thousands of years of successive selective breeding, and even then, we normally require huge amounts of post-harvest processing to get anything pleasantly edible. The leaves, too, are not very nutritious, and, to deter would-be herbivores, grasses have an internal equivalent of spiked railings – sharp opaline crystals called phytoliths are distributed throughout their tissues, enough to feel gritty in the herbivore's mouth, and to distinctively scratch and slowly wear down the enamel. All in all, for an animal to be a grazer is to subject itself to a hard diet of poor nutrition and constant abrasion, slowly eroding its teeth.[22]

We needn't even look down a microscope to see the effect that grasslands are having on the anatomy of the animals here. A lifetime of biting and chewing even soft food can produce plenty of wear and tear on teeth. With grasses, this is hugely increased, but natural selection, in its pig-headed approximation of wisdom, has responded not by giving up. The resource is there to be exploited and so there is some benefit for some creature, however hard the work is, to obtain it. Grazers are developing teeth that continue to grow, no matter how worn down they become. Such teeth, with high, flat crowns, plenty of hard enamel and cementum, and limited or no roots, are called 'hypselodont'. At its extreme, there are creatures whose teeth grow continuously, even as their gums recede with a lifetime of grit and grass, a strategy only known in woolly rhinos and notoungulates among larger animals. In the early Oligocene, when grasses are yet to spread widely in North America, horses are housecat-sized browsers, feeding on the leaves of the struggling broadleaf forests. As the plains and prairies open up in this transition to an icehouse world, they will adapt to become a creature of the open space, a long-limbed runner, a high-toothed grazer, a herding herbivore. From teeth to limbs, many creatures in open, grassy habitats have independently arrived at very similar solutions in both locomotion and diet. The driving forces are complex and mixed – the openness, body size, and the hardness of the ground can all influence the development of this morphology.

Antelopes, American pronghorns, deer and some SANUs – all will converge on a new way of living; the archetypical grazer.[23]

The unique mammals of South America will, in time, fall. When North America and South America join together 2.8 million years before the present, as the Isthmus of Panama rises out of the Caribbean to separate the Atlantic and Pacific, northerners will move south, and southerners will move north. This event of mass migration in both directions, beginning about 20 million years before the present and continuing until the isthmus is fully closed by 3.5 million years before the present, is known as the Great American Biotic Interchange. In truth, the movement will favour the northern species, although the reasons for this are not well understood. Of native South American animals, only the North American porcupine and Virginia opossum have truly thrived across North America, with armadillos also found across the southern, desert regions.[24]

Even in South America, the northern species will win out. All the great pouched carnivores will disappear, all the SANUs. In the modern day, all that remains of the native South American mammal diversity are 101 known species of opossum, six species of sloth, four of anteater and twenty-one of armadillo. The giant ground sloths will last tantalizingly long – surviving until perhaps only 4,000 years before the present in the Caribbean. Only 8,000 years ago in Brazil and Argentina, ground sloths were the largest tunnelling animals the Earth has ever seen, excavating vast networks of burrows in which whole families lived, burrows which still survive today. *Toxodon*, a notoungulate something like a hornless rhinoceros, and certain litopterns like the camel-like *Macrauchenia* also lived on until within 15,000 years before the present. The extinction of the Caribbean ground sloths and the last South American Native Ungulates happened, perhaps not coincidentally, about the same time that humans reached their habitats.[25]

Where their ecological and anatomical convergence to forms from elsewhere in the world has hampered our ability to place them in a mammalian family tree, these late survivors provide an opportunity for scientists to identify what they are. Preserved in the dry and the cold of Patagonia, some of the latest-surviving SANU fossil remains still contain molecules of connective tissues – strings of collagen that, like DNA sequences, can be used to find out how species are related. By comparing these sequences of amino acids, the identities of these uniquely South American forms have been laid bare, and we now know that they are most closely related to perissodactyls – the group of hoofed animals that today comprises horses, rhinoceroses and tapirs.[26]

As with any answer, that relationship produces a bigger question. Many lineages from across the Paleocene and Eocene have been claimed as relatives of perissodactyls, from places as far apart as North America, Europe and India. Given that the earliest perissodactyls lived in Asia, what does this say about global migration? In the separated world of this part of Earth history, where the continents were about as distant from one another as they ever were, how did the ancestors of all these families disperse so rapidly, so globally? Or, perhaps, are these all merely early convergences, similarities drawn between the same solution to an identical problem?[27]

Journeys themselves cannot be fossilized, but the destinations of those who made the trip are revealed by where their descendants ended up. Whether island-hopping or ocean-rafting, and whichever route is taken, throughout the history of the Earth, life has travelled, dispersed, and thrived in new surroundings. What has begun in Tinguiririca will soon spread worldwide, as the grasses of the South diversify to create the largest expanses of life on the planet, from the great plains of North America, to the Eurasian steppes, to the African savannahs. From bamboo forests to chalk meadows, the age of grasses has begun.

# 5. Cycles

*Seymour Island, Antarctica*
Eocene – *41 million years ago*

'Eppur si muove' / 'And yet it moves'

– Galileo Galilei

'They went darkling through the dusk beneath the solitary night'

– Virgil, *Aeneid* (tr. David West)

The beach is filled with the shouts of seabirds, the older ones insistently calling for their mates, the young pretenders eyeing up possible nesting sites. Littered with the unicorn horns of turritellid sea snails, spiralled *Polynices* gastropods, and the smooth hooded platters of *Cucullaea* clams, the shingle has been turned into an exceptionally crowded breeding ground. Painting stones white with guano, excrement infuses everything with an acrid and ammoniac smell, the phosphates seeping into the sand and changing the very chemistry of the rock that this will become. Pebble nests have been constructed in every nook, the smaller birds preferring to nest within crevices or sheltered by vegetation, the larger birds by necessity out in the open. A large sheltered inlet on the leeward side of a long, thin peninsula, near where a stream cuts a torn-paper precipice in the sandy bank down to the river estuary, this is an ideal location for raising chicks. Around the beach, the slopes are steep and densely forested; a hanging wood of scale-barked southern beeches, *Nothofagus*, pours down the hillside. Interspersed among them are tight-packed conifers – monkey-puzzles, cypresses, celery pines – and all are garbed in epiphytes, those plants that grow entirely on the surfaces of others. Vines and lianas, ferns and hair-like mosses,

*Anthropornis nordenskjoeldi*

set off by the complex, show-off inflorescences of the proteas, form a cloudy green palette. The humidity in the ocean westerlies has turned to rain on hitting the narrow strip of land thrust out into the Southern Ocean. This is a temperate coastal rainforest, where every surface is a collage of greens. Even halfway up a tree, a plant can grow roots out into the air, collect its own compost from falling leaves, and still suck up enough moisture to make a living. The fallen and rotting branches on the forest floor attest to the maturity of this woodland, an ancient, undisturbed place, of plants climbing over one another to reach the low-hanging sun.[1]

The continents may dance around one another, the global climate may warm and cool, but there are unalterable astronomical constants that define the physical world in which organisms must live. Sunlight comes from so far away that, to a near approximation, sunbeams all hit the Earth from the same direction and with the same energy. Where it lands, however, makes a big difference to the intensity that is felt at the surface. If the ground is directly facing the sun, the heat is concentrated on a smaller area, warming the environment more. If Earth falls away at an angle from the sun, such that a viewer on Earth would see the sun low in the sky, the sunbeams are spread over a wider area, and it is colder. This is, crudely, why it is colder both at dusk and dawn than midday, and colder at higher latitudes – those parts of the world further from the equator.[2]

By itself, however, that does not explain the existence of seasons. The annual rhythm that life settles into anywhere on the planet is a particular consequence of Earth's early history. Careless collisions in the crowded solar system knocked the north–south axis off kilter. Without a lean, our orbit would be uniform, each day unchanging, our progress around the sun unmarked. With the tilting of the Earth, the year has had a meaning. For six months in turn, each pole faces or is shaded from the sun, with an endless summer day and continual winter night. The tilted waltz of the Earth defines the seasons, and the inhabitants of the high latitudes must either migrate to avoid the changing conditions or stay and cope with them. In the modern day, one continent of our icehouse planet sits rooted to the bottom of the world, frozen all year round, and on land almost nothing remains all winter. But in the Eocene, life is different at the poles, here on the northern peninsula of West Antarctica.[3]

At the onset of the Eocene, the world warmed at a rate that was almost unprecedented, caused by high carbon dioxide and methane concentrations. Although there is uncertainty, it is thought that the carbon dioxide levels were up to about 800 parts per million: more than twice that of the modern day and four times that of the nineteenth century. Already a warm part of Earth's history, the Paleocene-Eocene boundary is marked by what is called the Thermal Maximum, a peak in both temperature and carbon dioxide levels. A huge influx of carbon dioxide and methane – about 1.5 gigatons over a period of 1,000 years – is the biggest that the world has ever seen, and will only be exceeded in rate by our own post-industrial era. Temperatures rose by at least 5°C. Exactly where this carbon has come from is unclear, it's been so sudden, but lying in the rock record are hints that in the deep sea, solid crystals of methane – a more potent greenhouse gas even than carbon dioxide – dissolved in oceans that were beginning to warm after a bout of intense volcanic eruption in Greenland. The warming of the seas intensified the dissolution, in a vicious cycle of warming begetting warming.[4]

Around the world, ecosystems responded. Mammals throughout the northern hemisphere became smaller; the amount of heat produced by a hot-blooded creature scales with mass, but the amount of heat lost scales with surface area. Smaller animals have a high surface area relative to their weight, so in an excessively hot environment, they are less likely to overheat. In the seas and on land, everything from minuscule plankton to giant herbivorous mammals either went extinct or rapidly evolved into new forms. The Eocene, the 'dawn of the recent' is, in many ways, the time in which the modern world was born, the basic structure of global biology moulded in the heat of the greenhouse world. By now, in Seymour

Island, the highs of the Thermal Maximum have receded, but average global temperatures are still far higher than in the modern day. The equatorial regions are not much warmer than today, with average land temperatures in island India very similar to those in modern-day hot-humid ecosystems. At high latitudes, though, the story is very different. This is not our icehouse world; the poles of the planet are not whitened by snow. Water is not locked up in mountain glaciers or endless sea ice, so the sea levels are 100 metres higher than in the present, and the climate on all continents is, from a human perspective, rather hospitable.[5]

Even Antarctica, the forgotten continent, the implicit exception when a modern-day species is described as having a global distribution, is warm, with summer temperatures reaching 25°C. The seas are a balmy 12°C. The entire continent is covered with a lush closed-canopy forest and filled with the shrieks of birds and rustling undergrowth. But the Earth keeps spinning, and the physical laws that define the relationship between living beings and the soils and seas in which they live still exist. Antarctica remains rooted to the southern extreme of the world, locked into a cycle of endless summer days and eternal winter nights. The same rules of sunlight, the same rules governing the flow of air and water around the planet are in force, and dominate the ecology of this polar rainforest.[6]

The beach is difficult to access from the steeply forested slopes, so it is isolated from predators. Not only are the seabird nests sheltered by geography, they have the protection afforded by numbers. This colony is one of the larger ones in the region, housing up to 100,000 birds. And these seabirds are iconic. Even if the weather seems unusually warm, there is no mistaking this for anything other than Antarctica. No bird better evokes an entire continent than penguins do Antarctica. With New Zealand, this is their ancestral home, the penguins from Seymour Island are among the first to make their mark in the fossil record. The colonies cover the beach, in a strip over 400 metres long. From up on the sandbanks, they merge into a scrum of black, yellow and white, a buzzing and chuckling mass where individuals shimmer into anonymity.[7]

Up close, the scale of these birds is even more shocking. The smallest of the birds, the dolphin penguin *Delphinornis*, was about the same size as modern king penguins, but it is entirely overshadowed by most species here. All these birds are members of the family of giant penguins, much larger than their more diminutive modern-day cousins. Some, such as Nordenskjöld's penguin *Anthropornis nordenskjoeldi*, stand at on average 165 centimetres, about the height of an average human. In this mixed breeding ground, they are generally largest, although there are a few big females of Klekowski's penguin that reach 2 metres in height, with a weight of nearly 120 kilograms – the proportions of a large rugby player. The spear-shaped beaks of these penguins are disproportionately long compared with modern penguins, and can be up to about 30 centimetres in length. Alongside these giants are seven other species of penguin, all larger than most modern penguins. For a single colony to exhibit such high species diversity, especially among those that are feeding in functionally the same way, is unusual. Ordinarily, species coexist only where their ecological niches are distinct enough that they are able to divide up the resources of the environment to avoid competition – so-called niche partitioning. Here, though, the bounty of the oceans is a big enough draw that, faced with the choice of living in a poorer site or competing for space in a crowded bird metropolis, the penguins have built a diverse society.[8]

They have already adapted to a marine life, with dense bones to overcome buoyancy, and a more waddling gait, although they still retain their inner toes, which later penguins will lose. Their wings are looser, more like a guillemot's, not yet the rigid flippers for flying underwater that later penguins will adopt, and their feathers are less densely packed, not yet adapted to the extreme cold. Those that are not milling on the shingle are floating in the bay, readying themselves to head out to the fishing grounds. There they will hunt herring, wrasse and hake, marine catfish, and the whetted snouts of knifejaws, swordfish and cutlassfish. Nautiluses, shelled relatives of octopuses, squid and cuttlefish, bob in the shallows, a fairly rare sight in high latitudes. Above all, the waters are filled with relatives of cod. Plankton is plentiful, and the fish use this excellent source of food as a marine nursery, a schoolyard for sprats.[9]

The peninsula lies in the Drake Passage, that region of raised sea-bed where the outstretched fingers of South America and Antarctica have only recently lost contact. It is a meeting of continents, a meeting of oceans, and a place where open-water creatures thrive in great numbers, a pelagic paradise. Here, cold water rises from the depths, sometimes bringing with it deep-dwelling oddities such as the big-eyed beryciforms: shiny slimeheads and fangteeth. The cold currents also bring nutrients and dissolved oxygen to fuel the community from the level of the seabed to the surface. They plough into the shallower water in the Passage, before turning, heading northwards, and circling round again, as they have here for some 20 million years.[10]

The presence of this upwelling is down to local geography, but the engine driving the conveyor belt is the sunlight falling on the equator, thousands of miles away. Equatorial air is heated fastest, and so it rises into the high atmosphere, pulling humid air from the tropics behind it. As the air rises, it cools and is pushed to the north or south by the still-rising air beneath it, before sinking down to create a cycling cell within the atmosphere that defines the tropical region. The movement of air at the edge of these cells pulls with it more poleward air, setting up two further rings of atmospheric cells that define the temperate and polar regions.

Add in the rotation of the Earth under these packets of air, and the result is the trade winds, the strong east-west winds that blow at sea level throughout the tropics. The same is true of the polar air at 60 degrees latitude, which is heated by the sun, rises and moves north and south. The air that moves polewards descends rapidly, and rushes east-west under the same Coriolis force. The air that moves towards the equator collides with the cool air moving away from the equator and is dragged down with it. While in the polar and equatorial regions the surface wind moves westwards, in these intermediate latitudes it moves eastwards. The Southern Ocean around Antarctica is therefore governed by westerlies, winds that blow from west to east.[11]

In the modern-day Southern Ocean, the constant westerly winds impart momentum to the surface water that is not blocked by any continent, causing a flow of water moving continually in an eastward circle as fast as friction allows – the circumpolar current. But the Earth's rotation affects the flow of water as much as the flow of air, and just as the spin of a roundabout sends riders outwards, the ocean water is forced towards the wide equator. The winds set up by the sun and the motion of the planet combine to drive water away from Antarctica, and the fertile waters of the deep well up, allowing life to bloom in the polar seas.

That bounty of fish brings in the predators, and penguins are not the only ones to exploit this cold sea. Darting among the penguins on the foreshore are little charadriids, members of the plover and lapwing family who feed on the insects that the massed animals attract, while further up the estuary are poised ibises, probing for molluscs and crustaceans in the tidal mudflats. Riding on the edge of the wind on long, narrow wings are the masters of the oceanic sky, small, tube-nosed albatrosses and petrels, and the huge, false-toothed odontopterygians. These birds, dwellers on clifftops further along the coast, all exploit the westerly winds that blow around the southern hemisphere to fly long distances without expending effort. Their most obvious feature is their white-rimmed wings, spanning in some cases over 5 metres, that are much longer than deep, and adapted for fast, wind-assisted flight, like that of a glider. Their size prevents them from taking off from the water, so they have become surf-runners, swooping low over the wave-tops to snatch fish from the surface at high speed, facing into the oncoming wind. Slippery fish and squid are tricky to keep hold of at the best of times, let alone on the wing in an Antarctic gale, and to that end, adult odontopterygians have heads dominated by jagged, bread-knife jaws. Their piercing eyes are set high in their small skulls, and their beaks are kingfisher-like, a long spear. From beneath the beak, bony spikes project, a crocodile smile approximation of teeth that grow directly from the bone, only appearing at adulthood. Like almost all seabirds, odontopterygians have a long lifespan, and raise only few offspring at a time. The toothless juveniles simply cannot catch their own food efficiently, and so they must be cared for for over a year, with each parent sitting in shifts while the other skims the surf.[12]

In the modern day, albatrosses voyage the oceans in a great circle, following the westerly winds by day, sleeping on the sea at night. The Eocene albatrosses and odontopterygians perhaps do the same, certainly disappearing out to sea for long stretches at a time. While flying, they use their scimitar wings to dynamically soar, barely flapping, constantly falling with the wind into slower, lower air, then turning and using their momentum to rise faster than before, a microcosm of the atmosphere.[13]

About the only thing that the birds have to fear on the water is the remarkable diversity of sharks. At least twenty-two species live here or regularly visit to feed on the booming fish population, dividing up the prey species and hunting grounds between them. Inshore, where the clear ocean shades into the shallow, milky tea of the estuary, the water suddenly foams as a spiky grey snout flashes back and forth, a waving, toothy plank, before quickly submerging again. A sawfish, *Pristis*. Sawfish, a type of shark with a snout resembling a horizontal chainsaw, are unusual visitors to Antarctica even in the warmth of the Eocene summer, usually restricting themselves to tropical and subtropical waters. The abundance of food at Seymour Island has proven too much of an attraction, and this sawfish has presumably followed the eastern coastal waters of South America to reach its destination. The saw acts as both locator and capturer of food, with thousands of sensitive ampullae along its length detecting changes in electric fields. Because vertebrates control muscle movement using flows of charged calcium ions, if a herring so much as twitches, the sawfish will know, swiping its saw through the water at high speed, hacking at the seabed with the edge, or pinning its prey down with the flat as it manoeuvres the fish towards its mouth.[14]

The floating penguins flap in concern as a burst of mist explodes from the water. The roiling body of a sea-serpent, a monstrously long beast, straight out of a sailor's tale. This is the 21-metre body of a basilosaurid, a member of the so-called 'emperor lizard' family. In a categorical error of judgement from early scientists, unforgivingly preserved by the strict rules of naming, this is in fact a whale. The first whales evolved far away in the island Indian subcontinent on the Tethyan shores only a few million years ago. *Pakicetus*, one of the earliest, was a long-legged amphibious predator and scavenger, a seal-wolf, dense-boned with high-set eyes for hiding in water, perhaps to ambush prey. Since then, they have fully committed to life in the water, and basilosaurids are the first to be unable to return to land.[15]

The animal blowing in the shallows has changed its body substantially to adapt to its new ecology, with flippered hands and a fluked tail. The nostrils have retracted to the top of the head, but the skull has not yet become the scooped-out, telescoped structure of modern whales. The support of water allows an animal to grow much larger without needing to worry about being crushed under its own weight, and, freed from the need to move on land, the hind limbs have been reduced to tiny external flippers, barely useful even for turning. Their inner ears are becoming more and more sensitive to low frequencies so they can hear better underwater. The cochleae are growing longer, more tightly coiled, and with thinner walls, all of which help in listening to the deeper notes that travel long distances through water. There is not yet, however, the fatty 'melon' structure – the bulging forehead that toothed whales, including dolphins, use to enhance their echolocating shouts. Basilosaurids can listen to the music of the oceans, but they have not yet learned to sing.[16]

Into the river and upstream, the channel narrows and deepens, snaking between the richly forested slopes. In this drowned valley, sunk by the sea rises of the Thermal Maximum, a lumbering, woolly creature strides down the bank, sending helmeted frogs sliding into the water. Horse-lipped, with a slight tapir-like trunk, barrel-bodied but with slender, five-toed legs, it wades into the estuary water like a brown bear after salmon, setting the gathered ibis to flight. With two protruding upper teeth and large hidden tusks, it nibbles on the soft sedges and rushes of the water's edge. This is the astrapothere – the 'lightning beast' – *Antarctodon*, a clue to the shared biological history between South America and Antarctica.[17]

Geographically, Antarctica is a crossroads, a connection between the several continents that once made up the supercontinent Gondwana: South America, Africa and Australia. Indo-Madagascar has defected, and India is now colliding with Asia, a tectonic bulldozer piling into the northern continent with mountain-building force. Nonetheless, Seymour Island is part of a chain of connectivity, the West Antarctic peninsula reaching an arm out towards very similar forests in Patagonia, only recently separated as the Weddell Sea flooded across the isthmus 10 million years before. Over the high East Antarctic mountains, Australia lies not far offshore. The animals and plants of Antarctica – the southern beeches, the penguins – are part of a broader Gondwanan flora and fauna. This biota forms a bioprovince extending across all the southern continents. *Antarctodon*'s lightning beast relatives are one of the native ungulate groups we met in Tinguiririca, and there are others in Seymour Island as well.[18]

Despite the piled debris, the centuries-old mattresses of conifer needles, the hovering epiphytes and fungus-infested logs piled catawampus among the living trees, the forest slopes are not entirely impenetrable. A gap in the trees marks the easiest route up the hillside, a well-worn trail made by generations of three-toed footprints. The camel-like *Notiolofos*, a litoptern, has broken ground through the forest. About the size of a small dromedary, it browses on low-hanging leaves in the southern beech forest. Living in an environment that fluctuates more within a year than over long timescales, the anatomy of *Notiolofos* has not changed in millions of years. At the anatomical scale, evolution has stood still, leaving *Notiolofos* a well-rounded generalist, able to cope with variations in the environment reasonably well while not being particularly specialized to any. This stasis in the face of chaos has been dubbed the 'plus ça change' model. In wild environments such as tidal zones or polar regions, versatility is a valuable trait. Constancy breeds specialism, but in evolutionary terms this is complacency. No environment stays the same for ever, and if your niche disappears, extinction follows.[19]

Deeper in the forest, a huge and recently fallen monkey-puzzle, which must have been 30 metres high in life, is now propped at an angle, held up by the density of vegetation. It is quickly rotting, with mushrooms sprouting along the sides. The invisible hyphae, the root and communications network of the fungus, have penetrated the bark and are forcing their way inside the dead wood, prising apart its cells one by one. Decay is rapid in this humid environment. In a cavity in the fallen giant is what looks like a child's football in green. The outside is constructed of large waterproof leaves, wrapped stylishly and tightly into a sphere. Through the entrance hole, the inside is lined with mosses and the young plants of spring, close-cropped and dried into warm hay, as soft and dry as a woollen slipper. There are no primates here, the tree as yet puzzles no monkeys. This nest belongs to a relative of what the Spanish-speaking people will come to call the monkey of the mountains, a tree-dwelling marsupial, the monito del monte.

Monitos are appealingly large-eyed, fluffy, nocturnal opossums about the size of mice, with grasping hands and a coiling, fattened tail, hairless on its underside, which, counterintuitively, serves the dual function of aiding climbing and storing fat for winter. They live like dormice, sleeping through the day and throughout the cold months. Seymour Island is home to two early species, but these include one that weighs about a kilogram, more than twenty times the size of the present-day species. Indeed, the presence of the monito del monte in South America in the first place may be thanks to Antarctica. Marsupials are split into two major groups, the Ameridelphia and the Australodelphia. Ameridelphian marsupials include the living opossums and several extinct groups, including the sabretoothed carnivorous thylacosmilids. As the name suggests, they are endemic to the Americas, particularly South America. The australidelphian marsupials include all the forms from Australia and the nearby landmasses; kangaroos and koalas, wombats and Tasmanian devils, numbats, sugar gliders, quokkas and quolls. They also include the monito del monte, which is in the present only found in the high-altitude Valdivian temperate rainforests of Chile and western Argentina. Indeed, modern monitos cannot live anywhere else, as they rely on a particular plant to survive. The quintral is a type of mistletoe, a parasitic plant whose seeds are spread by monitos, and is a key part of the *Nothofagus* forest ecosystem. The deep association of monitos with this ecosystem adds further fuel to the biogeographic puzzle. Did the ancestor of the monito del monte move back across from *Nothofagus* forests in Australia? Did the Australian lineage move across and then diversify? The other marsupials at Seymour Island are all part of the American marsupial grouping, throwing no light on this puzzle, the answers to which are hidden under the kilometres of Antarctic ice that will encase this rainforest.[20]

For now, though, somewhere within that rainforest lurk some elusive birds. The ratites – relatives of ostriches, emus, cassowaries and kiwis – are another classically southern group, with members in all the southern continents. The relationships among ratites are not dependent on continent, though. The two New Zealand ratites, the kiwi and the moa, are not close cousins. The kiwi's closest relative is instead the extinct elephant bird of Madagascar, linked by their adaptations to nocturnal foraging, with poor eyesight, an excellent sense of smell and whiskered heads, with feathers closer to shaggy hair than the hi-tech flight feathers of the odontopterygians. For something like these unusual Madagascan flightless birds, navigating the dark forest of the Eocene Antarctic winter would not be a challenge, but we don't really have any concrete evidence about the Seymour Island ratite except for a single ankle bone with a distinctively ratite appearance.[21]

Alongside these, a third group of giant, flightless birds lives in the forests along the riverside – phorusrhacids. Long-limbed but heavyset, and with shrivelled wings, over half of their skulls are made up of their beaks, deep, narrow and rectangular, with can-opener hooks on the end. Those at Seymour Island are known as brontor-nithines, 'thunderbirds', and are likely to be found scavenging, or lurking and ready to ambush along the litoptern paths that trace the island. Along with the gastornithines of Europe and the mihirungs of Miocene Australia, all of which are carnivorous relatives of waterfowl, these thunderbirds are the last hurrahs of large land-dwelling dinosaur predators. Later, in the Miocene, a lithe, 3-metre-tall phorusrhacid, *Kelenken*, named for a demon in Patagonian folklore, will have a skull 71 centimetres long, most of which is a deep, bladed beak the size and shape of a bush axe.[22]

Phorusrhacids have excellent eyesight, so as the Earth continues its path around the sun, and the seasons change, the darkness will not be an encumbrance. The alteration to the environment, though, is all-encompassing, as Seymour Island transforms from the mid-night feasts of summer to the dark season. The sun sinks lower and lower as it wheels around. Tomorrow, it will fail to rise, the start of a night that will last three months.[23]

Although the sun itself will never appear, the winter sky will still change daily. The day will be brightened with sky-bent light as the sun skirts beneath the horizon. In a cycle of twilight and night, the ordinary rhythms of life cease. When the changes in daylight become less distinct, the body clock, the internal circadian rhythm, cannot be maintained. In humans unused to the polar night, this causes stress, a sort of perpetual jetlag, where the body cannot match expectation with the external validation of reality. For some polar animals, the cycle simply stops, and life follows internal needs. Animals sleep when they are tired, and wake when refreshed. Others maintain their daily routine in the absence of days. Not all is constant; plankton still rise and fall in the ocean according to the phases of the moon, but for many the winter is a pause. The plants themselves stop respiring, slow down their metabolism. Conifers may keep their needles, but many plants, including *Nothofagus*, drop their leaves, and a forest holds its breath. In their branches, the monitos del monte simply hibernate to avoid the winter cold, snug in moss-ball nests. Larger animals cannot easily do this because of energy requirements, and so the *Antarctodon*, *Notiolofos* and ratites must venture out to find food.[24]

In the darkening wood, nocturnal animals come into their own, as do those adapted to the half-light, so-called crepuscular animals. The twilight of this final day draws out a whiskered head, superficially beaver-like, although much smaller, from a burrow among the root system of a cypress. Its large eyes show that it is well adapted to using what few photons descend from the polar night sky. Gondwanatheres like this are found from India to South America, and are one of the more ancient lineages of mammals, a holdover from the Mesozoic. Their front limbs are sprawled out to the side, but their hind limbs are held underneath, giving it a curiously combative, sumo-wrestler posture as it crawls hesitantly towards a *Nothofagus*, attracted by the sweet-smelling leaf fall. The gondwanathere is not alone in seeking out the southern beeches. Rather than set seed reliably every year, and risk constant predation, the trees normally don't produce any seeds at all. In a so-called 'mast' year, as it has been this past summer, each tree sets seed at once and in huge number, a coordinated dump of food for seed predators. With food usually in short supply, there are never big populations of those that eat *Nothofagus* seeds, and so the number of seeds released is far higher than can be consumed, ensuring that some seeds survive to become saplings. Exactly how this cunning behaviour is coordinated is unknown – do the trees communicate using hormonal signals, or are they all responding to some stimulus from the environment, some clue that it is time to mast? The gondwanathere, along with opossums, monitos and birds, is after the *Nothofagus* nuts that have not yet been eaten, a crop of little pursed cups containing seven or eight seeds still littering the ground. They are easy to locate, and it picks them into its mouth and chews with great gusto, jaw moving forwards and backwards in a gurning bite.[25]

Life evolves to fit the world in which it finds itself, but geography, of ocean currents, the position of the continents, wind patterns and atmospheric chemistry defines the parameters of that world.

Seymour Island is diverse because of the accumulated consequences of the physical state of the planet. That resource availability brings in animals and plants in great numbers, driving competition, adaptation, specialization and speciation. The climate also defines life's limits. The dark winters are still cold, and this excludes many species from living here. Unlike at Gargano, where the island rule benefited a medium size – giants of small creatures and dwarves of larger ones – in the polar regions the extremes are favoured. There are two ways to survive the cold. One is to hibernate, as the monitos and other small creatures do, to modify internal physiological processes such that a winter can be endured. The other is to increase in size, reducing surface area relative to volume, and keeping warm through bulk. A medium-sized animal cannot do either, and so, in the Eocene of Seymour Island, there is nothing between the sizes of rabbit and sheep.[26]

That pressure will only increase, for the years of plenty for Antarctica are almost at an end. In the high peaks of its centre, the mountaintop snows last all summer. For now, the cold will remain at high altitudes, but as the Earth cools once more into the Oligocene, the ice will descend, spreading across the entire continent and forcing out almost all plants and animals. It will begin gently, with glaciers flowing to the east of the West Antarctic Peninsula and calving short-lived icebergs into the Weddell Sea. As India collides with Asia, the Himalayas are now beginning to rise, causing freshly exposed rock to weather, react with carbon dioxide, and drawing it into the earth. As carbon dioxide levels decline, the ice grows. As it does, the white surface reflects more sunlight back into space, reducing the heat that the land absorbs and facilitating the growth of more ice. Patterns of air flow and rainfall will change, ocean currents will reorganize, temperatures will drop, and the Antarctic rainforest species, one by one, will find themselves living outwith their natural tolerances. Even generalized *Notiolofos* will find itself unable to survive. Every species has its limit.[27]

Exactly when and where the Antarctic biota died out on Antarctica itself is not known; we have only fragmentary records from the

Oligocene onwards. Unlike the early doomed human expeditions to the Antarctic interior, there is no diary, no record of the dates and places of species death. Those records that do exist are buried deep beneath the ice sheet, only very occasionally breaking the surface. By the inland Beardmore Glacier, *Nothofagus* shrublands will survive until the Pliocene, but of the lush Eocene vegetation, only a few hardy mosses, lichens and liverworts, a pearlwort and a hairgrass will make it all the way to the present. The ecological province that Eocene Antarctica epitomizes remains, in *Nothofagus* forests scattered around the southern fringes of Australasia, South America and Africa, but it has been thoroughly altered since the time of the polar rainforest. Of the animals, only the emperor penguins, thanks to their social huddling behaviour, exceptionally strong mate fidelity, and a suite of heat-retaining characteristics, will stubbornly hold on as the last permanent inhabitants of a land they and their kin have called home for tens of millions of years.[28]

In the early winter sky above Seymour Island, the sun sets for the last time for three months to come. On the edge of the Antarctic winds, the birds wheel through the night, navigating, perhaps, by the bright stars, or the magnetic field swirling from iron deep in the earth.[29] The sky seems to rotate around the South Pole, as the constellations steer through the sky, and the tilted Earth rolls through its seasons. Below, wakeful under the stars, thunderbird and lightning beast crackle over the newly frosted ground.

HELL CREEK

Laramidia

Appalachia

*Western Interior Seaway*

● Chicxulub Impact Site

# 6. Rebirth

*Hell Creek, Montana, USA*
Paleocene – *66 million years ago*

'As doors to the next world go, a bog ain't a bad choice'

– Ransom Riggs, *Miss Peregrine's Home for Peculiar Children*

'Trümmer von Sternen:
aus diesen Trümmern baute ich meine Welt'

'The debris of shattered stars:
From this debris I built my world'

– Friedrich Nietzsche, *Fragment of the Dionysos Dithyrambs*
(tr. John Halliday)

The world has ended. Two years ago, a piece of rock at least 10 kilometres long appeared high in the sky to the north, travelling southwards and westwards at thousands of metres per second. Almost immediately after lighting up the stratosphere, it collided with the shallow seas at Chicxulub, in the Yucatán of modern-day Mexico. The crust shattered and melted with the impact, and hot magma splashed high into the sky. In the cool air, the droplets of rock solidified, raining hot glass spherule bullets over half of North America over the course of three days. The accompanying pulse of heat burned forests, globally killing two thirds of tree species, down to the last specimen, and causing deforestation as far away as New Zealand. Seismic vibrations rang around the planet, and on the opposite side of the Earth from the impact side, ridges in the Indian Ocean cracked open. Shockwaves annihilated nearby ecosystems on land, while massive tsunamis churned the seabed. Rearing over

*Baioconodon* sp.

100 metres high, the waves crossed the gulf in under an hour, and flooded not just the coasts but far inland, destroying established communities throughout the Caribbean region. Across the shallow seaway that covers part of North America, a standing wave sloshed back and forth as if it were merely a bathtub. Beneath the meteorite-punched hole, 100 kilometres in diameter, oil long buried in the earth beneath the impact site was instantly incinerated. The resulting fires threw smoke and soot into the atmosphere that, spread by high-altitude winds, quickly concealed the Earth in a particulate blanket. In the months immediately following, rainfall declined to a sixth of what it had been. The sky darkened and, without light from the sun, plants and phytoplankton stopped producing energy. They have not yet started again. In some places, the temperature has dropped by at least 3 or 4 degrees, and globally, the average temperature on land has dropped below freezing. After two years of darkness, two years without photosynthesis anywhere worldwide, two years of rain infused with nitric and sulphuric acid entering the oceans, populations have failed. Warm-adapted species could not survive, and large-bodied herbivores and carnivores alike, deprived of a reliable supply of food, have starved. Decomposers have taken over, with fungi digesting the remains of dead and dying communities under the day-black sky. For three quarters of species on Earth, every male, every female, every adult and every child is dead. It is the winter that lasts a generation.[1]

The start of the Paleocene, an epoch born in fire, appears in the fossil record like a glitch in a CCTV recording, a few juddering frames of static after which the picture returns and everything has changed. A layer of iridium, a chemical element found in high concentrations in meteorites, is found all around the world, dusted into rocks laid down 66 million years before the present, an alien signature of the death blow. It has been estimated that only an eighth of the Earth's surface was sufficiently high in hydrocarbons to produce the sooty cloak that caused the extinction winter, but that bad luck changed everything. Centimetres below, in those immediately older layers, are the remains of the world of dinosaurs – plant-eaters like little frilled *Leptoceratops*, the dome-skulled *Pachycephalosaurus*, toothless *Ornithomimus* – alongside their *Tyrannosaurus* predators. Azhdarchid pterosaurs, the largest-ever living flyers, bigger and lighter than Orville and Wilbur Wright's early aeroplanes, glided overhead. Leviathan reptiles seethed in the nearby seas. Centimetres above the iridium layer, we find a motley collection of small- to medium-sized mammals, eating roots, tubers and insects. Alongside these are a few crocodiles, maybe a turtle. Up to three quarters of all plant and mammal species, and all dinosaurs except a few groups of bird, are gone, and there are new organisms in their place. The transition is so quick as to beggar understanding, and indeed it baffled early scientific efforts at comprehension. It took over a hundred years of geological research before the Paleocene was even recognized as having taken place at all. Eventually, it was inserted at the beginning of the Eocene as an in-between stage joining the world dominated by archosaurs – pterosaurs, dinosaurs and the relatives of crocodiles – to that of early horses, primates and carnivorans.[2]

H. G. Wells wrote in 1922: 'There is a veil here still, over even the outline of the history of life. When it lifts again, the Age of Reptiles is at an end. [ . . .] We find now a new scene, a new and hardier flora, and a new and hardier fauna in possession of the world.'[3] To meet these new and hardy plants and animals, and to understand how they inherited the Earth, we must go a little further away, in both time and space, from the destructive collision itself. After the end of the world, there is only one place we can go. We must cross the river. We must go to Hell.

Thirty thousand years after the asteroid struck, the air is full with the unmistakeable sensation – too intrusive to be a mere smell – of bracken and bog, at once damp and exhilarating. Below, the suppurating suck of the damp ground. As a tropical storm rolls through, the rain is not so much falling as infiltrating every aspect of the environment, a pervasive, osmotic wetness. On the distant hills to the west, an oily brushstroke of grey on green is devoid of life; the rain has caused yet another landslip, clearing the latest generation of infant trees. It is as if the hills have given up hope and are sinking slowly into the sea. This is the western edge of the Western Interior Seaway, a warm and shallow sea flooding the low-altitude centre of North America and splitting it into two smaller landmasses, Laramidia in the west, made up of the later Rocky Mountain region, and Appalachia in the east, comprising everything from Florida to Nova Scotia via Tennessee. The sea has been retreating for the last few million years, and Laramidia and Appalachia are beginning to connect at their northern extremes. Nevertheless, the North American continent remains mostly divided by this shallow and productive sea. In the flatlands of the eastern Laramidian coast used to be a gently meandering river system, bounded by tall forests and inhabited by enormous beasts, known in the modern day as Hell Creek. On the day the world went dark, as the glowing red glass rained down, a flood of intense heat burst from the south, a huge infrared wave. Those forests burned, and nearly four of every five large plant species here were wiped out for ever. Deep-knitted roots, the maintainers of the earth's integrity, could no longer hold on with their bodies above in cinders and ashes. With no trees to clasp the hills or drink deeply from their rivers, the storm now saturates the soil, and water, the eternal architect of the Earth, levels hills into plains. The dense basement rock does not permit the water to percolate down, so the water table has risen. Rivers become bogs, peat mires, ponded floodscapes. Around their edges, on patches of higher ground, the surviving tree species have begun to spread from their refugia, and now build sparse woods, interwoven with channels of dingy swampwater.[4]

It is as if life on Earth has been reset. Lichens, algae, mosses and especially ferns spread across the new landscape, recapitulating the early evolution of plants, the circumstances asking the world to choose again its inhabitants. After disasters, it is the opportunists that rise first, and among plants ferns are some of the greatest opportunists of all. Able to cling on in nutrient-poor soil, quick to grow and versatile, fern spores germinate and succeed where others do not. Worldwide, there is a spike in fern populations, as they

throw their distinctive spores into the wind, each cell a cheap invest-
ment in new real estate, a foothold on a devastated landscape,
gaining a quick win while others suffer. These are the disaster taxa,
the pioneer species, the modifiers of the environment that make
the world more habitable. Sometimes these are species that shore
up the environment, for example by developing more fertile soils,
creating conditions that other less adaptable species can thrive in.
Other times, the successful species are more actively competitive,
simply fast-growing species that take rapid advantage of free
resources. They exclude others for as long as they can, but ultimately
succumb as slower-growing, more risk-averse species out-compete
them. Whatever the mechanism, succession ultimately restores the
ecosystem to its former diversity. The fern spike will be intense and
short-lived, the millennium-scale boom and bust of the evolutionary risk-taker, but recovering the pre-extinction diversity will take
nearly a million years. In geology, time and distance are inextricably
intertwined; here, the fern spike lasts for a single centimetre of rock
record, a layer of spores and clay.[5]

From a small slope, the fern bog seems to extend for miles, but
this relative high ground is a refuge for other plants too. A few
swamp cypresses, *Glyptostrobus europaeus*, mark the edge of the
mire, sitting like lifeguards in half a metre of duckweed-coated
water, their knock-kneed roots arching from the surface to help
them breathe, while just inland, young, lanky dwarf redwoods,
*Metasequoia occidentalis*, are climbing to the sky. Most of the trees
here are proud, straight *Populus nebrascensis*, cousins of aspens and
poplars. Among these are many *Artocarpus lessigiana*, early relatives
of jackfruits and plane trees, whose duck-footed leaves seem oddly
appropriate for the weather.[6]

You can tell a lot about a landscape from its leaves. The main
organ of food acquisition and respiration, they are a lung and intestine
rolled into one, and are vulnerably held at the plant's extremities.
This poses challenges. Too dry an environment, and the stomata
will leak water too rapidly into the air. Before the evolution of $C_4$
photosynthesis, before the origination of succulent plants, there are

still a few ways to combat this; leaves can reduce their number, or their size. The wax produced by all leaves becomes denser, thicker, to prevent water loss. If the world is rainy, accumulated water would break the leaf or provide a haven for fungal infection, so leaves adapt to have gunnels that channel the rainfall down to the end, where, like the lip of a jug, there is a 'drip tip' to guide the water to the floor without breaking the leaf. Measure the proportion of leaves that have drip tips in a place, and it will give a reasonable guess of the local rainfall. *Populus*, the aspen, has one; the leaf itself a raindrop shape. *Platanus raynoldsii*, the plane, has three for every leaf.[7]

As if held to a deadline, the rain abates suddenly, and the trees seem to relax, their branches sighing upwards as the weight of the weather is lifted. They continue to drip, and the water that runs off and soaks into the soil takes with it some of that water-resistant wax. Each type of wax has a chemical composition characteristic of the leaves from which it drips, and will give the soil a signature of the plants that once shaded it. Flowering plants produce more wax than cone-bearing plants, and both have wax made of longer molecules than the wax of mosses. In more arid environments, the wax molecules are longer, which prevents the loss of water into the dry air. That chemistry is retained even as the soil hardens and mineralizes into rock and, to a point, can reveal which plants were once present. Long after death, incorporated into the bedrock, their combined and dappled chemical shadows will remain.[8]

It is not just the plants exploiting this island in the marsh. One tiny beast, *Mesodma*, dominates this ecosystem, making up nearly three quarters of the community here. With its large front teeth, square jaws and clambering movement, you might mistake it for a wood mouse at first glance, but *Mesodma* is no rodent. Deep inside that mouth is a tooth unlike that of any mammal of the modern day. Resembling half of a circular saw embedded in the gum, the massively enlarged premolar fulfils a similar purpose. The serrations extend in fluted channels down to the gumline, and this arrangement makes for a rounded blade custom-made for blitzing woody stems. *Mesodma* is a multituberculate, part of a group that has been diverse since the Jurassic, mostly mouse-sized and ranging from seed-eaters to those that feed on fruit or stems, from burrowers to climbers.[9]

Not many mammal groups – of those known from the Cretaceous – have survived, and none is unscathed. In the southern hemisphere, monotremes, which include modern platypuses and echidnas, have barely survived, but they are evolutionary endurance athletes. There may not be many, but they will limp through to the modern day, never diverse, never common, but always there, though invisible to the fossil record. The pouched metatherians, the progenitors of marsupials, used to be common all over North America. Now, only a few survive here. Eventually, they too will be restricted to the south. Two other unusual insectivorous mammal groups, the sprawling symmetrodonts with their piercing triangular molars and ankle spurs, and dryolestids, animals like spineless hedgehogs, may have survived. The only symmetrodont is a controversial specimen named *Chronoperates*, the time wanderer, a late Paleocene tooth from a Mesozoic group, out of place in the geological record. A dog-sized dryolestid, *Peligrotherium*, the 'lazy beast', is known from the early Paleocene of Patagonia, while another, *Necrolestes*, the 'grave-robber', is from the same part of the world, but much later, in the Miocene. A mole-like burrower with a sensitive snout, *Necrolestes* is so highly specialized that exactly which group it belongs to remains uncertain, while *Peligrotherium* has been thought to be a placental mammal by some. For both to be dryolestids requires nearly 40 million years of missing evolutionary history, a large but not insurmountable gap. Perhaps, like the monotremes, these groups survived but were not preserved by virtue of their ecology.[10]

When it comes to preservation, all environments are not equal. To become a museum specimen, the dead must resist decay, be covered in sediment, and avoid being eroded, metamorphosed or sunk beyond the reach of chisel and awl. The mammals of Hell Creek have an advantage over the birds in this respect – their teeth. Coated in protective enamel, teeth are physically and chemically hardier than bone, and preserve at a far higher rate. Mammal teeth, and in particular molars, have distinctive patterns of cusps and basins, with various types of ridges connecting and dividing them and running around borders. The shape of a single lower molar can identify a species precisely, although it is less easy to use teeth to work out how species are interrelated; convergent adaptation to similar diet to some extent overwrites the features that linger from family history.[11]

For many families, genera, species, the Paleocene is not an end, but a beginning. All over the world, a recovery is taking place. By necessity, this means diversification of those few surviving lineages, the origin of whole new groups, where a species can become an order. Bony fish, lizards, marsupials, many types of bird are all diversifying in a world where niches are vacant and ripe for the taking.[12]

As humans, we are eutherians – in self-important Victorian terminology the 'true beasts'. Our eutherian kin were diverse in the Cretaceous, and survived the mass extinction event. Ancestrally insectivorous, like so many mammal groups, they are defined by their descendants, the placental mammals. Although diverse in the northern hemisphere during the Cretaceous, they are also known from the island continent of India, now about as distant from other landmasses as it will ever be. Even with this wide distribution, more than half of eutherian families were entirely wiped out. Only three groups of eutherians made it through for any length of time: the mostly predatory cimolestids, the jerboa-like leptictids, and the placentals. We do not know exactly how many placental lineages survived to leave descendants; most estimates are around ten. Neither do we know directly what the anatomy of the surviving placental lineages was like, because there are none known from before the extinction event. It can be inferred backwards that their ecology was that of a small, nocturnal insectivore. Placentals are found in the fossil record only from this point onwards. This, almost immediately following the cataclysm, is their dawn.[13]

Across a narrow rivulet, a cluster of ferns creaks and opens, tearing the orb web of a spider, as a slender animal the size of a cat, followed by two kits, emerges to drink from the water. Perhaps they would be better called calves, but it is difficult to know; it does not yet make sense to talk of cattle or dogs, of monkeys or horses. None of these groups exist yet, but this is where, or at least when, they begin, in the world recovering from the wreckage of the Chicxulub impact. This is one of the earliest placental mammals. Names lose tangibility in the depths of the past, and our language has no description for the young of common ancestors, in the time when these groups began to split from one another. Somewhere out there are tribes of organisms, the gathered remnants of wider distributions, separated by calamity, never to meet again on the branches of the tree of life. These two young *Baioconodon* are brother and sister, but perhaps, when grown, one will migrate away to pastures new. Perhaps their children will never meet, their communities never mingle. Speculatively, these may be ancestors, respectively, of bats and horses.* At some point in the genealogy of those animals, their lineages must converge on populations of ancestors, and for most placental orders, the Paleocene is their cradle.[14]

Where *Baioconodon* and many of the other enigmatic mammals of this time fit in the mammal family tree is uncertain. To reach across this stretch of bog and pull a clump of hair, extract some DNA, would fulfil the fantasy of many a palaeontologist. But 66

---

* Speculative because species are not, of course, descended from individuals, but from populations. Within those populations, though, if two siblings find themselves on opposite sides of a split, in different gene pools, their separation would be one divergence among many. As for bats and horses, they are surprisingly closely related, within the placental 'superorder' Laurasiatheria. Some suggest that bats, perissodactyls (including horses), and carnivorans form a tight-knit group entertainingly known as Pegasoferae (the 'savage winged horses'), but laurasiatherian interrelationships are notoriously hard to pin down. Creatures like *Baioconodon* are somewhere near the base of the placental and laurasiatherian radiation, but claiming with certainty that one species is a direct ancestor of another is neither fashionable nor wise.

million years is too far a stretch, and *Baioconodon* would surely startle and flee.[15]

For now, we must remain content to watch and hope for a better
day with better evidence. Their anatomy is too non-committal, too
similar to and yet too distinct from too many living orders to be
placed with confidence. The facial resemblance is outwardly like a
giant, snub-nosed hedgehog, or something like the Madagascan
mongoose relative, the fossa, but this is to superimpose too much
of later groups on them. They are an unspecialized, Platonic placental, a lump of living clay from which all others are stretched,
pinched and pulled into shape. This is true to some extent of many
of the Paleocene mammal families, who we now lump together into
a group we call 'Condylarthra', known to be a wastebasket, a museum
cabinet with a brown and peeling label where we throw all the early
mammals we don't know how else to file. Within that hodge-podge,
*Baioconodon* is part of the 'bear-dog' family of arctocyonids – perhaps
a wastebasket itself.[16]

Although their ancestors were insectivorous, deep within the
cells of these animals, the genes that code for the enzymes required
to break down the hard outer exoskeleton of insects are being
turned off. Even where an animal is adapted to eating insects, it
might be easier to avoid competition and try a new, less common
diet, however challenging, such as plants. If insect-digesting
enzymes are unused, there is no longer an advantage to maintaining
them. There is no mechanism checking the instructions to make
chitinase as they are passed from parent to child, and in an evolutionary game of Telephone, the information contained there slowly
becomes useless. The remnants of these genes can be found in
humans, in horses, in dogs and cats, a vague genetic memory of an
insectivorous past, and one in which, intriguingly, the losses appear
to have happened independently of one another.[17]

Scampering past the *Baioconodon*, a *Mesodma* climbs, squirrel-like,
along an aspen branch, looking for food. Clambering headfirst down
a woody vine, it rejects some dark, low-hanging berries among the
dagger-leaves of *Cocculus flabella*, the moonseed. A *Procerberus*, a

slightly larger animal sheltering under the vines, something like a large and aggressive shrew, barks in alarm and scurries away through the vegetation, its hiding place located. Moonseed fruit is not good food for a *Mesodma*; it grows quickly, climbing over the larger trees, using their height to reach the sun, so is plentiful, but the seeds themselves are toxic. The neurotoxin can paralyse a mammal, but is less effective on birds, a resistance the plant exploits to disperse its seeds. The birds that live in these sparse woods are ground-dwellers resembling quail or tinamous. For the tree-dwellers, the burning of the forests was truly devastating. It is possible that, within every family of birds, only those that nest on the ground have managed to survive. Their fragile bones mean that their record is fragmentary in more ways than one, but the earliest known birds from the Paleocene fit with this theory. They are all rock-nesting seabirds, inhabiting the western Atlantic seaboard, over the Western Interior Seaway on the island of Appalachia.[18]

Nobody is certain what permitted creatures like the arboreal *Mesodma*, *Procerberus*, or the earliest placental relatives to survive where so many others went extinct. Some aspects of their life history are likely to have helped; smaller animals need less food to survive and, like the ferns, reproduce quickly and have many offspring, a scattergun approach that is more likely to bear fruit in an unpredictable environment. Earlier reproduction helps with adaptation, as less time is needed for any individual to survive before it replaces itself in the population many times over. Living underground in burrows, where the temperature is less variable, would have been a protection against all of the searing heat, the fallout and the meteoric impact winter, and was almost certainly a contributing factor to the survival of many animals. During the 1960s, when nuclear weapon testing was heavily underway in the Nevada Desert, the burrows of kangaroo rats, only some 50 centimetres below the surface at their deepest, were isolated enough to mean that they survived and thrived despite the blasts of atomic bombs.[19]

Being aquatic was perhaps also a protection – turtles, as well as salamanders and other amphibians, fared relatively well, but, even then, several aquatic animals such as crocodile relatives and other marine reptiles were either destroyed or lost almost all their diversity. Today's remaining crocodilian diversity is spectacularly low compared with their counterparts from the Cretaceous. Far from all being semi-aquatic ambush predators, Cretaceous crocodiles included the agile and cat-like *Pakasuchus* of Tanzania; the fully marine thalattosuchian family with their flippers and shark-like tails; and the pug-nosed *Simosuchus,* a clove-toothed burrowing herbivore from Madagascar, only the size of an iguana, that, despite the advantages of that lifestyle, still didn't make it. Life history matters, and even the apparently indiscriminate death brought by a meteoric collision affects some ways of life more than others.[20]

Sometimes, being common is enough. Before the extinction, the rivers of the Hell Creek area played host to at least twelve types of salamander. Only four of those species survived – the four that between them had made up 95 per cent of the pre-extinction salamander population. Their abundance made them more resilient to population crashes. The water in Hell Creek still holds their gelatinous egg-masses, but they are now almost exclusively one species. In the fresh but low-oxygen water, bottom-dwelling fish like bowfin, equipped with gills but able to breathe air, or guitarfish, who bury themselves in the mud and hunt for clams – a slow life – are the major inhabitants. Freshwater turtles hoist themselves onto logs. Somewhere out there, predatory lizards called champsosaurs and large alligator-like crocodilians lurk, mostly submerged in the water, snapping at fish. Of the crocodylomorphs that lived in the latest Cretaceous, the survivors around the world are those that are adapted to wider ranges of salinity, the ones that live in and around the liminal parts of the world, where salt water meets fresh. Again, versatility breeds survival.[21]

On the water's surface, duckweed floats in mats, while the overlapping circles of *Quereuxia* pads, bobbing gently on the water surface, are beginning to bud and flower. Damselflies hover precariously, a wing-length above the water. Half-hidden among the trees is the real herbivorous specialist. *Mimatuta,* another mammal with the placental strategy of long-term internal gestation, is the size of a fox terrier, and has wandered out of the boggier ground to gnaw on the exposed root of a ginger. Long-tailed and low-slung for its size, from its neck backwards it is not unlike a small brown badger, half crouching and thick-set. Its head is more domed than a badger, though, and its jaws are deeper than *Baioconodon*; this is an animal used to chewing. So the family of periptychids is adapting to a life of herbivory. The relatives of *Mimatuta* are growing larger, developing bulbous teeth ideal for crushing tasty roots deep in the forest, somewhat like wild boar. *Mimatuta's* face is densely coated with vibrissae, whiskers, sensitive to food in the undergrowth.[22]

After the dinosaurs disappeared, the newly enlarged placental mammals like *Mimatuta* are the biggest on land. *Mimatuta* is an entrepreneur, exploiting plants now unchecked by the competition of *Triceratops* or *Pachycephalosaurus*. The sharp taste of the ginger is meant to put off predators, but *Mimatuta* presses on regardless, as do beetle larvae. Close by, the leaves of a laurel are scribbled with apparently random pale lines, initially resembling snail trails but within the leaf itself. These little tunnels indicate the presence of minute larvae that dig into and feed on the leaf tissue. The thin skin of the leaf is a transparent window to the activity within, as the larva sweeps its head back and forth, wriggling in its little shaft. It is a leaf miner, the silky caterpillar of a gracilariid moth. The young larvae hatch on the side of the leaf, and, after their fourth moult, pierce and enter. There they remain until they are ready to pupate, wrapping themselves in silk, and emerging as minuscule flutterers.[23]

Insects do not preserve well in the fossil record; they are too small, too dust-blown, to be preserved in all but the finest sediment. In Hell Creek, the river system and marsh are too poor in resolution to preserve them, but in the breathless depths, it does preserve those tell-all leaves, sunk to the bottom of the mire. With them will be preserved all the activity of their minute predators, boreholes, excavated tunnels between the veins, galls, and wounds from piercing, sucking bugs. No type of plant is immune; thousands of leaves of cycads, ginkgos, conifers, ferns, even the floating pads of *Quereuxia* are damaged. A distinctive semi-circular nibble here and there, the only evidence of Hell Creek's butterflies that will survive the ravages of geology. But the extinction has killed more than the vertebrates. Even the insects, though abundant, are not as diverse as once they were. The different numbers of ecotypes – ways of eking out an existence – have declined. Where host plants went extinct, larvae could not feed, and followed. Of the specialized insects, 85 per cent were lost, and it was the generalists that survived. The beetle larvae feeding on the ginger are not fussy eaters, and are still here; that gracilariid only lives because the laurel did.[24]

In the complex game that is an ecosystem, every player is connected to some, but not all, others, a web not just of food but of competition, of who lives where, of light and shade, and of internal disputes within species. Extinction bursts through that web, breaking connections and threatening its integrity. Sever one strand, and it wavers, reshapes, but survives. Tear another, and it will still hold. Over long periods, repairs are made as species adapt, and new balances are reached, new associations made. If enough strands are broken at once, the web will collapse, drifting in the breeze, and the world will have to make do with what little remains.

So, after a mass extinction event, a turnover happens, with new species appearing, the web self-repairing. Where exactly *Mimatuta* and *Baioconodon* came from is not entirely clear. They do not have any obvious ancestors in the Late Cretaceous, and so we are forced to ask whether they simply evolved too rapidly for the frame rate of the fossil record to keep track, or whether they arrived from somewhere else, somewhere unpreserved by the fossil record, already partially on the way to their omnivorous niche, but hidden by the erratic preservation of environments across geological space and time. Did they evolve off-camera, in some Cretaceous cradle, separating from one another and only becoming quite so distinct after an opportunity arose to expand their range, both in geography and ecology?

These are unanswered questions about the enigmatic beasts, and only some of many, often difficult to answer with certainty. It is as if in death they have crossed over Lethe, the river of forgetfulness, the memory of their ancestors erased by the passage of time.

The mammals of Hell Creek, and of the earliest Paleocene worldwide, have always held an almost mythological appeal. The species of *Baioconodon* lingering by the water's edge was originally given the name of *Ragnarok*. This is taken from the apocalypse of Norse mythology, the end of the world predicted by the three crones weaving at the loom of fate, constantly making and unmaking, threading the world and letting it come apart.[25]

Others are taken from more recent mythologies; another of the earliest Paleocene mammals has been called *Earendil undomiel*. In J. R. R. Tolkien's Arda mythology, Eärendil is the voyager, the morning star that heralds oncoming joy, a reference to an Anglo-Saxon poem which uses this image to describe John the Baptist, in Christianity the herald of Christ. By the vagaries of taxonomy, the specimen named *Earendil undomiel* is now considered to be a species of *Mimatuta*, a close relative, perhaps a descendant, of the *Mimatuta* here. *Mimatuta* itself has a Sindarin elvish etymology, meaning 'jewel of the dawn'. The names given to these species evoke morning for a reason. The periptychids and arctocyonids of the early Paleocene may or may not leave descendants that will last until the modern day, but they are the ecological pioneers of the Age of Mammals. Where they lead, other groups will follow. Looking back, it is easy to see these mammals as the heralds of later exploration, the physiological boundary pushing that culminates in the bizarre shapes of bats and whales, of armadillos and elephants.[26]

There is a world to conquer, damaged ecological webs to repair, and although dinosaurs will remain more common even to the present day (there are still twice as many bird species as mammal), it is the mammals that will, in general, sit on top of the food chain. In all of mammalian history, it is now that they begin to reach new heights of diversity in number of species and disparity of anatomy. For us, periptychids and arctocyonids are part of this new mammalian fauna.[27]

At the moment the Chicxulub meteor struck, all primates, flying lemurs, tree shrews, rabbits and rodents had yet to diversify. They were united within one, perhaps two or three species, common ancestors to all. Our ancestors are here, and they contain within their genetic code the essence of what it means to be a primate. The ancestors of some of the largest land mammals that will ever exist, the 17-ton rhinoceros cousin *Paraceratherium*, are here, and are the same individuals whose descendants will miniaturize and fly as the smallest, the bumblebee bat. The range of anatomical forms will expand rapidly, exploring the different possibilities of being a mammal, before eventually specializing into the groups we know in the present. It seems an almost biblical promise: your descendants will reach all corners of the Earth and beyond it.[28]

But to look forward too far is to be teleological; the world of Hell Creek does not depend on the hazy future that we call the present. Many species here will not have such genealogical success. The family of *Procerberus*, the cimolestids, will move into new niches, diversify, and survive for another 30 million years before they too succumb to extinction, last being known as a few semi-aquatic otter-like forms in the early Oligocene of Europe. Likewise, the multituberculates like *Mesodma*, for now the ubiquitous rulers of North America, will last until the late Eocene, with one genus, *Ectypodus*, still being found in the Arctic right up until the group disappears for ever after 120 million years of existence. Extinction is an inevitable part of life, but only rarely do mass extinctions occur, only rarely is the established order overturned so quickly. For now, multituberculate, cimolestid, periptychid: all are part of the group of mammals exploring this broken Earth, this world whose diversity has been destroyed by devastating happenstance and rapid climate change. The full recovery of the Earth's natural cycles will take several million years.[29]

Already, recovery is underway. Even at the site of the asteroid impact, life has returned to a highly productive ecosystem. In what will become Colorado, further south along this self-same coast, the site known as the Corral Bluffs records in detail the subsequent recovery of the globe. The earliest communities, like the one at

Hell Creek, show a fern-dominated landscape, with a few disaster taxa making up the majority of life. Within 100,000 years of the mass extinction, the number of mammal species will double. Within 300,000 years, niche specialization has begun to occur, and alongside the new and hardier fauna of the Paleocene, new types of plants besides ferns and palms are becoming significant parts of the eco-system. The earliest trees of the walnut family and the earliest bean pods will soon arise – their nutritious seeds offering protein-rich supplements to the diet of herbivorous mammals whose recent ancestors had mostly fed on insects. Warmth will return, and, with it, the forests will from pole to pole spring to life once more.[30]

Even in the Norse myth of Ragnarök, hope prevails. Although the Earth has been burnt by the fire demon Surtr, although most of the gods are dead, it is not an end. Under Yggdrasil, the great ash that links all worlds together, there is light. The woman Lif and the man Lifthrasír, whose names mean 'life' and 'life of the body', emerge from some underground shelter, the only humans to sur-vive. A new age begins, with new gods, and new worlds. After death, life; after extinction, speciation. In the Laramidian swamp, the spider sets a new strand of silk aloft. *Mimatuta* chews lazily on a fresh flower. It is spring.

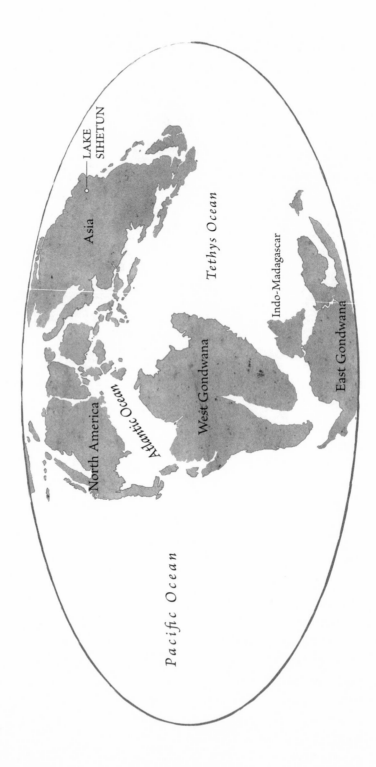

# 7.  Signals

*Yixian, Liaoning, China*
Cretaceous – *125 million years ago*

'Zanio in xochitl tonequimilol, zanio in cuicatl ic huehuetzin telel
a in tlalticpac' / 'Only flowers are our adornment, only songs
turn our suffering to delight on earth'

– Nezahualcoyōtl (tr. John Curl)

'Det som göms i snö kommer fram vid tö' / 'What is hidden
in snow is revealed at the thaw'

– Swedish proverb

By the lakes of Liaoning, near the restless volcanoes, golden ripples spread as the dark night recedes. Across the expansive water, on a narrow patch of sand, an early-rising pterosaur slowly bows its head, the smooth reflection showing off its vulture mane. A mouth full of needles meets the empty water and submerges, sending undulations over the mirror surface. Teeth enmeshed, it separates minute shrimp from the night-cold lake, the air gradually warming, cricket sounds receding with the advancing dawn, the sleeping forest transforming into a lively bazaar.

It is a fresh, spring day in the Early Cretaceous, and this particular pterosaur, one of the ctenochasmatid 'combjaw' family, a flying reptile about the size of a crow, is concentrating too hard on its task to worry about presentation. It has arrived early, trying to beat the rush to the lake edge. Like all combjaws, *Beipiaopterus* uses its comb-like mesh of teeth for filter feeding in much the same way as the beaks of modern flamingos or the baleen of toothless whales. It stands at the side of the water on all fours, wings folded closely, both to keep them

*Oregramma illecebrosa*

out of the way and to avoid losing too much heat through their thin membranes. Compared with birds, pterosaurs have large, long-snouted heads, seemingly out of proportion to their bodies and mounted on fairly long necks, and so in general they are not natural floaters. They can swim well enough, and indeed get most of their food from the water, but they are too front-heavy to glide with the grace of a duck. Instead, this one leans forward, dipping its mouth into the water, and resting on hands set a third of the way down its wing. The rest of each wing is held tight to the body, the fourth fingers that support the flight membrane pointing backwards like a pair of ski poles. From them, the wing membrane itself droops to where it attaches at the ankle, over webbed feet.[1]

The cracking of branches and the sound of branches scraping against skin reveal a departing pod of gigantic titanosaurs. All around Lake Sihetun run coniferous forests, a thousand ancient cypresses keeping the forest green through the seasonal snows, pinned into the knee-high undergrowth, needle straight. As their seismic steps pass through, the crushing herd creates clearings, opening space in which ferns, slender horsetails and other young plants can thrive. The vistas of the Cretaceous are the poster-children of the ancient Earth, the heyday of non-avian dinosaurs. Dinosaurs are certainly the biggest creatures around – and these sauropods are among the largest beasts ever to live on land, the titanosaur *Dongbeititan*. Their long, thick, muscular necks reach over 17 metres in height, with each adult weighing several tons. The titanosaurs live in nomadic pods, moving around in search of fresh food sources to sustain their massive bulk, and travelling between areas as the seasons turn.[2]

Their footfalls are wide, leaving behind crescent moon footprints in the firm ground. As they walk, their necks are held aloft, barely straining to reach the high-held trees. Contrary to popular belief, they will not rear onto their hind limbs, as their flexible spines prevent them from stably lifting their forefeet off the ground at the same time. Instead, they swing each foot forward one at a time. As they build up speed, their gait suddenly shifts into a fast amble, rolling with each movement, first the left feet swinging, then the right. From the front, they look almost like knuckle-walkers – and in a sense, they are. Titanosaurs like *Dongbeititan* do not have digits on their forefeet. The same bones that in pterosaurs lengthen and strengthen into a wing are in sauropods reduced to almost nothing, a simple vestige at most. Instead, *Dongbeititan* walk on clawless, fingerless knuckles.[3]

Sauropod dinosaurs, although they are large herbivores, are not simply reptilian versions of creatures like elephants. Some of their traits are similar through necessity, with all large herbivores having to be bulk feeders – a 30-ton sauropod in the Early Cretaceous would have to consume at least 60 kilograms of nutritionally rich understory plants, or even more canopy vegetation, every single day. Many traits, though, are not shared, usually down to physiology. Compared with elephants, sauropods have extremely light bones, with their vertebrae also surrounded by extensive air sacs, a feature that may have helped them get so big. Several also have showy displays that are more impressive than those of many mammals; the South American dicraeosaurids have a row of large spines protruding all down the back of their necks, a mane of spiked keratin thought to serve both display and defensive functions. Other sauropods, like *Saltasaurus*, have armour within their scaly skin. But dinosaurs, unlike mammals, also have excellent colour vision, and so many sauropods are boldly patterned with blotches and barring, eye-catching visual signals to their kin.[4]

Lake Sihetun is a caravanserai of activity, the hubbub of the day barely subsiding at night. The lake is productive, with turtles and pterosaurs jostling for space with aquatic lizards, bowfin and lampreys, snails and crustaceans. In an undulating upland landscape, the lakes are the main source of drinking water for life on land. Within the cypress wood, following the thaw, and left behind by the departing pod, lies the 20-ton corpse of a titanosaur, broken and large as a fallen tree. Cold, open and reeking, it bears the traces of scavenging, scratches and tears, and is scattered around with broken feathers and a lost tooth from some lion-sized, bipedal theropod predator.

Theropods are the group of dinosaurs that includes, for instance, tyrannosaurids, dromaeosaurs like *Velociraptor*, and indeed, birds.[5]

Breaking through the half-light of the early morning, the first changes to the rhythm of the night are in the sounds, as birds and insects begin to call. Songbirds will not arise until the Eocene, in Australia, so the dawn chorus is not yet filled with the melodious, complex tunes of perching birds. The chirruping of insects has been a feature of landscapes the world over since the Triassic, when the crickets first began to rub their wing cases together. Several groups of insect have modified their hard exoskeletons into musical instruments, the ridged 'file' and smooth 'plectrum' rubbing together, making sound by stridulation, the same way as a young child running a stick along a row of railings. By the Jurassic, the idea had evolved independently in many groups and become sophisticated; certain katydids at the time are known to have sung not with a coarse rasp but with a pure, single tone. Cricket or katydid, grasshopper or beetle, each does it in a slightly different way. In the Cretaceous, the high-pitched chirp of grigs is interspersed with the gentle rasp of longhorn beetles. The air is alive with these displays, the insects eager to find a mate, advertising their sexual availability and location into the ether – the best way to ensure mating success in a crowded ecosystem. As daytime arrives, every living thing is sending out signals, whether loud and clear, intended for all to hear, or coded to members of their own species.[6]

A month ago, the nearby morning ground was coated with hoar, but the lake remained unfrozen, warmed by the heat of the underlying volcano. Now, the signs of spring are all around. As the world awakes from winter dormancy, it seems that every animal, every plant is engaged in gossip, reacquainting with one another after a period of torpor. Among the tall cypresses are shorter trees and shrubs: cycads, with wide-splayed leafy crowns thickening the understory; mossy and budding ginkgos, their new leaves triumphantly emerging like trumpet bells; fading red cones of yews; and the jointed scaffold poles of low and brushy gnetophytes, relatives of mahuang and Mormon tea. All of these plants are gymnosperms;

the name means 'naked seed' because their seeds are exposed on the surface of specialized leaves, and they are a group that has dominated terrestrial life now for 180 million years. Often, the seed-bearing leaves of the gymnosperms are modified into bright cones, standing out yellow and pink against the dark needles and leaves to attract beetles, scorpion-flies and lacewings.[7]

An old cypress, striped and red-wounded from peeling bark, oozing yellow sap, grows perilously close to the water's edge. It leans out raggedly over the water, a drooping branch gently stroking the surface with the wind. In the water, in the shadow of the cypress, are little stems bearing long, pointed seed pods and brush-like clusters of yellow filaments, bunched into the air. Beneath, in pale green, wispy leaves wave, and fresh stems grow towards the surface, ending in rounded, underdeveloped versions of the shapes above. This unassuming water plant is part of a sexual revolution, and one which will change the appearance of the Earth's ecosystems for ever. It is among the first flowers on the planet. The flowers, emerging lily-like from the water, are bisexual, with both male and female tissues on one single stem, unlike most gymnosperms. The yellow bristles are the male stamens, covered in pollen. Above them, the female carpels, in which seeds are developing into a peapod-like casing only a couple of centimetres in length. In these early days, the flowers are not yet overly showy; these flowers have no petals. It is strange to think of flowers without petals, but there are many that have evolved in the modern day, from brightly coloured examples like the Australian bottlebrush and most members of the anemone family to the inconspicuous flowers of grass. This plant, *Archaefructus*, grows in water about 30 centimetres in depth, and its leaves have little bladders at their base in order to help the scrawny stems float to the surface. Only the flowers are held above the water, in order to aid in pollination. Around Lake Sihetun, several aquatic plants have this newfangled way of developing seeds, and it is thought plausible, though by no means confirmed, that flowering plants may have had their origins in fresh water. Not long after *Archaefructus* and its kind grow in Lake Sihetun, the first water lilies and hornworts will be found across the world – in what will become Portugal and Spain. Once seeds begin to grow fleshier and more nutritious, plants begin to co-opt vertebrates into dispersal. About a quarter of angiosperm species already use multituberculates, reptiles and perhaps birds to disperse their seeds.[8]

Small birds call as they stand among the cypress needles, while others uncertainly glide in and out of its branches. Strikingly adorned with a jay-like feather crest and a mottled black spotting on their throat and outstretched wings, the final artistic flourish is a pair of exceptionally long streamers from the otherwise short tail. It is as if the sky is filled with children's kites. This is *Confuciusornis sanctus*, the Sacred Confucius-bird, and its tail feathers serve two entirely ornamental purposes. The first is as a display – the males are typically larger and have long tails, their display a demonstration of fecundity, a dance to impress the watching females, which lack the ornamental feathers. They are mesmerizingly thin and practically weightless – unlike most feathers, which have a cylindrical central shaft or 'rachis', the ribbon feathers are semi-circular in cross-section, open and light, and about the same thickness as a strand of spider silk. Although they can be a centimetre wide and more than 20 centimetres long, at their thinnest they can be only 3 microns thick, thinner than a fine droplet of mist. The early morning sunlight passes through them, reddened by the thin tissue, so that each bird appears to be trailing streams of smoke. The second purpose is as a distraction to aid in escaping predators. *Sinocalliopteryx*, the Chinese prettywing, is a giant dromaeosaur – an ostrich-sized theropod dinosaur that regularly preys on small birds. The thin feathers are easily detached, and, if caught in the snapping jaws of the prettywing, will be left behind. It is not just predators that can take the feathers, though. A feather protrudes from a scar of sticky resin on the cypress trunk – a clumsy Confucius-bird, clinging to the tree, has left its best clothes behind.[9]

Where combjaw pterosaurs like *Beipiaopterus* are specialist filter-feeders, rather like flamingos, the Confucius-birds are opportunistic. Sometimes they can be seen diving for fish, hunting for shiny silver

wolf-fins, a name that belies the tiny, minnow-like fish with spark-
ling oval scales. Other times they will snap at insects on the water's
surface or in the air. Frogs, wary of the flying hunters, croak their
mating display from the safety of half-submerged cypress roots.
Advertising your presence to potential mates means advertising
your presence to your predators as well, and, most of the time, animals want to remain undetected. During mating season, however,
there is nothing for it but to sing, and risk the consequences, perhaps compensating by being cautious elsewhere in life. Spring is in
the air, and for the song of the frog and the dance of the Confucius-
bird, the lake edge is an exceptional place to display.[10]

Ferns, slick with dew, rustle at the brush of an enormous leg. The
pterosaur startles and takes flight, crouching low before pole-
vaulting forward over its wings, and catching itself in the air. It
flies low over the lake and rises high, mane still puffed with alarm.
The owner of the leg, approximately the height of an Asian elephant and about 8 metres in length, is an adult *Yutyrannus*, the
Beautiful-feathered Tyrant. Like its more famous younger cousin
*Tyrannosaurus*, it is a tyrant lizard, a predator on two legs, tail and
body balanced see-saw style. Small, three-fingered hands are held
close to the body. Unlike *Tyrannosaurus*, which will live in the
warmth of Hell Creek nearly 60 million years in the future, at the
end of the Cretaceous, *Yutyrannus* is a true northerner, adapted to
the mild summers and harsh winters of Yixian. Here, the snows last
through the winter – the volcanic hilltops around these woods will
still be snow-capped for months to come – and even a large dinosaur
needs a feather coat to keep warm. Mottled brown and white, the
light plays off it, dappling and disrupting the outline, disguising
even the largest of predators. Large dinosaurs are not particularly
noisy creatures; their vocal organs are much simpler than those of
birds, so they cannot produce the complex trills and flutters of
songbirds. Bigger animals generally don't vocalize as much, and
hisses, wing-beats or clacking of jaws is common. Crocodilians and
the bigger modern-day dinosaurs like ostriches or cassowaries do
perform low grunts with their mouths closed, and *Yutyrannus* is

similar, throat swelling and falling back with a rumble. Modern crocodilians and birds, however, use different organs to produce sound – the larynx and syrinx respectively – suggesting that vocalization evolved independently in each group. Exactly how Cretaceous dinosaurs produced sound is unclear, but no non-avian dinosaur syrinx has been found. The penchant for visual display, at least, will persist in dinosaurs until the modern day – no group of vertebrates has the variety, the detail and the vibrancy of colour and shape as birds. Indeed, reptiles, from birds to lizards, have colours that humans cannot see, patterns that fluoresce under ultraviolet light. Given that this seems to be an ancestral trait, it is possible that non-avian archosaur display, including pterosaurs and dinosaurs, extends beyond the human visual spectrum. *Yutyrannus*'s concession to fashion is a striking pair of feathery crests above its eyes, perhaps a disruption pattern, disguising the black dots of eyes with a cut-across streak of colour. Other dinosaurs use colour to hide. Prey species, like the quill-tailed *Psittacosaurus*, a dog-sized early relative of *Triceratops*, beaked and frilled, need to hide in the crowded forests in which they live. Their dark backs and pale undersides mean that in a world where the light all comes from above, the shadows cancel out their colouration, remove the contrast, so that they appear flat and almost invisible. Pay particular attention, though, and they can be spotted by distinctive horizontal black stripes on the pale inside of their hind limbs. These partially function as camouflage, like okapi stripes, but perhaps also have the added bonus of disrupting insect bites as zebra stripes do, by preventing flying insects from being able to easily judge landing distances at close range. The inner thighs of *Psittacosaurus* are otherwise vulnerable; they are thin-skinned and scale-free, and when the summer's heat comes in, the woods will be teeming with biting horseflies, gnats and midges.[11]

The *Yutyrannus* is not here to feed. Instead, it wades a few steps into the lake, and draws a deep draught from the surface, eyes always alert, before lifting its head to complete the swallow. This is repeated several times, and the sun is by now well above the horizon. As the day progresses, there is always something emerging from the trees, always another creature arriving at the still, mossy lake in the centre of the woods. The flattened shell and long neck and tail of an *Ordosemys* turtle draws circles in the surface, as little pterosaurs whir among the choking clouds of midges that now hang over the water, snapping them out of the air. Here, a shimmering and iridescent dragonfly, hunting wasps and horseflies on the wing. There, a cluster of blood-red snails hanging from plants like berries above the shallows. Above, sparrow-sized *Eoenantiornis*, 'dawn opposite-birds', scout for insects among the branches of a ginkgo, while the languid 7-metre wings of the pterosaur *Moganopterus* beat through the sky.[12]

Life abounds, from the leaf litter to the skies above the canopy. The invertebrate life, having mostly slept through the winter, is now well and truly awake. Running around the fallen twigs are cockroaches, laying eggs in well-hidden crevices in rotting wood or bark. Rammed into the cracks of a tree trunk is a little leathery capsule, a deep brown structure, keeled like a peapod, bulging at intervals with sixty or seventy cockroach eggs. The egg case is a protective measure, but cockroaches have a particular enemy in Cretaceous Liaoning. Humming through the air are tiny hatchet wasps with waists so thin that, at speed, the creature seems to be a pair of separate bodies, the one tailing the other. *Cretevania* is a parasitoid, an animal that reproduces inside another animal, always killing it in the process. Specifically, the female *Cretevania* seeks out egg cases, and then lays one of her own eggs within each cockroach egg, injected by her syringe-like ovipositor. The nutrients intended to nurture the growing cockroaches will instead go towards the rearing of more hatchet wasps. Such relationships are surprisingly stable, and hatchet wasps and ichneumonids – another family of parasitoid wasps known from Lake Sihetun – will fulfil the same ecological roles for the next hundred million years or so. The same is true of the bloodsuckers, horseflies and midges, which will merely adapt to new hosts – horseflies predating horses by some 70 million years.[13]

Where plants and vertebrates of the Cretaceous are substantially different from those we are familiar with in the modern day, the insects and other minibeasts are, in large part, very recognizable. The cockroaches and wasps are brightly marked, black and yellow or black and red, a warning of danger, or poison, or simply unpalatability. These colours are found in creatures that want to be seen, that are advertising their unsuitability as food, just in case of a mistaken crunch from a bird's jaws. Black and yellow are high-contrast and stand out against the green of the leaves even in creatures with no colour vision, deterring even individuals unfamiliar with the danger. There is a continuity here, with the same signals that caused a dinosaur to think twice about messing with a wasp as those that cause a modern-day picnicker to do a double take. The warning colouration of insects uses a shared visual language that has persisted for more than 100 million years.[14]

As with sauropod scales, the feather coats of even large theropods are marked with various colours. The real outlier in dinosaur fashion is a large, sloth-like dinosaur, *Beipiaosaurus*. A little smaller than an ostrich when fully grown, *Beipiaosaurus* is a therizinosaur, a 'reaper lizard'. Like all reapers, their most distinctive features are their long forelimbs tipped with scythe-like claws. *Beipiaosaurus* is one of the earliest; later species will take the claws to extremes, such as *Therizinosaurus*, whose claws were up to half a metre in length. These claws are not weapons exactly, and are instead usually used for grasping vegetation, adapted to the same feeding style as giant ground sloths and gorillas, large arms drawing food to the mouth. A *Beipiaosaurus* is densely feathered, with a tasselled appearance. A short, pale undercoat of downy feathers covers the whole body, but around the head and neck is a brown bush of long, thick, stiff feathers, several centimetres long and somewhat like the spines of a porcupine.[15]

As it passes an aged cycad, glossy with green moss, the *Beipiaosaurus* leans towards it and rubs its side against the rough bark, scratching off a few untidy older feathers from its dullish, downy coat. Unlike lizards and many other dinosaur groups, maniraptorans (including birds) do not shed their skin in large sections, as this would interfere with the feathers. Instead, they shed small patches at a time like mammals, the skin continually growing and flaking off dandruff. The winters in Liaoning are cold but the summers will be warm, and excess feathers will be an encumbrance.[16]

Disturbed by the sudden shaking of theropod fluff, a flurry of unseen lacewings burst from the cycad, each of their wings a perfect mimic of a cycad leaf. Deceptive signals are found in insects too, often disguising themselves as parts of the plants on which they live. The great masters of such hiding places are the family of stick insects called the phasmatids. A growing sapling at the base of the cycad is covered with early stick insects, dark stripes on their elongated bodies and wings mimicking the plant's veined leaves. As a family, phasmatids have mimicked stems and stalks since the Jurassic, and are now coming to mimic leaves and flowers too. They hide in plain sight, inhabiting the gymnosperms of Lake Sihetun.[17]

But disguise is not the only display option open to insects. Lacewings here are as common, as large and as varied in colour as butterflies. In some cases, without an expert eye and the knowledge that butterflies have yet to appear on Earth, you would not be able to tell the difference between a fluttering kalligrammatid lacewing of the Cretaceous and a butterfly of the twenty-first century. In particular, they have unusually wide wings for lacewings, and, in species like *Oregramma*, have converged on the same solution as butterflies will find to the risk of predation – eye-spots. These dark circles surrounded by bright colour are hidden at rest, but revealed when the insect is startled, causing any would-be predator to double-take and think twice, momentarily, about attacking. It is commonly thought that eye-spots and similar patterns serve to mimic the enemies of a predator, so that, for example, a butterfly might mimic the eyes of a hawk to scare away a songbird. Perhaps, preserved in the wings of ephemeral *Oregramma*, are some of the last mirrors to the gaze of a non-avian dinosaur.[18]

As the summer comes, lacewings will dance over the lake's surface, rising and falling as if they have not yet mastered staying in the air, teasing the water's surface and the fish that lurk underneath.

Now, they are hatching from the undersides of leaves, each little larva possessing a serrated blade that it uses to saw the egg open from the inside. Those that have already hatched are beginning to develop a form of camouflage that is distinctively lacewing, and very difficult to detect. Rather than simply resembling the surroundings, several families of lacewing start life as what are known colloquially as 'trash bugs'. That is, they collect items from their environment – fern spores, sand grains, shed exoskeletons of insects and so on, and pile them on their backs. This pile becomes a coat, carried round by the larva, and making them essentially indistinguishable from the collection of entirely harmless debris that coats the forest floor.[19]

Away from the congested vegetation, from the bracket fungus and moss, the air cools and the canopy thins. Preferring the open, a small, bipedal dinosaur, *Sinosauropteryx*, slinks across the ground, head and tail held flat and low. It moves only a few metres at a time, stopping periodically, its tail instinctively rising and falling. In look, it is a sepia stereotype of a silent-film prisoner, with its russet and white striped tail and bandit mask over the eyes. The stripes act to disrupt the outline of the animal, disguising the prominent, giveaway tail and eyes of a theropod predator. Its dark back and pale underside help to soften its three-dimensional appearance, helping it to hide even in the open, giving another element of surprise. A shaking *Prognetella* shrub attracts the dinosaur's interest. Inside, a furry, gerbil-sized animal, *Zhangheotherium*, cowers in the shelter of the branches. Holding itself side-on to the dinosaur, it squeaks a warning – and it is not as defenceless as it might seem. Protruding from its heel is a spur, a sharp spike of keratin, which, if it manages to get a kick in accurately, will deliver a dose of venom, enough to hurt but not kill a *Sinosauropteryx*. Therian mammals, marsupials and placentals, have lost this structure, but male platypuses and echidnas are still venomous, and perhaps all non-therian mammals like *Zhangheotherium* have venomous spurs. Having been spotted, this *Zhangheotherium* has adopted a strong defensive posture, and is well protected by a cage of twigs. It has not been quick enough this time, so, recognizing a lost cause, it turns away from the sheltering mammal and disappears into the brush, looking for other small animals.[20]

It is in the scattered light of the more open ground that *Sinosauropteryx* is most at ease, with plenty of hiding places and cover to spring from, and space in which to run quickly. Further to the north, where the woods thicken at Lujiatun, there is more competition with woodland dinosaurs – other lithe, hunting theropods like wide-eyed troodontids and, at night, the mammalian predators such as *Repenomamus*, a badger-sized carnivorous mammal, the largest in the Cretaceous world, known to catch and kill baby dinosaurs.[21]

Although some mammals are awake during the day, it is when the sun sets over Lake Sihetun that they truly take over. Nocturnality is an unusual ecological choice, and in the Cretaceous mammals are rare among vertebrates in this respect. Little else is really active at night, perhaps only some small, predatory dinosaurs. Ectotherms – creatures such as lizards or amphibians that rely on external sources of heat to maintain their core temperature – sleep, cold and inactive. Mammals and their kin are different. Even back in the Permian, the distant predatory relative of mammals, *Dimetrodon*, might well have been nocturnal, and, since then, mammals have become experts at life in dark places, perhaps several times independently. Their eyes are large and well adapted to the low light, better at gathering light of any kind than differentiating between colours. Ancestrally, the tetrapod condition seems to be tetrachromatic – to have four different pigments to detect colours in the eye. Most tetrapods today still have them, but mammals are mostly colour blind.[22]

During this foray into the nocturnal world, there is no need to distinguish colour, only to concentrate on any bit of light, and so the pigments decay from lack of use or are lost. Metatherians, including marsupials, have three-colour vision, having lost one pigment, while eutherians, including placentals, are down to two. Even in the modern day, almost all placental mammals are dichromatic – they have cells detecting red light and cells detecting blue light. Two diurnal groups of placental mammal that depend on being able to distinguish ripe and unripe fruit – catarrhines (monkeys from Africa and Eurasia, including us humans) and one species of howler monkey – have recovered a green-sensitive pigment by duplicating and modifying their red-sensitive pigment. The similarity in the DNA sequences that control the red and green pigments, and the fact that they are next to one another on the X chromosome means that copy-errors occur very frequently, causing red-green colour-blindness. About 8 per cent of human males are dichromats of one kind or another, much higher than in other catarrhine species. Compared with birds, which can see a little into the ultraviolet part of the spectrum, we mammals are all colour blind. Our biology in the modern day, our poor colour vision, is a direct consequence of our reliance on scent, our abandonment of vision, our ancestral journey into the night.[23]

For those that cannot generate their own heat, a bask in the sun is required to get up to speed. A little willow-lizard, the gecko-like *Liushusaurus*, darts out of a thin crevice, its wide body held flat against the sun-warmed rock to absorb the heat of the sunlight. Its back is pale and camouflaged, its belly dark and, at first glance, spiny, but this is a toothless warning. Any predator brave enough to risk a bite would discover that those spines are just a trick of the light, a warning colouration, dark in the middle and light at the side, giving the impression of sharpness, producing enough doubt in the predator's mind for the lizard to escape.[24]

A series of heaped piles of decaying vegetation marks a sauropod nursery. The young of *Dongbeititan* are minute compared with their fully grown parents, and the developing eggs, about the size and shape of a cantaloupe melon, number up to about forty at most. Ancestrally, dinosaurs laid soft eggs like turtles, but over time several families have independently evolved harder, calcium-rich eggshells. In a herd of 17-metre giants, duck-sized sauropod young would be crushed, and a pod simply cannot stay in the same area for long if their food plants are to recover. So it is that the nomadic adults lay their eggs, scrape earth over them with their enormous hind limbs, and cover the nests with vegetation – the rotting plant matter generates heat and keeps the eggs warm. The nests are prone to raiding, particularly by snakes, but the sheer number of eggs across all nests means that a substantial clutch will still emerge. Once hatched, the precocious infants roam the plain together until they are big enough to join the adult caravan themselves. Lizards here do the same. Where a stream flows to refill the lake, a heavy, moss-covered boulder is occupied by a gaggle of wet, green crocodile-lizards. None is yet an adult, the oldest being two or three years old, and the youngest barely a year since hatching. Where their size cannot help them, these small reptiles are banding together to bring the benefits of community. The more eyes a group has, the better the chance that danger is spotted early, and all can scarper into a crevice. Until they are adults, the best chance of survival for these lizards is in a group.[25]

Not all parents are so carefree. A round, fort-like, earthen nest is dotted with oval blue eggs, turquoise gemstones set into a ring of dust. Around them stalks a grey-black, turkey-sized dinosaur, a *Caudipteryx*, flashing flamboyant arm feathers and performing choreographed bows, lifting a round fan of black and white banded feathers at the tip of his tail. He is guarding his brightly coloured eggs, and waiting for another female to approach and engage with him in their ritualistic courtship dance. The eggs that are already fertilized have been laid, with the more pointed end downwards, in a circle, and partially buried by their mothers. The nests of *Caudipteryx* are communal, with a single male guarding the eggs of several females. The eggs' mottled colouring, distinctive to each mother, provides an additional signal to the father. Any female that can produce the complex pigments protoporphyrin and bilirubin to the extent that she lays strongly blue-green eggs must be a healthy, successful feeder, and the father can expect that any offspring hatching out of those eggs are likely also to be successful. Oviraptorosaurs are caring parents, but brighter eggs in living dinosaurs elicit a bigger caring response from the father; one of the few examples of sexual selection occurring after mating has already taken place. Once the eggs have been laid, the *Caudipteryx* father will sit in the middle of the circle, crouching low over them and covering them with his wings to keep them warm until they hatch.[26]

Taken together, this temperate forest, lake and shrubland ecosystem is a bustling metropolis of creatures from the top to the bottom of the food chain. Insects and birds pollinate the plants, including the innovative angiosperms. Other plants, like the gnetalean *Prognetella* drop parts of their body – peduncles – into the water where they are spread by water currents. Regular rain, warm summers and cold winters help to support an exceptional diversity of life.[27]

This diversity is supported by high primary productivity, driven by the area's fertile, volcanic soil, continually augmented by nitrogen-rich ash from regular eruptions. But the source of life in this northern land also threatens death. Lake Sihetun lies within a crater, filling the collapsed caldera of a low-lying – and, for now, dormant – volcano. The volcanic area is large; the lake is deep and covers about 20 square kilometres. All around, eruptions still take place, and when they do, pyroclastic flows arrive, or, more subtly, heavy gases – carbon monoxide, hydrogen chloride, sulphur dioxide. All are toxic, and all displace the air, as they drift down hillsides and collect in these geographic bowls. Any creature trapped within the invisible gas cloud will be suffocated, including most of the water's inhabitants.[28]

Bodies that are washed into the lake will sink, and when the ash blows in and settles, the entire community memorialized in fine silt, a marvel of preservation.

These lakes have a very low sedimentation rate – the fine silts that drift out of the water and settle build up at a millimetre every two to five years. Because there is little decay at the bottom of Lake Sihetun, the exceptionally fine ash preserves everything from bones and cartilage to feathers and hair, even down to individual melanosomes, the subcellular packets of pigment that colour these organisms. By the distinctive shapes of melanosomes that contain reddish or blackish melanin, the colouration of these creatures will be preserved. Present, too, are the stridulation structures, iridescence of feathers and other physical markings so that warning signals, camouflage and sexual display flash out long after their bearers have died.[29]

For most of the fossil record, this sort of information is highly incomplete, the behaviour absent, the interactions among species difficult to reconstruct. At Lake Sihetun and the other localities of the Yixian Formation in north-east China, the brightness and diversity of life, its clamour and colour and conflict, leap from the golden siltstone canvas. As with the immortals of folklore, who spotlessly let the years tumble past,[30] so it is with the landscapes of Cretaceous Sihetun. This is a world that has been preserved in exquisite detail, where even the transience of a song, a startling wing-flap, is made solid and lasting. As the Confucius-birds and kalligrammatids, the first flowers that bloomed and huddles of infant lizards emerge from the rock, it is as if they have just been resting, awaiting a chance to sing and bloom once more.

Boreal Sea

Ural

Greenland

North America

Baltica

Laurentia

L-B

○————— NUSPLINGEN

Iberia

Atlantic Passage

Tethys Ocean

Adria

South America

Africa

Gondwana

L-B London-Brabant
Sponge reefs

# 8.   *Foundation*

*Swabia, Germany*
Jurassic – *155 million years ago*

'You do not have to travel to find the sea, for the traces of its
ancient stands are everywhere about'

– Rachel Carson, *The Sea Around Us*

| 'Nami kaze no / ari mo arazu | 'The waves before the wind |
|---|---|
| mo / nani ka sen / ichiyō no | rise and fall, and yet what am I |
| fune no / ukiyo narikeri' | to do in this drifting world |
| | with a boat of a single leaf?' |

– Ichiyō Higuchi, 'Koigokoro' / 'Loving Heart'
(tr. L. Rasplica Rodd)

Wave crests appear and disappear from view with abandon, throwing points of light haphazardly into the air. The very reflection of the sky from the warm water is blinding, and the shoreline, some kilometres off, is scarcely visible. All around, small, white bodies are plunging from the skies, throwing up enormous splashes every time they enter the sea. After each splash, there is a pause of a moment or two, before a shiny, fuzzy head and a needle-toothed, keratin-tipped smile emerge from the depths. More often than not these mouths are empty, but from time to time one will break the surface with a little fish trapped in its jaws. *Rhamphorhynchus* is a truly marine pterosaur, one of several closely related species that diversified among the bays and cliffs of tropical Europe. These seas are its ancestral home, a place that has brought it and its kin millions of years of evolutionary success. Floating low in the water, it shakes its head, turning the fish around in its jaws until the tail flaps

*Rhamphorhynchus muensteri*

limply. A tilt and a flick, and it swallows the fish whole, throat distending. Lifting them from the water, it unfurls its wet wings, no mean feat with their long, rigid flight fingers. Waiting for its moment, it times its flaps to launch just as it tops the crest of a wave, gaining height, ready to dive once more. Under the water, other plunging *Rhamphorhynchus* push through the water with webbed feet and snatch at the shoal, the fish parting in panic, divided and defeated, not only by the flock, but also by what swims beneath.[1]

A shoal of fish will not rise to the surface and make itself vulnerable without being pushed from elsewhere. Predators from below have corralled and shaped the shoal into a tight, panicking mass, a bait ball, trapped against the deadly air. Fast-moving shadows belie the presence of ichthyosaurs. Like *Rhamphorhynchus*, these are land-dwellers that have adapted to a life at sea, but unlike *Rhamphorhynchus*, these live beneath the waves. With the opportunities afforded by the marine realm in a world of higher sea levels and flooded continental margins, around the world, many four-limbed beasts have given up their life on land and entered the ocean. In the modern day, there are few fully marine tetrapods. Only whales, the closely related manatees and dugongs, and sea snakes have entirely embraced the ocean. All other marine tetrapod groups, from seabirds to seals, saltwater crocodiles, polar bears, marine iguanas, sea otters, and even sea turtles, have a lifestyle that requires a return to land to reproduce.[2]

In the Mesozoic, there are far more groups of fully marine reptile. The fish-like ichthyosaurs and long-necked plesiosaurs are the best known, but there are others. Patrolling the open water between the tropical islands and slinking into lagoons and bays are geosaurines, smooth-skinned crocodiles as large as an orca. Life in the open seas has changed these crocodilians almost beyond recognition – legs are now flippers, the dense bony armour has been discarded, and even the tail has a vertical shark-like fluke. *Pleurosaurus*, whose closest living relative is New Zealand's superficially lizard-like tuatara, also resembles a sea snake, with a long, undulating body, a flattened fin-like tail and short limbs pressed hard against its streamlined sides. Hunting many of the other marine reptiles are the pliosaurs, short-necked and large-headed versions of plesiosaurs who appear to have had a general diet of anything that moved. The marine reptiles of the European seas coexist by adapting to different diets, with some specializing as hard-object feeders, some hunting large prey, and others feeding on fast fish and squid-like prey. Even with this diverse assemblage, the Jurassic is a time of recovery for marine reptiles. They were severely affected by the mysterious Triassic-Jurassic extinction, the cause of which is hotly debated. The leading candidate is runaway climate change thanks to the release of gas as magma rises to the surface, sulphur dioxide and carbon dioxide bubbling out as from a soda can. After the ocean acidification that followed, the range of forms displayed by marine reptiles, and their functional variety, are now in the middle of what will be a 100-million-year-long recovery.[3]

Of all the extinct worlds that the Earth has housed, the world of pterosaurs and marine reptiles, of the seas and islands of Jurassic Europe, was among the very first to be pieced together. The first description of a pterosaur fossil, written in 1784, interpreted it as a swimming creature with its wing fingers as long paddles. Because the scientific community of the day was yet to accept extinction as a real phenomenon, the pterosaur was assumed to be a living creature, restricted to some remote and unexplored environment. The rationalization that followed assumed that pterosaurs were modern creatures of the deep sea. With the turn of the nineteenth century, fuelled in large part by the discoveries by Mary Anning of further extinct marine creatures in the cliff-falls of the Dorset coast, the evidence built in favour of extinction. Ichthyosaurs and plesiosaurs, so unlike the marine organisms of the modern day, and yet so frequently found, gave scientists grounds for a vision of a past crowded with a fauna alien to the modern day. Those particular dead individuals that will eventually be carefully prepared in the back room of Anning's white-washed Lyme fossil shop are already buried in the sea floor to the north, with 40 million years' worth of slowly settling sand and silt piled on top, but their descendants are now piercing and snapping at a bait ball. The shimmering mirrored surface of uncountable fish bodies flexes and turns with the attack, breaking into tori and changing direction all together, their only defence numbers and confusion, and the hope that the predators will be sated. With them driven to the surface, this will only last so long. Attacked from below and above, their eventual annihilation is inevitable.[4]

Jurassic Europe is an archipelago. A series of islands up to about the size of modern-day Jamaica, separated by warm, shallow seas, the flooded margins of continents which here and there dive into deep oceanic trenches. The nearest continental-scale mainland is the unflooded west coast of Eurasia. In the Jurassic, the world has been building through a total greenhouse state, with temperate climates reaching into the polar regions. The seas have been ever rising, increasing the area of the seabed available for inhabitation by marine animals, producing a series of species-rich marine communities throughout the world.[5]

The particular richness of the European archipelago is a result of its status as an oceanic crossroads. It is a series of ribboned landmasses set within stripes of shallow inland sea on the continental margin between Asia and Appalachia. White beaches of fine sand, with briny, still lagoons, are rimmed by reefs. Forests of conifers hang down almost to the sea, stopping only where the tide runs in and out over smooth mudflats. Some islands, like the Massif Central, are flat, former mountain peaks eroded over a hundred million years. Others are on their way up, built by tectonic activity and the rising of reefs. To the south, the island continent of Adria lies in the warm and humid Tethys Ocean that separates Europe from Africa. Eastwards, the Tethys expands into its widest point along the southern margin of Asia, where a plunging oceanic trench runs from Greece to Tibet and beyond, splitting the northern world of Laurasia from the southern Gondwanan landmasses. To the north, the seas narrow into a pair of channels around the landmass of Baltica, before diving into the cooler, less rainy, Boreal Sea. In the west, the landmasses that will become North America are already separating from Gondwana, giving birth to a narrow but growing sea passage, now a mere offshoot of the Tethys but one that in time will be large enough to deserve its own name: the Atlantic. In places, the continental waters of Jurassic Europe are deep compared with seas on continental shelves today, with trenches up to about 1,000 metres from surface to seabed. Mostly, though, the seas are barely 100 metres deep, housing a huge diversity of animal life.[6]

At the meeting point of three oceanic systems – the proto-Atlantic passage, the Tethys and the 'Viking passage' to the Boreal Sea – Europe is a choke point for underwater currents. Like the Gulf Stream that warms northern Europe today, the ocean currents act as a feedback system that evens out global temperature differences. Some 15 million years before, these seas would have been far, far warmer. The narrowness and shallowness of the straits around Baltica are such that, with so much tectonic activity, the corridor between the Tethys and the Boreal Seas was closed by a rift rising in what will become the North Sea. Cutting off the route by which warm water could flow from south to north, the Boreal Sea became isolated, cooled and froze, throwing the Middle Jurassic Earth temporarily into an icehouse world. Now, with continents beginning to separate once more, the currents are beginning to flow again. The Europe of the Late Jurassic, 150 million years before the present, is a lush greenhouse on land, and a swirling meeting ground for hot and cold in the seas. As tropical and polar air mingles in the Boreal seaway, it brings stormy weather to northern Europe.[7]

Shelly plankton and other invertebrates grow and die, their calcium carbonate shells accumulating on the ocean floor. As sea levels fall and as the Tethyan trench drags Africa towards Europe, the calcium-rich seabed will be raised into the towering limestone of the Swiss and German Jura. These will one day become the sources of two of the great rivers of Europe, the Danube and the Rhine, that carve their path through the tectonically lifted beds of the ancient seas. Most of the geological periods are linked through their etymologies to a place, and the mountains of southern Germany and Switzerland are the reference point for this one. In the Austrian

Tyrol, a golden-topped stake protrudes from the mountain. It has been hammered by geologists into a specific point in time, marking the definitive boundary below which is Triassic and above which is Jurassic. The Alpine region of Europe is the 'golden spike' for the period, these seas the definitive Jurassic waterpark.[8]

Not far away from the frenzied activity at the surface is a world of calm, a shimmering crystal structure in the deep dark sea below. High-piled tubes, icy, like frozen lace, are stacked tens of metres high, each a brilliant white net of woven glass strands. Built on top of one another, some are knobbled like melting candles, a devotional altar that disappears into the blue-black haze in all directions. Although they are fixed in place, these are animals, growing on the skeletons of their predecessors. These are the reef-builders of the Jurassic, the glass sponges. Sponges are some of the simplest animals, at least in terms of their tissues. Organized into only two layers of tissue, a layer of cells with hair-like structures called flagella flaps wildly, sucking water into the centre of the sponge, and filtering out detritus from the water column as food. An exhaust hole at the top, the osculum, sends the water out again, the whole system functioning like a jet engine, while also able to detect blockages. Supporting the tubular structure are spicules, generally minute structures made from calcium, silicon or a modified form of collagen called spongin, and ranging in shape from unassuming hearts to spiky shapes resembling throwing stars, spears, grappling hooks or caltrops. Each cell is semi-independent, and a single sponge blurs the line between individual and colony. If you were to put one in a blender, it would re-aggregate – a different shape, but still a working organism, a functioning sponge.[9]

Glass sponges go one step further. The cells that make up their supportive tissue fuse together, channels opening in a way that allows the flow of the internal cell fluid, the cytoplasm, from cell to cell. In a very real sense, glass sponges come close to being a single-celled organism, with the 'syncytium' leaving very little to distinguish the functioning of a glass sponge from that of a highly complex single cell. This interconnectivity means that a glass sponge can easily send electrical signals throughout its body, allowing it to respond quickly and effectively to stimuli and to change the rate at which water is filtered through its body – impressive for a creature that otherwise lacks a nervous system. The oddities of glass sponges do not end there. Their skeletons are built with silicon, but the four- and six-pointed spicules that form the mesh structure of their support, anchoring the animal to the seabed, are enormous. A single, star-pointed silicate crystal in some species can reach 3 metres in length. Those species that form reefs mesh the spicules together into a solid scaffold, robust enough to remain for decades. In fact, this fused skeleton is exactly what remains of the glass sponges after death, and their piled bodies provide future generations with a perfect frame on which to set down their own roots. They are ideal colony-builders. Since they obtain their food from a simple, water-filtration system, cleaning the detritus of other organisms from the water column, they don't need to be close to the surface, as do corals, which have a symbiotic relationship with light-hungry algae.[10]

The Late Jurassic planet is similar in temperature to climatologists' optimistic predictions for the end of the twenty-first century, warmer than the pre-industrial level by about two degrees. There are woodlands, not ice, at the poles, and there are extensive deserts near the equator, but in the highest mountains glaciers can still be found. There are scattered coral reefs throughout this archipelago, but they are more common on the steeper slopes in other parts of the world. Rarer still, but nonetheless hidden in corners of Europe, are reefs being built by oysters, mounds of shells anchored to those of their ancestors. This, though, is an age dominated by sponge reefs, their spicule skeleton more resistant to high temperatures and acidic seas.[11]

Glass sponges – hexactinellids – need clear water. Sponges are filterers of seawater, and are peppered with minute holes called ostia. In a single day, a sponge weighing a kilogram can pump 24,000 litres of water – more than a typical power shower could pump in the same time – and extract most of the bacteria in the water for food. Silty water gets in the way of these pores, and so most sponges can close them when they need to avoid being clogged. Glass sponges, though, cannot.[12]

Because of this sensitivity to particles, they need to live in still waters, far from the murky outflows of rivers. While corals grow in the shallows, below the storm base, where the water is calmer, the sponges grow tall in the dark, tens of metres tall and spreading for kilometres in each direction. Each coiled and ridged mound has grown over a period of thousands of years from a tiny initial colony, circular, symmetrical. The spicules of the first colony-builders are still there, sunk into the soft seabed, blown over and buried with sediment, but remaining solid and giving a firmer foundation than the ocean floor on which new sponges can grow. Sometimes, this growth is in a motte-like mound; occasionally, the sponges lean out over 20-metre precipices. As each colony grows, it encounters others, joining into a conurbation. These tall 'bioherms' are diverse places – in this location, on the sea floor of what will become the Swiss-German border, there are about forty species of sponge, all living and growing together.[13]

Bioherms are built quickly. Over the course of a century, one will rise by up to 7 metres, and expand across the sea floor, following the contours and topology of what already exists. The prevailing current flows from east to west, draining the Tethys Ocean through the islands of Europe and out, through the Atlantic passage. Each herm casts a shadow of still water behind it. There, it is easier to settle, to grow and to build, and so reefs form in elongated lines, each mound baffling the seabed currents like a windbreak. The growth is like that of a city, and life flocks to the reef, as it creates crannies and cubbies for other life forms to thrive in. Sponges are extraordinarily efficient at capturing nutrients on which others will ultimately feed, and so these become diverse metropolises, fringing the northern margins of the Tethys. From Poland in the east to Oklahoma in the west, they cover a length of about 7,000 kilometres of ocean floor. At three times the length of the Great Barrier Reef, these silicon constructions are the largest biological structures ever to have existed.[14]

In the space above the sponge altar, ridged, lacquered coils rise and fall, or jet about at some speed. From each spiral shell, tentacles coyly poke. Ammonites, with perhaps the closest that any non-vertebrate fossil gets to celebrity status, are iconic inhabitants of the Mesozoic seas. While most of the early forms were fairly small – millimetres to centimetres in diameter – others will get very big indeed. By the Late Cretaceous, just before ammonites become one of the many casualties of the Chicxulub impact, the largest known ammonite individual, part of the species *Parapuzosia seppenradensis*, will have a shell diameter of about 3.5 metres. For most of their evolutionary history, though, ammonites are not these armoured krakens, but a common and diverse collection of shelly cephalo-pods, the group of molluscs that includes octopuses, squid and nautiluses.[15]

The shells of ammonites are artistic wonders. As the animal grows, it continually adds new living chambers at its opening, an aragonite shell forged from calcium and carbonate ions foraged from the raw seas and secreted, exuded, to form a rigid, ridged fort-ress. Inside, it is a smooth refuge. The angle at which each chamber connects to the previous one and the size to which it is built is different from species to species, but each follows a logarithmic spiral, with truly bizarre shapes emerging from this simple rule. The classic 'snakestone' shape, a tight spiral on a flat plane, is most common, but others coil helically like a snail, and still others, in the Cretaceous, will develop extraordinary shapes where the spiral is loose and uncoiled, each turn separated from its inner predecessor. The strangest of these are uncanny, 2-metre paperclip shells, arms waving gently from the opening as if to deny any notion of the ridiculous. It is in the detailed insides that an ammonite's real beauty can be found. Within the growing chamber, the place where the ammonite secretes itself, the architecture of the shell is revealed. Each chamber is secured to the previous one by way of a convo-luted suture, a complex fractal dovetail standing out against brilliant nacre.[16]

A series of muffled booms passes through the water, and the reef jerks back and forth for a few seconds. Ammonites, like all cephalo-pods, are unable to hear sound except for a brief period after hatching, but they have pressure-sensitive organs, a series of little fluid- and hair-filled sacs called statocysts that distort with pressure and can detect particle movement associated with low-frequency sound. Now, statocysts pick up the swell as the shockwave passes through the ocean. At the meeting-place of continents, the jostling of plates builds up tension; released, the sea floor appears to boil with a submerged earthquake. Thrown up by the jolt, disturbed white sediment blooms like smoke. The bottom of the reef is churned to opaqueness. Though the epicentre is likely to be miles away, the effects are felt far afield. Even now, the wave is passing through the seas of Europe largely unnoticed until the seabed rises towards the land. There, the tidal wave will be lifted, slosh unstop-pably against the tropical islands, and wreak destruction. Tsunamis are faster in deeper water, and the shallow carbonate shelf of Europe is not particularly deep.[17]

At the surface, the *Rhamphorhynchus* take flight, launching them-selves into the air on their fingerstrut wings. Seen from the sky, the islands of Europe rise darkly forested above the sunset sea. The island of the Massif Central, an ancient highland region about the size of Hispaniola, can just be made out on the western hori-zon, silhouetted against the sun, where stagnant pools along its coast recover from the burning heat of the day. This is an archipel-ago as dense and as bustling as the Caribbean, where life thrives in rainforest and on the hot sand between land and sea. On the tidal flats of a smaller island of the Jura, between the spike fields of man-grove roots, a family of *Diplodocus*-like sauropods make their heavy way. For a large, bulky animal like a sauropod it is easier to pass along the beach than navigate a forest. It also makes it easier for them to be found by others. Harrying the sauropods, advancing and cautiously slinking back, are megalosaurid theropods, the largest carnivores of the Jurassic. *Megalosaurus*, the namesake of the group, is one of the three definitive dinosaurs, used in 1842 to define the name Dinosauria. They are the first group of dinosaurs to become large-bodied predators. Though slenderer and longer-snouted than the later *Tyrannosaurus*, they are built on the same essential plan: small, reduced arms, walking on two powerful hind limbs. They are beach-combers, scavenging on the carcasses of sea creatures that wash up on the strandline – sharks or plesiosaurs, large fish or crocodilians. When a pod of sauropods migrates past, though, the younger and weaker members of the pod are an attractive opportunity.

Relatives of the predator *Allosaurus* and little sickle-clawed dromaeosaurs are found on some small Swiss islands, and stegosaurs daunder around the dark forests of London-Brabant and Iberia. Those coasts, though, are not where the *Rhamphorhynchus* are flying. On taking off, most of the pterosaurs bear north for a few kilometres. They hang in the air like gulls, flapping only when necessary as their soaring path takes them a short distance towards one of the smaller islands, Nusplingen, a place bustling with coastal wildlife.[18]

At Nusplingen, the air tastes of salt and stone. A deep, clear lagoon is surrounded by the breaking of waves where sponge reefs, tectonically lifted, are among the first parts of the Alps to emerge out of the surface of the sea. The lagoon is a two-pronged inlet on the eastern edge of a small island, capped with a forest of cycads and the tall, angular araucarian relatives of kauris, Wollemi pines and monkey-puzzles. It is tinder-box dry, the summer climate similar to that of the Mediterranean, the heat sparking occasional wildfires. The upper portions of the beach are pocked with sticky, resinous cones, dropped from the branches of araucarians and powdered with broken fragments of shells, some fine enough to be sand. The clean white of seashell sand is darkened at low tide by iodine-scented thickets of bristly, bladdered seaweed, before giving way to an exceptionally pale blue. Even at high tide, the rim of seaweed does not extend far; from the shoreline, the sea floor quickly plunges away into lightless depths of over 100 metres. At the bottom, the motionless water has become starved of oxygen, but the lagoon is for the most part a still haven for diverse creatures. The seaquake has disturbed this stillness, having shaken the edge of the atoll, dislodging parts of the exposed reef, and sending boulders into the depths. This quake was a minor one, but even on the leeward side of the resultant tidal wave, the shaken sea has littered the beach with molluscs, brachiopods and other littoral creatures. Eventually, the stillness of Nusplingen will be shattered for ever by a sudden rising of the sea floor of the entire Swabian Jura, bringing the island down upon itself as Europe begins to convulse into being.[19]

Once the *Rhamphorhynchus* have landed, their long tails are no hindrance to walking on the beach. They walk upright on their forefingers, wings carefully folded away. Nusplingen itself is only big enough to support a small population of pterosaurs – *Rhamphorhynchus* among them, but also two species that are more unusual – at least for now. Early pterosaurs resemble the general biology of *Rhamphorhynchus*, but by the end of the Jurassic a new order from within the pterosaur ranks will have all but entirely replaced these early models. *Pterodactylus* and *Cycnorhamphus* are part of the sleek new look, the future of the pterosaur lineages. With very short tails, long wrists and often flamboyant crests, pterodactyloids are the avant garde.[20]

Picking their way eagerly through the detritus of the wave, several *Cycnorhamphus* bicker over a particularly appealing crustacean. Standing upright, supported by their long forelimbs, they snap and shake their heads, none yet committing to blows. Of the three species of pterosaur that inhabit Nusplingen in any number, this is a particularly unusual one. It is a combjaw, but gone is the typical ctenochasmatid array of needle teeth; *Cycnorhamphus* has a paltry few left right at the front of each jaw. Behind these stumpy teeth is a smug grin. The bones of the upper and lower jaws lose touch with one another to form a round, nutcracker gap. Were it not for the stiffened plates covering that awkward hole, the jaws would look like a pair of coal tongs. Grabbing its prize, the triumphant *Cycnorhamphus* anchors the unfortunate prey in the gap, and bone and sheath deliver a crushing end.[21]

Juvenile *Rhamphorhynchus* – pterosaur young are informally known as 'flaplings' – don't come out as far as the reef. They are too small to hunt for fish themselves and, like many vertebrates, pterosaurs do not invest much in their offspring. This means that, in at least some species, they emerge from the eggs with wings and a backbone that are immediately capable of supporting full independent flight. The flaplings, with short faces and only very little teeth, must seek out their own food, and restrict themselves to land, agilely snapping at insects until they are old enough to venture out to the fishing waters with their relatives. By then, their faces have matured and lengthened, and those little jaws for crunching beetles have grown into spike-filled fish-catching machines. The distinctive tail vane – thought to help stabilize flight – can be taken as an indicator of age, too. Starting out as roundly elliptical, it changes through diamond and kite shapes, finally becoming an inverted triangle. The development of pterosaurs from newly hatched infant to full adult is not as rapid as in birds, which often reach adult size within a year, before a sudden slowing or cessation of growth. *Rhamphorhynchus* grow slowly and continuously, with a more gradual transition from juvenility to adulthood. Because of this, they can take at least three years to reach full size, while capable of flight almost that whole time, a pattern more like their 'reptilian' relatives.[22]

The last of the fishing flock returns to the island as the light begins to fade. Carelessly swooping low over the lagoon, perhaps hoping for a final snatch at a squiddish *Plesioteuthis* lurking close to the surface, a *Rhamphorhynchus* stutters and stalls as if realizing its mistake. Too late; a spray of water hides a huge black mass. Futile wing tips beat the salt water desperately, until stillness once more takes over. There are risks even for adult pterosaurs in leaving the island. In the lagoons of Nusplingen and Solnhofen live large and heavily armoured fish, *Aspidorhynchus*, which lurk, their pointed snouts just out of view, ready to flick their tails in a powerful leap to snatch passing pterosaurs from the wing.[23]

Much safer, then, to stay on land, holding onto the trees like the *Rhamphorhynchus* flock; while they are perfectly capable on the ground, they are still more at home in vertical settings or wandering the beaches like the pterodactyloids. Even once the diurnal inhabitants of the island have gone to sleep, they have left their mark on the ground. Their footprints remain along the tideline. The splay-fingered, web-toed footprints of *Rhamphorhynchus* contrast with those of the pterodactyloids, which walk with their hands held sideways. Here, the skittering marks of landing pterosaurs, where the claws have dug into the sand as they landed hind-feet first, a jump, skip and hop to a stop. There, the drag marks of a horseshoe crab carapace, hardly different from those of the modern day, or the regurgitated beak and discarded shell of a belemnite mollusc.[24]

The *Plesioteuthis* cephalopod that escaped the jaws of *Rhamphorhynchus* also has a meal on its mind. Although a relative of sedate octopuses, *Plesioteuthis* are active predators. Swimming at high speed, this one chases a small ammonite, grabbing it with suckered arms. Crushing the shell with its pointed jaws, it leaves tiny pits on the surface, cracking it open to remove the soft parts from their mother-of-pearl home. The body of the ammonite is sucked down, and left to digest, but this presents the predator with a problem. Within an ammonite head are two hard, calcified jaw parts. Digesting these is, in fact, chemically impossible for a coleoid, which, unlike humans, have alkaline stomachs. The easiest way out is the way that they came in, and so the hard remains are vomited out once more, sinking to the sea floor as a sticky, mucous mass. Fossil vomit has a particular name – regurgitalite. It is considered a 'trace fossil', the fossilized result of behaviour, of something other than the body, such as burrows, footprints or faeces. Ammonite faeces – worm-like strands, coiled by the time they settle on the seabed – are some of the most common fossils in the limestones that make up the Jura.[25]

By the entrance to the lagoon, a log floats in the waves, gently tossed like an uncertain rowing boat. Once part of a fat-trunked araucarian conifer, its thick bark has protected it from the ravages of the ocean. As it heaves with the waves, shining stalks, colourful medusa-hair projections, break the surface and roll under once more. Beneath the surface, the stalks end with feathery parachutes, furling and unfurling, out and back again, as they coil food into handkerchief mouths. Sea lilies, or crinoids, like the members of this *Seirocrinus* colony, are echinoderms – relatives of starfish and urchins – and they passively feed on plankton and floating biological detritus as they float through the water. About fifteen of them have attached to this log, trailing behind it like the landing parachutes of a space shuttle, where the resistance to flowing water is lowest. Their stalks are stacked rings of hard calcium, each stalk supporting a feather-duster fringe that cleans up the oceans.[26]

Like reefs, the floating logs become islands of diversity in an otherwise barren ocean. As they are travelling at only one or two knots at most, it is hardly difficult for creatures to hitch a lift. Joining *Seirocrinus* on these driftwood paradises are all manner of molluscs as well as more active swimmers. Small fish follow these communities on their voyages as an easy source of food, the shellfish and echinoderms filtering out nutrients from the water, and the fish feeding on the bodies that they build. Even in total isolation in the middle of the ocean, a log, for as long as it survives, can support a thriving community.[27]

The rafting colonies of sea lilies and their hangers-on are extremely long-lived, with some being twenty years old, and the sea lilies are correspondingly large; some of the stalks on *Seirocrinus* sea lilies can be up to twenty metres long – about the same length as an adult fin whale – with a crown a metre in diameter. Modern-day raft communities last up to only about six years, and the biggest echinoderms, which are not rafters, are starfish only about one metre across. Eventually, either the log sinks under the weight of new colonists or, having soaked for too long, it disintegrates. The presence of the oysters preserves the longevity of the system, helping to seal cracks in the bark and preventing water from permeating the internal structure too quickly. Even unsealed, a large log like this could last a couple of years, but the adult sea lilies attached are a good ten years old. This is in part because in the Jurassic oceans, there are no wood-boring predators. Shipworms, the scourge of mariners during the age of sail, will not appear until the Cretaceous.

Their appearance will make this way of life impossible, something that can and will never be replicated in quite the same way again; wood just doesn't float for as long as it used to.[28]

Although *Seirocrinus* colonies have been found as far away from Europe as Japan, this fragment of an araucarian is probably more local, having drifted here from the islands to the east, or the Asian west coast. Those islands in the west of Europe's archipelago are botanical gardens, homes to diverse forests, with no one tree kin to its nearest neighbours. On the larger continental landmasses to the east, there are wide-spreading forests, dominated by araucarians, and this is where much of the floating wood in the sea originates. The island communities are distinctively similar to one another, containing different species from the forests of the east. There is an invisible line dividing these biomes from one another, a sea border that prevents migration and maintains difference. The most famous example of this kind of invisible border in the modern world was described by the co-discoverer of natural selection, Alfred Russel Wallace. During his time in the Indonesian archipelago, he noticed that all the islands to the east of Borneo and Bali were distinctively Australasian in their species, and different from the typically Asian creatures of the western islands. This division reflects the connection between the landmasses during the last glacial maximum; Borneo, Sumatra, Java and Bali were all connected by land bridges to Asia, while Papua and other eastern islands were connected to Australia. The 'Wallace line' is one of the invisible borders that layers geographic history and ecology together, separating biogeographic realms from one another. Many other geographic features act as these borders in the modern day, from the Himalayas to the deserts of North Africa. In the Jurassic, the European islands mirror the island divisions of modern-day Indonesia.[29]

These seas form a world dominated by meeting places and divisions. Pterosaurs hunt and are hunted at the boundary between sea and sky. The world's currents flow around continents that are beginning to go their separate ways. Diversity is recovering after a worldwide change that is marked by a mass extinction. The Jurassic is best known for its dinosaur inhabitants, for plesiosaurs and ichth-yosaurs. It is here that past life began to be understood, where the earliest three genera first used to define what 'dinosaur' meant, *Iguanodon*, *Megalosaurus* and *Hylaeosaurus*, roamed. But they could not have existed without a firm ecological foundation. The island diversity of the European archipelago was built from the sea floor up.

The rise of sponges and corals forms diverse reefs and islands, built on the remains of their ancestors. Onto these bare and emerged islands life has alighted and clung, a merging of east and west, of north and south. Trees grow, built of sunlight and the mineral skel-etons of those former reefs. In death, commandeered by lilies and oysters, trees ride the currents of the world. In ecology, nothing is completely isolated. Everywhere, always, life is built upon life.

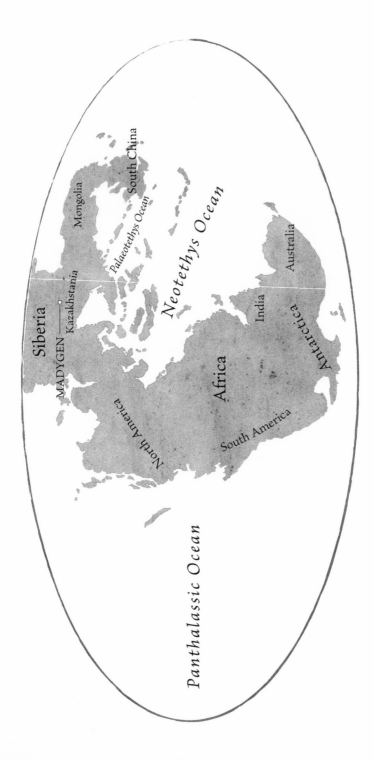

Siberia

MADYGEN

Kazakhstania

Mongolia

Palaeotethys Ocean

South China

Neotethys Ocean

Australia

India

Antarctica

Africa

North America

South America

Panthalassic Ocean

# 9.  Contingency

*Madygen, Kyrgyzstan*
Triassic – *225 million years ago*

'I live on the mountain
no one knows.
Among white clouds
eternal perfect silence.'

– Han Shan (tr. J. P. Seaton)

'Isn't it astonishing that all these secrets have been preserved for
so many years just so we could discover them!!'

– Orville Wright in a letter to George Spratt, 1903

It is cool in the shade of the *Baiera* tree, its ribbon leaves lumines-
cent as they frame an inverted triangle of afternoon sunshine, steep
forest slopes rising either side of a mountain glen. The tree canopy
below contains clues to features otherwise unseen. In the distance,
a gap in tree cover marks a lake edge, while a ragged line of darker
vegetation traces the route of the narrow river that has excavated
this valley. Mosses grow along the ground, where thick, black soils
form a soft, fragrant carpet. To modern ears, the silence in this forest is unnerving and unnatural. There is no birdsong, for this is
before birds. Only the sounds of wind, water and insect wings disturb the air. To modern eyes, this forest is deep and exotic. Even the
densest and most diverse forests in the modern world show the
marks of thousands of years of human management, but these
woods are truly pristine, every surface given over to lichens and
ferns and mosses, trees rising through the thick-trunked debris of
their fallen, rotting forebears.[1]

*Sharovipteryx mirabilis*

The rich earth is the result of accumulated decay, years upon years of leaf-fall, but the plants that grow from it are less familiar, for this is before flowering plants. The forests of Central Asia are a mixture of ginkgophytes, seed ferns, cycads and dark-leaved swathes of *Podozamites* conifers. Deciduous, broadleaved trees with branches spread wide, *Podozamites* cover the land, in places so dominant in the canopy that little else grows tall. They form monocultures that are common throughout the temperate world of eastern Laurasia, having spread from China not long ago. Here, these conifers ripple over the slopes of the low mountains of Madygen, in the Triassic of Kyrgyzstan.[2]

For a cone-bearing tree to have broad, veined leaves rather than needles is, in the modern day, rare. There are exceptions that diversified and still coexist alongside the angiosperms of the present – the kauri pine, yellowwood and nagi, for example – but in Madygen, they are only outnumbered by the smaller seed ferns. Here, the open *Podozamites* canopies allow light through, supporting a lower understory of others who have managed to get a foothold in the Alpine landscape.[3]

The trees lean into the rugged valley, one of many that lie parallel to one another in this basin and range system ranging from gently undulating glens to entirely impassable cracks. Streams splash into pools that overflow in occasional waterfalls, cascading, tumbling, growing into rivers that eventually trudge through the floodplain into the oil-smooth lake of Dzhaylyaucho. Though only about 5 square kilometres, Lake Dzhaylyaucho has a pleasing flatness among the forest slopes that rise a few hundred metres above its surface. In the misty distance, where the lake can be seen spilling out, refreshed, on a journey to the coast that has some 600 kilometres to go, the clouds break up the jagged horizon. Occasional peaks emerge unsupported in the sky, while white vapour hides in unseen dips in the forest, or streaks across the flatter margins of the lake. The air is not too humid, and with rainfall evenly spread throughout the year, warm summers and snowy winters, the climate is ideal for a stable and diverse ecosystem to develop. Away from the sporadic cliff edges, the forest closes in, and the ground is littered with the debris of abundant life. Long, thin cupedid beetles clamber over the rotting humus – they are specialists at consuming soft, decaying wood and the fungus that causes that decay. Indeed, insects as a whole are exceptionally diverse here; of the 106 insect families known to have inhabited the Triassic world, ninety-six are known from Madygen, with over 500 species of insect counted to date, including the earliest known weevils and earwigs in Earth's history. Many plants here are robustly defended against insect feeding; it is thought that the hairy leaves of cycads evolved as a deterrent against insect predation. But as much as the insect population consumes, there are also those that feed on them.[4]

Here and there are scattered cobble-sized blocks of limestone, eroded from above and carried down the flow, reminders of the ancient oceanic past of this landscape. Within a few pieces, traces of fossil shells can be seen – the coils and shapes of the long-extinct creatures of the Carboniferous seas, well over 100 million years old even now. As with so many ranges, these mountains are built from the deep sea, and though this is over 200 million years before the present, the complexion of the Earth is still contingent on an ancient history. Dark, fragile, fissile shales, layered fine as paper and scattered as scree around the steep sides of the valley, are from the soft undisturbed mud of the sea floor. Thick, pale beds of limestone, rough-skinned with weathering, are dense concentrations of the minute shells of marine creatures that lived in the Devonian and Carboniferous Turkestan Sea, an extension on the western edge of what will one day become the Pacific Ocean. A shelf of volcanic basalt rock tells where the tectonic conveyor belt was pulling the ocean floor under another plate, and where, throughout the Permian and Triassic, the risen seabed has been eroded. Dragged out of the flow by occasional floods, the ancient rocks of the mountain streams are quickly overgrown by plants that love the wet spray – soft mosses so deep as would sink an unwatchful step, shining flat liverworts, and the overhanging coiled sprays of ferns.[5]

A shadow streaks across the canvas, instantaneous, joyous and unmistakeable, gone as soon as it appears. The archosauromorph

*Sharovipteryx mirabilis* is almost unique to Madygen. When still and holding onto the vertical trunk of a tree, it is indistinct, another browny-green shape among many others, but in full gliding pose, silhouetted against the bright sky, it is a stopping of time, an indelible afterimage that lasts far longer than the experience.

Four limbs splayed wide, with thin sheets of skin stretched taut between its hindlimbs and tail, and a second, smaller sheet of skin connected to the forelimbs: the triangular profile of a *Sharovipteryx* in flight is a startlingly efficient and manoeuvrable gliding posture; the same wing shape is used in aeroplanes from modern fighter jets to Concorde. Compared with the apparatus of modern-day gliding animals, *Sharovipteryx* is hi-tech. It must glide at quite a high angle, chest thrust into the wind, to provide any lift, but minute movements from the knees change the shape of the main delta wing, and so alter with high precision the direction of flight.[6]

A moment after passing by, its path takes it blustering into a tree trunk, and on it clings, legs wrapping around the tree as a young child holds onto its parent, knees bowed out. Its elegance in flight far outstrips its appearance on the tree, with its membrane retracting and limbs thrown all directions like a collapsing deckchair. Those knees, so useful in the air, look continuously poised for a frog jump, held out as the feet are moved up the body to maintain a hold on the tree. The belly of the creature is slightly concave, so as to more tightly hug rounded branches.[7]

*Sharovipteryx* is almost unique,* but in the experimental times of the Triassic, even more distant relatives are taking to the air. Several species of reptile around the world are flying on paragliders supported by extremely long, hinged ribs. At a time when only insects are truly flying, all these gliders are in the front line of vertebrate evolutionary innovation. Before too long, more lineages of

---

* To date, there are only two species within the family Sharovipterygidae, both of which are hindlimb gliders. One is *Sharovipteryx mirabilis* itself, from Madygen in Kyrgyzstan, and the other is *Ozimek volans*, a slightly larger animal from the Triassic of Poland with the same long legs and light bones.

archosaur will take to the skies – first pterosaurs, and then at least three groups of dinosaurs. It will be about 170 million years before mammals finally fly as bats in the late Paleocene or early Eocene.[8]

In fact, along with birds and flowers, the Triassic is almost beyond the existence of mammals at all. Mammalian diversity, the ancestors of everything from you to a platypus, a wombat or a manatee, is contained at this time in, depending on who you ask, only one or a few species worldwide, from this point in time back to the beginning of life. *Adelobasileus* is an early mammal (or at least a very close relative) contemporaneous with the site at Madygen, but it lives far away in what will become Texas. A Triassic naturalist would hardly give it a second glance, perhaps noting the odd bony casing of its inner ear, but otherwise regarding it as a small if unusual member of the cynodonts. The cynodonts are, in a manner, a stepping stone in what is, looking back, the path of mammalian evolution. As such, they share several of the features that we think of in the modern day as exclusively mammalian. The Triassic immediately followed a devastating mass extinction, and the cynodonts diversified in its aftermath in much the same way mammals did in the Paleocene.[9]

Even Madygen's own fairly anatomically conservative cynodont, *Madysaurus*, is like a mammal in many ways. A hard palate separates its breathing from its feeding. Its teeth are differentiated into slicing incisors, stabbing canines and grinding premolars and molars, rather than the uniform row of identikit teeth that characterizes most other vertebrate groups. Its hairy hide covers skin with oil-producing glands. It is an egg-layer, but, unlike a platypus or an echidna, will probably not feed its hatched young on milk – mammary glands seem to have developed at a later stage of cynodont evolution than that which led to *Madysaurus*, perhaps initially as a way of keeping their thin-shelled eggs from drying out.[10]

What with gliding *Sharovipteryx* and the strange and novel physiology of cynodonts, there is a lot of anatomical experimentation going on all around the world. Ask any vertebrate palaeontologist about which geological period includes the weirdest beasts, and the overwhelming majority will plump for the Triassic. While the innovations of the cynodonts persist in humans, the Triassic is peculiar in having such a range of disparate forms, many of which have not survived. In no group is this truer than in the archosaurs and their kin, of which *Sharovipteryx* is one. In the modern day, archosaurs include birds on the one hand and crocodilians on the other. In the past, even excluding the undoubted otherness of dinosaurs, the diversity was much higher; archosaurs also include pterosaurs and, in the Triassic, several forms that explore the anatomical and physiological limits of the group, and begin their rise to ecological dominance.[11]

In parts of what will become Europe, there are semi-aquatic creatures called tanystropheids. Many of this family are enormous beasts, up to 5 or 6 metres in length, and they all are found near bodies of water. They hunt squid and fish, aided by necks that are half of their overall length – that is to say, up to 3 metres long – which allows them to separate their inconspicuous feeding apparatus from the giveaway gigantic body that might alert their prey. In shallow, muddy water, they ambush faster-swimming prey with a sudden sideways swipe of their heads, propelling their bodies forward with gigantic froglike kicks. Unlike plesiosaurs and ichthyosaurs, their feet appear to be able to move on land, and they have strong pelvises that suggest that they supported their weight from further back on their bodies than might be expected for a walking fishing rod.[12]

Sharovipterygids are not the only bizarre Triassic reptiles in these woods. The signs of activity are everywhere, with scrabbling bank-climbing footprints leading into the seed ferns. Paths scratched in the mosses on tree trunks mark the presence of drepanosaurs, another of Madygen's evolutionary oddities. Living somewhere above, these squirrel-sized, tree-climbing reptiles begin here; *Kyrgyzsaurus bukhanchenkoi* is the oldest known species of drepanosaur – a group that will later spread around the northern hemisphere. With folded skin like that of an iguana, and a pendulous throat sac, *Kyrgyzsaurus* is not an elegant creature. In many ways, drepanosaurs are the Triassic equivalents of chameleons. *Kyrgyzsaurus* has a short, delicate, triangular face armed with a battery of minute teeth with which it snaps at its insect prey. For a drepanosaur, it is fairly average in size – larger species are as big as cats – and it is adapted to life in the trees. Drepanosaurs have grasping hands and feet with opposed digits that allow them to grip branches tightly. In several cases, their tails, long and flattened from side to side, act as a prehensile fifth limb, with the final vertebra modified into a claw with which to gain additional purchase on slippery bark within the canopy. *Drepanosaurus* itself, the namesake of the group, has an enormous thumb claw, as big as the rest of its digits combined, with which it presumably prises open bark to find its food living underneath.[13]

As the river bounces onwards, the limestone cobbles disappear under gravelly curves cut into muddy banks. Overspilling water begins to pool into ever wetter ground. Here, earth and plant begin to merge into peat, sucking in more moisture and condensing more and more into unconsolidated coal. Dissolving the lime along the way, the water grows in minerals and loses its oxygen. Never hurrying, the ageing river exhales into the waiting lake. There, beneath the water, worms perforate the mud in complex, branching burrows, making their homes only as far as the sun's warmth penetrates.[14]

The lake, so clear from high above, is invisible from its shore. A marsh of horsetails, *Neocalamites*, stands in a wall 2 metres tall in shallowly submerged clay. Thick, prickly stems, divided into bamboo-like segments, bear leaves at the joins. Beyond the stand of horsetails, where the water becomes too deep, its surface tension is stretched from beneath, but rarely broken, by dense and floating lime green carpets of crystalwort. Under the surface is a reflection of a forest that provides a haven for the larvae of the hundreds of species of insect inhabiting this land, and the egg-masses of *Triassurus*, the earliest salamander. On the water-lapped shores, damp mosses, too, are the breeding grounds of swarms. Massing a thousand strong are clouds of shrimp called kazacharthrans, with large, heavily armoured heads, shaped like the cross-section of an apple. From the front of this shield, Chinese dragon whiskers flick and sense their surroundings. Combined with their awkward, undulating swimming, when their legs are hidden beneath, they are fairly similar at first glance to tadpoles, and indeed they are relatives of so-called tadpole shrimp of the modern day. They flutter after the detritus that floats in from the river, or the eggs of flies that have been laid on the still surface, and are unique to Central Asia.[15]

During the spring, when the rivers are in full flow, there is plenty of food to go around, but when things get tighter, the kazacharthrans face competition for food from not only swimming arthropods, but also static colonies. What initially looks like a rock coated in some kind of algal slime is in fact a colony of animals called bryozoans – moss animals. Every microscopic individual in a bryozoan colony is a clone of the original animal that settled on the lake floor, and is a hermaphrodite, both male and female. Their skeletons, unlike those of other colonial animals like corals or glass sponges, are not even mineralized. Instead, they are made of jelly-like proteins, so that the colony has a very wobbly texture to it. Madygen, being continental, gets cold in winter, and conditions can be very volatile. Now, in the height of summer, the bryozoans are making proverbial hay. They are producing special, chitin-insulated clusters of cells, called statoblasts, and releasing them to land elsewhere. These statoblasts are biological insurance policies against a harsh winter. If the lake freezes, or if the water level drops too far, the colony will die, but the statoblast will survive, and will be primed to open when conditions improve. The microhabitats formed in the liminal regions of freshwater lakes are hugely important for maintaining the diversity of the whole ecosystem, from the plants to the apex predators.[16]

Growing ripples on the water's surface mark the passing of just such a creature, the biggest animal known from Lake Dzhaylyaucho – as long as a river otter – and a survivor from another time. The mountain isolation of Madygen means that many of its creatures – including all of its vertebrates – are endemic, not known from anywhere else in the world. All the rest have reasonably close relatives elsewhere, but in the case of *Madygenerpeton*, none of its kind has survived so long. In the modern day, the four-limbed vertebrates of the world can be divided into amphibians, which still breed in water as is ancestral for tetrapods, and amniotes, which enclose the developing embryo in a series of membranes, whether in a shelled egg or within a womb. Both amphibians and amniotes are equally distant relatives of *Madygenerpeton*, which is a chroniosuchian – the name means 'time-crocodile'. With backs armoured by interlocking plates, chroniosuchians have followed a crocodile-like lifestyle in the waterways of Asia for 30 million years, but their time is almost up. The crocodile lifestyle is a successful one, however, and where chroniosuchians are laying down the baton, it is being picked up by giant amphibians like *Mastodonsaurus*, a 6-metre salamander-like animal with a flat, almost perfectly triangular skull so shallow that the two largest lower teeth, conical and sharp as an awl, protrude through special holes in the top of its snout. Phytosaurs, another group of Triassic archosaurs, are so superficially like modern crocodilians that, were it not for their nostrils being positioned far back on their snouts, it would be very easy to confuse the two groups.[17]

*Madygenerpeton* has adopted a particularly aquatic existence. Its interlocking plate armour, a bony brace for the body, is more flexible than that of its ancestors, allowing a sinuous spine. Weighed down by this additional armour, it can lie low in the water, its small, alligator-shaped head barely emerging from beneath the surface. Floating weeds tangle with its rough, knobbled surface, broken through by little raised eyes and nostrils. The recent end-Permian mass extinction struck as chroniosuchians were beginning to diversify, and it has nipped them in the bud. Those that survived have merely stuttered along until now, and *Madygenerpeton* is thought to be the last remaining species. Silently, it melts back into the crystalwort.[18]

Deep at the bottom of the lake, too far down to concern those at the surface, are coelacanths and lungfish, but also a surprising inhabitant for a mountain lake so far inland – sharks. Wait for ever at the surface and you would probably never see one, but occasionally their leathery egg cases, spiralled and pointed like an elongated lemon, wash up on the shore. Finding the egg case of a shark far inland in the high mountains is like finding the carcass of an ibex on the ocean floor; until they were discovered in Lake Dzhaylyaucho, the only sharks known to lay eggs were from that unfathomably different marine realm. Here, in Madygen, there are two types of egg-laying shark, the most common of which is a humptooth shark. Hybodonts, as they are properly called, have distinctive long, curved spines at the leading edge of each fin, and the Madygen hybodonts are, among sharks, small and slow, more like a dogfish than a great white.[19]

For a long time, nobody knew anything about how hybodonts reproduced, beyond speculation related to our knowledge of other, rather different shark groups. The discovery of egg sacs at Lake Dzhaylyaucho changed all that. At least for one species of hybodont, the mountain lakes of Central Asia provided a nursery for their young, a place where adults would congregate and mate. In the shallow water, among the plunging stems of the horsetails, infant sharks hatch and start their slow life. As they grow, they move away from the shoreline, and begin to inhabit ever deeper water. Exactly where they go after this is entirely unknown. Many hybodonts are freshwater all their lives, but others are marine. Lake Dzhaylyaucho is so far from the sea that the simplest explanation for their appearance there is just that the adults are somewhere deeper in the lake, far below the surface, away from the outflow of the rivers and the percolating sediment. The other less likely explanation is that, like sockeye salmon, this particular hybodont shark, the Fergana lance-shark, might be making a breeding journey from ocean to a protected inland nursery.[20]

What have certainly been revealed, indeed what are perhaps least avoidable along the shoreline of Lake Dzhaylyaucho, are the black flies. Despite the number of insectivores at Madygen, from drepanosaurs and sharovipterygians to fish and tadpole shrimp, there are large and diverse populations of true flies, the acrobatic dipterans. Only recently diversified, they are tricky creatures to swat out of the way, thanks to their clever engineering. Ancestrally, insects have four wings, and almost every group, from butterflies to beetles, crickets to bees, stick with that ancestral constraint. True flies, like fruit flies, houseflies or mosquitos, have taken this basic constraint and played with it. Their second pair of wings, the hind pair, is no longer used for generating lift. These wings have changed into club-shaped struts called halteres that are attached with a horizontal hinge to the body of the fly and vibrate wildly during flight. Whenever the fly changes direction at an angle to the hinge, the shaking bends the halteres at their base and turns them into a gyroscope. Detecting this movement, the fly's flight muscles automatically adjust and correct positions. In practical terms, this means that flies can pull off far more daring manoeuvres during flight than any other insect, which allows them to rapidly swerve away from danger, whether the open jaws of *Sharovipteryx* or a swinging newspaper, without spiralling out of control.[21]

On land, the most common insects to be found in the leaf litter at Madygen are cockroaches, but the entomologist's dream encounter here is surely with titan-wings, enigmatic insects thought to be relatives of grasshoppers. These are well disguised, holding themselves statue-still among the fern fronds. Having evolved in the Permian of Russia, they are not known from many sites worldwide, but Madygen is home to several different genera. They behave and look rather like mantises, but their scale far exceeds any modern-day mantis or grasshopper. The biggest wingspan of any insect in the modern day is that of the mariposa emperador, or white witch moth, which can reach 28 centimetres. Titan-wings can get bigger. The laconically named *Gigatitan* has individual wings that can reach 25 centimetres. Standing on four legs, this grasshopper cannot hop. Instead, the front pair of legs are held up, and bear sharp spines to help capture their unfortunate prey. Like modern grasshoppers, however, they can sing. Along the wings are the plectrum and file, which when rubbed together, produce a deep, baritone, bullfrog chirp.[22]

*Gigatitan* even outsizes several tetrapod inhabitants of these woods, including perhaps the strangest denizen of the trees around Lake Dzhaylyaucho. The creature known as *Longisquama insignis*, whose name means 'remarkable long-scale', is another bizarre lizard-like reptile, perhaps also related to archosaurs. This unique and tiny animal – it is barely 15 centimetres long – is a tree-climber with grasping limbs, but what makes it stand out, metaphorically and literally, are its utterly enormous scales, shaped like ice hockey sticks, that emerge in a ridge along its back. There are more than half a dozen of these protuberances along its spine, and each stands as tall as *Longisquama* is long. What these are for is not exactly known, but it is generally assumed that they serve the purpose of either display or camouflage; being so thin makes them unlikely to have any mechanical benefit. However, only one of these animals has ever been found, and the condition of the specimen does not help matters. As always, where there are reports of strange creatures in the forest, only more sightings can help to resolve the questions raised by the first.[23]

The forests and lake of Madygen are an important lesson in deep-time humility. The presence of creatures so hard to interpret as *Longisquama*, of lifestyles so distinct from those of their relatives in the modern day, like *Sharovipteryx* or the lance-shark, and the very local occurrence of species like *Madysaurus* or *Madygenerpeton*, the last of their kinds, remind us quite how much is still unknown about the life that once inhabited Earth. Madygen is a single data point and, as yet, there is little to compare it with. We cannot say how unique this community is, how far the sharovipterygids' wings took them, or what endemic delights existed in other inland landscapes.

Madygen and the Fergana basin as a whole tell a story of contingencies. From the basic scaffold of the tetrapod body plan, manifold variations emerge, but each is built using the limitations of its ancestors. Evolution is a process of adaptation within constraints, and the development of ways in which to break those constraints. From the wing-derived halteres of dipteran flies, first appearing in the Triassic, to the stretched skin of *Sharovipteryx* and its kin, new uses for old structures are changing the ways that animals navigate their environments. In fact, there are ingenious solutions appearing to evolutionary conundrums across the tree of life. The Triassic is a period of change and experimentation, a time on Earth when, to modern eyes, it would seem as if anything goes.

In part, this is probably thanks to the hangover from the mass extinction that occurred at the boundary between the Permian and Triassic periods, the worst extinction event ever to have hit the planet, in which 95 per cent of life was wiped out. After extinction events, the rate at which new species appear increases, and extinction becomes a temporarily rarer phenomenon. By the time of Madygen, the sparse landscapes across the Early Triassic world have filled in, and life is again gloriously abundant. Come the beginning of the Jurassic, the organisms that will come to typify the rest of the Mesozoic will have more or less risen to their ecologically dominant position, and the wild days of experimentation will be over.[24]

Throughout the Permian and Triassic, the area around Madygen has been a typical mid-altitude mountain range, rising slowly enough that erosion is keeping the mountain tops at a near constant height. Soon, they will begin to fall, and by the Oligocene, almost 200 million years into the future, the mountains will have given way once more to sea. In an unlikely mirroring, the modern site of Madygen, hidden in the northern foothills of the Turkestan mountain range, is a recapitulation of the Triassic topology. The modern mountain ranges of the Tien-Shan to the north and the Gissar-Alai to the south, built of that self-same Palaeozoic sea floor, descend into the huge Fergana Valley that forms the border region between Kyrgyzstan, Uzbekistan and Tajikistan. The modern plant communities are semi-arid steppes, filled with grasses, so the human inhabitants have historically been nomadic herders.

There is no separation of biology from history. Every living thing is the result of biological evolution, and is influenced by the lives of its ancestors. This can be anatomical, like the constraints on the different ways that a vertebrate can use its limbs. Or it can be geographic, like migration across the openness of the Pleistocene mammoth steppe. At the beginning of the Triassic, all the major continental tectonic plates were connected, forming the supercontinent Pangaea. The lack of major barriers between terrestrial communities meant that, after the dust of the Permian-Triassic mass extinction had settled, after the oxygen had returned to the deep oceans, and the flames had died down, the common survivors could spread relatively easily around the world, a homogenous fauna that only later became endemic in each place. By comparison, the ocean-separated continents at the end-Cretaceous mass extinction left behind faunas that were less globally similar.[25]

Contingencies in palaeontology extend into the geological record that remains. That Madygen, an inland ecosystem, will be preserved in such detail at all is lucky in the extreme. Inland ecosystems are generally not where sediment is being deposited. Wind, rain and the burrowing of roots conspire to weather exposed rock rather than make more. The history of terrestrial life on Earth is, in general terms, the history of the waterways, of rivers and coastlands, deltas and estuaries. Lakes are so infrequently preserved that the fossil record of lakes has been described as a 'megabias' – a pattern that prevents any long-term analysis because all the data are concentrated on a few isolated cases. When terrestrial sediments do form, they are often poor preservers of detail. Madygen is remarkable in its preservation, making its place in the ecological history of Earth clearer than most marine sites. Evidence of the swarms of insects on the floodplains around Lake Dzhaylyaucho is so plentiful that one expert on the formation describes certain layers of rock as being 'literally tiled with tiny, often hard-to-see wings', with over 20,000 specimens collected so far. We are constrained by what can be preserved, but Madygen breaks the constraints for a moment, and lets us see sharply what could never otherwise have been known.[26]

What exists now can only ever come from what existed before. For the Triassic, what had come before had been utterly destroyed. In the face of the removal of almost everything alive, there was little to work with, and yet evolutionary forces excel at breaking contingency, finding evolutionary loopholes and working with what remains to generate new wonders of diversity. Extinction and speciation usually go hand in hand, and the oddities of the Triassic inhabit a time when ecological options were open to a whole new set of surviving body types: ambushing tanystropheids with their implausibly long necks; drepanosaurs with coiled claw-tails; somersaulting flies. Somewhere on the slopes high above the lakeside, a *Sharovipteryx* shifts itself on the bark of a tree. A kick and a flick, and it launches itself into the unknown.

# 10.  *Seasons*

*Moradi, Niger*
Permian – *253 million years ago*

'Such rains relieve like tears'

– Mary Hunter Austin, *The Land of Little Rain*

'Water oozes up until our steps float
and then we are ankle deep in sky'

– Rachael Mead, 'Kati Thanda / Lake Eyre'

The winds have changed. Northerlies blow through a sparse dune field, whipping up ridges and bringing sharp shards of silicate hard and fast through the air. It is difficult to see anything distinctly. Out on the salt flat, there is no respite, no place in which to hide from the deep, red, piercing wind. A female gorgon tries to lift herself from the ground, shaking away the accumulating sand as she takes another step. To resist this ongoing burial is exhausting, but sandstorms are frequent enough that this will not have been the first she has had to endure. Her thick skin, scarred with age, provides some resistance to it, if not absolute protection. The rainy season is due any day now, heralded by the return of the north wind, but, until the sandstorm subsides, there is little to do but wander and wait. Her jaw is swollen, and she is limping. Ever since a flailing blow from a *Bunostegos* she had been hunting fractured her leg, it has never been the same. The injury is now healed, blood having flooded the broken area, helping the living tissue to knit together quickly – a handy physiological consequence of an active, hot-blooded lifestyle – but the strengthening of the new bone growth has healed as a lump, holding the fracture together from the outside, though weaker than it once was.[1]

*Bunostegos akokanensis*

For a member of the group Gorgonopsia like this, the apex predator at dusty Moradi, such injuries are occasionally severe but not uncommon. The swollen jaw, though, is more unusual. Looseness in her long, sharp left canine might have been the result of a new tooth coming through; despite having differentiated teeth like mammals – incisors, canines and postcanines – gorgons replace their teeth continuously, more like modern reptiles. Gorgons are active predators, and need a functioning pair of upper and lower canines in order to feed. To make sure that this happens, they alternate the replacement of their left and right canines and replace both upper and lower on one side at the same time. However, her right canines are themselves already mid-replacement, so a loose tooth on the left means something else. This is a more pressing development, an accident of cell division. Within her jaw bone is an odontoma, a cancerous tumour, pressed up against the root of her canine. The tumour is filled with miniature teeth, and as they develop, they are slowly eroding the nearby root. She shifts her jaw uncomfortably in her lips. The storm will soon be over.[2]

A burst of lightning briefly lightens the peaks of the Aïr Mountains as the storm thins, arousing her attention. She tilts her head forward, peering over her bull-terrier snout into the blowing sand. Below her sprawling feet is a dry lakebed, some 80 metres in diameter and identical in every direction – white, and paved with uneven geometric figures marked out by ridges of crystalline gypsum in the clay. Every year, this lake runs deep with fresh water. And every year, it runs completely dry, and the memories of that water are held in the light undulations of hardened mud. Even when the rivers from the mountains to the east flow into it, no water ever flows out. Some water percolates down through the soils, but mostly, the dry, hot air vacuums up the moisture through evaporation. It is an end-point, a playa lake with an entrance and no exit, a depression in an enormous landmass.[3]

Almost all the world's land, excepting islands at the coast, is locked up as a single supercontinent, Pangaea. There is land close to the north pole, land at the south pole, and a continuous band in between, striped with cool, temperate land, damp forests, and the great red western and interior deserts near the equator. The water cycle relies on the oceans as a source of cloud and rain, and when much of the supercontinent's surface is far inland, precious little rain can make it into the arid centre. However, when it does, it does so in dramatic amounts. Pangaea is roughly C-shaped, a great cup, open to the east around the equator, where the early Tethys Sea is forming. To the east of that sea, a great archipelago of ever-wet tropical islands – the continental mass that will one day become southern China and south-east Asia – acts as a barrier against the ravages of Panthalassa, the giant ocean covering the rest of the planet, bigger than the Pacific and Atlantic put together, covering more than an entire hemisphere. With large landmasses to the north and south, and barriers to the west and east, the Tethys embayment is not unlike an expansive and deeper version of the Caribbean. As those who inhabit the fringes of that sea today will know all too well, this geography is prone to storms.[4]

In the northern summer, the land in northern Pangaea heats up, while the wintering south cools. In between, the sea, heat-preserving water, remains roughly the same temperature. Being bounded by islands to the east, the Tethys has few strong currents, and so a pool of warm water – 32 degrees at the surface – has built up in this sea, shifting north and south as the seasons progress. Wherever the warm pool lies, air pressure is low, sucking air from the cooler half of the world and sending it on, freshly filled with evaporated Tethyan water, onto the summer coasts of the other half of Pangaea. The amount of rain that falls in the path of these winds has been estimated at up to 8 litres of rainfall per square metre every single day during the August peak. Pangaea is the land of the megamonsoon.[5]

Moradi lies in the southern half of Pangaea, just to the south of the lush, tropical region bounding the equator where the main rain belt falls. It forms the northern end of the southern desert belt, a world of purest drought. Sitting as it does on the boundary, about 2,000 kilometres from the nearest sea, Moradi gets both extremes. The land between the desert and the sodden tropics is warm, extremely dry, and sees exceptional rainfall for short periods of the year. When Earth turns its southern side to the sun, the mountains of the Aïr massif to the east capture the heavy rains, and life comes cascading down its slipped escarpments to replenish the parched landscape of the Tim Mersoï basin, splaying out as sloping fans of mud, and slowly swerving and curving into an ever-branching and ever-rejoining network of waterways.[6]

Away from the white flatness of the playa lake, the landscape rises slightly to a red-brown plain covered in loess. Here and there, small coniferous shrubs, voltzians, gather. The sandstorm has blasted them hard, and their long, ribbon-like leaves and many-needled dwarf shoots – the little budding structures that would one day have been new branches – lie torn and scattered, half-buried in the deposited sand. These voltzian conifers stand in sparse clusters along the branching river channels of Tim Mersoï, providing shelter and shade for the animals of Moradi.[7]

Gorgons are by far the biggest predators in the latter stages of the Permian, and the largest gorgons, known as rubidgines, are found only in the African part of Pangaea. Some rubidgines, like *Dinogorgon*, can have heads larger than a polar bear's and bodies to match. With powerful brows, short, thick tails, a pair of long canine fangs, and a deep, Shere Khan jawline, they carry an imposing authority. In overall aspect, a gorgon, pacing in the sand, looks somewhere between a big cat and a monitor lizard. Their feet are better than most at grasping, able to hold onto large and struggling prey just as a big cat would, but they are hairless and have a slightly sprawling posture. As a desert-dwelling carnivore, the Moradi rubidgine has a particular problem obtaining water. Some cool, shadowed pools are present all year round, and there small populations of lungfish, which can breathe air if necessary, and freshwater bivalve molluscs – *Palaeomutella* – cling onto life until relief arrives. During the summer the gorgon will, of course, drink from these pools as a supplement, but it is likely that, like other large desert carnivores, the majority of her water is obtained from meat and blood.[8]

The biggest prize of all the prey at Moradi is a charmingly

obtuse-looking animal called the Akokan knobblehead – *Bunostegos akokanensis*. A small herd of these are collected near a grove of trees between the dry remnants of two river channels. Here, a well-worn path in the bank has been hardened into incompressible lumps by their incessant footprints. With chunky, hairless bodies about the size of bison, they have short, thick tails and spade-like feet. The knobbles that give them their name are protuberant bony bosses – two or three at the front of their snouts, larger ones at each of the upper back corners of their heads, and one more over each eye. Down their backs runs more armour, with ridges of bony lumps – osteoderms – to protect them from gorgon attack. Size is this animal's main advantage, and all pareiasaurs quickly grow to their adult bulk. Compared with the other large tetrapods of Moradi, a *Bunostegos* quite literally stands out. Rather than sprawling, lizard-like, it stands tall, with its limbs beneath its body. It is the earliest example of a tetrapod that walks upright.[9]

For a big animal in such a landscape, walking upright is a useful adaptation. To find enough water and food to survive, herbivores need to be efficient at moving between whatever sources remain. If their body mass is placed over the limbs rather than between them, they need less energy to walk. In open, arid habitats, an animal's range is typically wide, and sources of food and drink are often far apart. For this reason, these creatures tend towards more vertical postures than those of their close relatives, and being large makes for proportionally more efficient fuel use. *Bunostegos* is the first to adopt this posture, but not all upright tetrapods are its descendants. Dinosaurs, which have upright hindlimbs and – where they are used for walking – sprawled forelimbs, and larger mammals, in which all four limbs are upright – are merely very distant cousins to *Bunostegos*. They are all 'amniotes', a name given to distinguish those animals that developed shelled eggs from amphibians; but they fall on distinct lineages that follow early divisions in the amniote family tree.[10]

It is in the Permian that the amniotes have begun to take over the land. The relative drought, or at least the extreme seasonality, of this period is a new phenomenon. During the Carboniferous, a group of amphibious animals had evolved a clever anatomy for their eggs. Being descended from aquatic fish, their eggs, like those of their ancestors, had the same basic salty chemistry as the sea. The proteins involved in development and DNA replication are adapted to work in a watery environment, and cannot function once desiccated. Although amphibians can leave the water, they cannot survive their early life without standing water. Amniota is the name given to the descendants of the species that first solved this problem by sealing off each egg, surrounding it with a series of membranes: a shell for physical protection; the amnion and chorion, two layers of protective sacs, in which the embryo develops, and finally, the allantois, which acts as a form of embryonic lung, bringing oxygen from the porous shell to the embryo so that it can continue to respire, and functions as a repository for the waste products of that respiration. These protective membranes maintain the internal chemistry of the egg as it develops, even in entirely dry environments.[11]

Over 30 million years around the end of the Carboniferous and the beginning of the Permian, the Earth's climate shifted from fairly humid to extremely arid. By the time the new dry world took full hold, amniotes were in their element. What had been a backup option in case of temporary drought allowed them to explore new niches and create new communities inland. Freed from the restriction of finding fresh water in which to lay their eggs, they settled the previously inaccessible deserts and high grounds of continental Pangaea. Where insects, arachnids, fungi and plants had gone before, each with their own version of a drought-resistant seed, spore or egg, vertebrates finally followed. In the modern day, the only non-amniote tetrapods are frogs, salamanders and the blind, burrowing caecilians, the so-called Lissamphibia. All others, including humans, are variations on the amniote theme. The amnion is familiar to us as the container of the 'waters' that break during human labour, the miniature ocean each of us creates to protect ourselves as we develop. The chorion and the allantois have joined forces to produce the familiar placenta. We still carry around the remnants of our ancient ecological traits. Our cells are unable to break the constraints of our most fundamental chemistry, and our bodies bear the legacy of the developmental loophole that allowed our ancestors to move onto land.[12]

Gorgons, our closest relatives at Moradi, lay soft-shelled eggs, as do the herds of *Bunostegos*. But unlikely as it may seem, there are a few hardy amphibians here too. Patrolling the gravelly central channel of the riverbed is an animal broadly resembling a large alligator in size and basic shape. Its small eyes bulge from the skull more goofily than an alligator's, and its nostrils, raised like little volcanic cones, are not at the tip of the snout, but halfway down it, while two long lower fangs protrude, bizarrely, through the top. *Nigerpeton* is a temnospondyl, more closely related to modern amphibians than to amniotes, but far bigger than today's diminutive amphibians. Even the giant *Andrias* salamanders, still found in a few rivers of modern-day China and Japan, only grow to a mere 180 centimetres in length, some 60 centimetres shorter than *Nigerpeton*. *Nigerpeton* and the other giant Moradi temnospondyl, its close relative *Saharastega*, may still be tied to the water for breeding, but they are nevertheless very much terrestrial animals.[13]

Perhaps the most comfortable inhabitant of the Moradi wet desert ecosystem is the species that was first found to live here – *Moradisaurus*, the 'Moradi lizard'. Not a true lizard, *Moradisaurus* is a captorhinid, or 'grabsnout', another type of early amniote that has no close living relatives. An important change in the diets of early amniotes, including grabsnouts like *Moradisaurus* was the adaptation to eating high-fibre vegetation. That fibre mostly consists of the carbohydrate cellulose, a molecule that is only digestible by an enzyme that vertebrates cannot produce. Humans, for example, cannot digest cellulose at all, so the energy we obtain from the plant matter we eat comes from other sources like starch or sugars. As a result, we mainly eat fruits and seeds, including grains and nuts, and tubers like potatoes or turnips. Where we do eat leaves or stems such as spinach, cabbage or celery, we do so for other nutritional reasons – vitamins, minerals and that indigestible cellulose fibre – than energy. Whenever there is an energy source that is inaccessible, the best thing to do is to collaborate with a micro-organism that can access it, just as corals collaborate with algae to digest light. A high-cellulose diet needs just such an association: *Bunostegos* and other pareiasaurs have a barrel-chested stomach filled with bacteria, in which to ferment. Others, like the smaller grabsnouts resembling *Moradisaurus*, help speed up the process by slicing plants with up to twelve separate rows of teeth. This new ecology opened up evolutionary opportunities elsewhere; Early Triassic droppings from other synapsids preserve eggs of a distinctive type of parasitic worm known to live only in herbivorous guts.[14]

In the riverbed, the heat-rippled shadows of the trees give way to dark reflections, and the sand foams. Some days ago, the monsoon hit the Aïr Mountains, and the rain has been making its way into this basin ever since. Now, the frontier has finally arrived, picking up the salty precipitations of the old river, tumbling thin ribbed leaves towards the air-filled lake. Now that water is filling the channel it is nearly impossible to think that the river was not always flowing, as the sudden blackness obscures the riverbed, and the oncoming water is endless. Opening out into the lake, the water forks over the flat, a tree drawn in black on white canvas, slowly locating the lowest part. As the branches expand and join together, the once bare clay of the playa lake swells. Crystals in the ground accept the water's bulk, and the cracked earth heals. In the slowing flow, the mountain dirt settles, and the lake becomes ever less salty, ever clearer. The banks of previous years are inconstant, ragged and horizontal, the consequences of a transient filling. Aridity-loving plants, growing near the margins of this lake, are over-brimmed, and hair-like strands of thin vegetation perforate the water's surface.

Against the sun, the water is a perfect mirror to the sky, and where there was once lakebed is now only the appearance of a mass of air, upended. No doubt, hundreds of miles upstream in the mountains, tempestuous rains are shattering leaves and destabilizing roots with deafening volume. But here, at the sandstorm's end, is serenity. A small, wedge-headed *Moradisaurus* waddles forward, its short trunk forcing it to throw its shorter limbs with each step, bending from side to side to get a bigger stride, feet twisting in the wet mud. It enters the water and then, as it floats on the surface, it uses its dangling limbs to punt itself along. Its tail has an abrupt break partway along its length, where a smaller, thinner tail emerges like a shoot from a tree stump; juvenile captorhinids, like modern iguanids, drop their tails if caught by a predator and later regrow them.[15]

For two months, the rivers swell, and the salt plains are entirely submerged. *Bunostegos* wander to the lake to wallow, their heavy frame sinking deep into the soggy ground. Coniferous plants, their tap roots dug deep to maintain a contact with the groundwater table, green and thrive, while the molluscs clinging onto life in the dark pools that remain take the opportunity to emerge and breed. Little lizard-like reptiles scuttle here and there. At the curve of the refilling river and by the filling lake, the inhabitants of Moradi emerge to drink deeply.

Plant flotsam from upstream gathers, caught against the edge of the channel, or washed by occasional surges onto shallow swales. Compared with the slow-growing plants that live all year round in Moradi, the outwash speaks of a lusher environment upstream. Also brought down, eroded from the sandy banks where they died, sharp features smoothed by tumbling water, are bones of *Bunostegos* and of *Moradisaurus*. In the dry lands, decay is slow. Perhaps some five or ten years have passed since the animals these once belonged to died. In that time, decay has broken the skeleton into isolated bones, and the skin, dried into taut sheets, catches sand in the wind. The sandstorms have scratched marks into the bones, and their surfaces have become chemically altered, forming a crystalline chrysalis. As the river returned and meandered, the bones, not yet fossilized, were swept into the current and buried once more in a new bank. The journey from living being to fossil is rarely a straight one.

Sometimes, however, it can be very simple. Meandering rivers erode banks and bars, but they can also interfere with burrows. At

Moradi, one such burrow, inundated by the flood, poignantly preserves four juvenile *Moradisaurus*, curled up cosily together, caught by the floodwater and buried unawares, perhaps while sleeping out the heat of the summer day. Now that the river has risen, there are new risks for any animal trying to drink from the flow. Where some curving bars catch light flotsam, perhaps a few bones, the rivers are wide and deep enough to carry bigger debris and even huge logs, washed down from the Aïr Mountains. At one such corner, a 25-metre-long log, almost a whole fallen tree, has trapped more and more woody debris, causing the beginnings of a logjam, a natural dam. Over its 100-kilometre journey, the more fragile branches have been broken off, leaving a large trunk. The log, in cross-section, shows almost no growth rings – the trees that were felled in this year's monsoon grew continually, a sign of constant, year-round rainfall. Now, the log blocks the flow, slowing the river during the rainy season, and providing shade for smaller animals after the flow trickles to nothing.[16]

Logjams tend to be important influencers of river ecology. When a flash flood flows through a logjam, it loses a lot of the energy that it contains as it fights past the uneven barrier in a burst of turbulence. Above and below a logjam, water slows. In this way, the disruption of the wood soothes the river and reduces the damage downstream. Upstream, the water is diverted onto a floodplain and leaves temporary pools, which might explain the persistence of amphibious animals throughout the year at Moradi. In a landscape like this, with a low gradient and slow water flow, the logjams have little impact on the geology, but in wider river systems, their sizes and effects can be hugely extensive. The largest logjam in historical times lasted for nearly 1,000 years in the lands of the Caddoan Mississippian culture, now in Louisiana. Known as the Great Raft, it at one time covered more than 150 miles of river, an ever-shifting carpet of trunks slowly decaying in the water, and was an important element of local folklore and agriculture, providing fertile floodwater and trapping silt for crops. It would still be here today if it had not been blown up to allow boats through. Once it was gone, the river flooded the land downstream, requiring further dams to be built, and changing the dynamics of water flow in the region.[17]

Moradi, as an ecosystem, is unusual even for the Permian. It is tempting to imagine a time period in the past as being homogenous worldwide. But planets are never just snow worlds, desert worlds, forest worlds. There is always variation, always provinciality. Across the world, species are distributed according to a combination of history and climatic tolerances. Moradi is fairly extreme in heat and aridity, so the creatures that live here are different from those other Permian ecosystems that are known from, say, the Karoo of South Africa, or eastern Europe. In those sites, the reconstructed climates are temperate, with cool winters and warm summers, and there, a very different selection of species are found. Therapsids, the group including gorgons, are present all around the world, and are extremely diverse, ecologically speaking. For instance, the superficially monkey-like *Suminia*, part of a particularly Russian group of therapsids, is the first animal to possess opposable thumbs, and the first tree-climbing vertebrate.[18] The difference at Moradi is perhaps down to the lack of *Glossopteris*, a type of seed fern with distinctively tongue-like leaves. Their forests dominate the south of Pangaea, and they are the preferred food of dicynodonts. In their absence, other herbivores, like pareiasaurs and captorhinids, have thrived.[19]

Extreme seasons pose a challenge to any community. In exactly the same way that the horses of Pleistocene Ikpikpuk ceased growing in the winter, *Bunostegos* stops growing every dry season. Its limbs are marked with growth rings, each one remembering a time of arrested growth, a drought survived. But although the lifestyle is harsh, desert communities that persist throughout the dry season are often incredibly diverse. In modern-day Namibia, a river flows through towering dunes, hundreds of metres high. At the low point, the water pools in a clay and salt pan – a playa lake called the Sossusvlei, a rare modern analogue for the wet desert of Moradi. Rainwater only flows through the dunes of the Namib Desert every few years, so the Sossusvlei stays mostly dry. Even so, there is enough groundwater that camel thorn trees, which have tap roots as long as 60 metres, can drive deep into the sand to extract it despite the high salinity, without changing the rate of water uptake throughout the year. In turn, the trees support a thriving community of lizards and mammals. Close by Sossusvlei, another playa demonstrates what happens when even that source of water runs out. Deadvlei is one of the major tourist attractions of Namibia, a strange landscape, below an invariably cloudless blue sky, of iron-orange dunes, pale white clay ground, and deadened, dry camel thorn trees, graphite black. Here, several hundred years ago, the dunes shifted into the path of the river, blocking it. Since then, the trees have stood as angular, sun-blackened monuments to a lost ecosystem, too dry to decay. But for the annual flood, but for the rains on Aïr, but for the monsoon, but for the shape of a continent, Moradi, too, would become truly deserted.[20]

Nevertheless, even reliable rains cannot protect against the change that is to come. The rock record of Moradi lasts until 252 million years before the present, then suddenly ceases, a 15-million-year hole in time. That gap in the archive is not all that is lost. The hot Pangaean wind is rising, and from the top of the Earth, the Arctic is about to send down a blast unlike any other. Siberia is about to erupt. When it does, it will expel 4 million cubic kilometres of lava – enough to fill the modern-day Mediterranean Sea – which will flood an area the size of Australia. That eruption will tear through recently formed coal beds, turning the Earth into a candle, and drifting coal ash and toxic metals over the land, transforming watercourses into deadly slurries. Oxygen will boil from the oceans; bacteria will bloom and produce poisonous hydrogen sulphide. The foul-smelling sulphides will infuse the seas and skies. Ninety-five per cent of all species on Earth will perish in what will become known as the Great Dying.[21]

As the skies go dark over Moradi, the megamonsoon continues unconcerned, but the water that it sends down from the Aïr is undrinkable, laced with arsenic, chromium and molybdenum. Deprived of a life source, the desert's derelict bones sink under the storm.

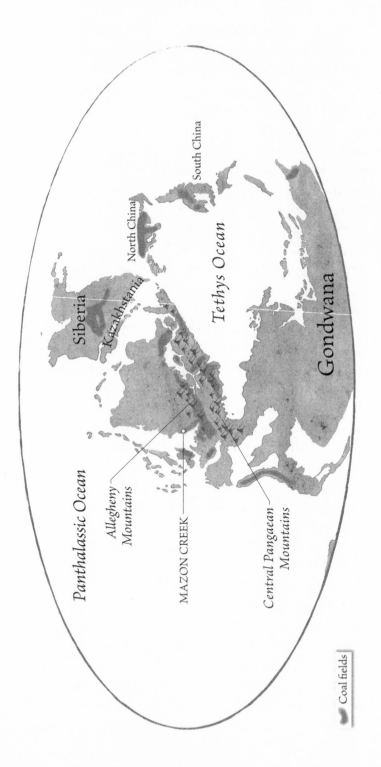

Panthalassic Ocean

Siberia

Kazakhstania

North China

South China

Tethys Ocean

Gondwana

Allegheny
Mountains

MAZON CREEK

Central Pangaean
Mountains

Coal fields

# 11.  *Fuel*

*Mazon Creek, Illinois, USA*
Carboniferous – *309 million years ago*

'I saw the Medullosae
with multipartite fronds
and watched the sunset rosy
through *Calamites* wands'

– Dr E. Marion Delf-Smith, 'A Botanical Dream'

'Toutes les saisons sont abolies
dans ces zones inexplorées
qui occupent la moitié du monde
et la parent de floraisons inconnues
et de nul climat'

'All the seasons are abolished
in these unexplored zones
which occupy half the world
and adorn it with flowerings
unknown and of no climate'

– Jean-Joseph Rabearivelo, from *Traduit de la Nuit*
(tr. Robert Ziller)

Crushing humidity and invigorating heat. An almost impenetrable mire of vegetation, sinking into still, still black waters. Proud, straight horsetails and sprays of tree ferns stand tall, clambering over one another to reach the sunlight. The air is intoxicating – the massed plant material all over the planet has pumped the atmosphere full of oxygen, with levels 50 per cent higher than in the modern day. On the western Pangaean coast, a river pours through a dense equatorial swamp, emptying a wedge of silt into a huge epicontinental sea. It is a far cry from the expansive Corn Belt landscape of modern Grundy County, Illinois. Where the Illinois River will in the modern day begin its journey through monoculture fields towards the breadth of the Mississippi, a river with no name

*Lepidodendron* sp.

enters the sea and deposits the eroded heights of the early Allegheny Mountains into a rich delta.[1]

Standing close-knit in the peaty mire is a large patch of trees, each no more than a couple of metres from its nearest neighbours and a relatively uniform 10 metres tall. Their trunks are crocodile-green, and textured with diamonds, overlapping like scales. Because each scale is slightly offset from those above and below, together they tessellate into a helix, giving the impression of coiling staircases, leading up into the dark fuzz above. For while the lower 5 metres or so are shiny, unadorned scales, from halfway up each tree to its top, each scale lends support to a single long, thin leaf, a dark, brushy bristle, which intermeshes with its closest neighbours and casts patches of darkness on the shallow standing water below. Narrow leaves, already dropped from the lower scales, float among the reflections. The trees do not block out the light as would a broadleaf forest of the present day, but they are not inefficient when it comes to catching the light. The light that shines through the sparse canopy can still be brought to use; in the scale tree, *Lepidodendron*, every part of its diamond-patterned side remains photosynthetic, the whole of the bark able to turn air and sunshine into new plant material.[2]

In the early evening, most of the light that penetrates beneath this bottlebrush canopy is caught horizontally, reflected at low sun off patches of deeper water where the trees do not grow and where the sky is open. Compared with the cool shade, the equatorial sun of Carboniferous Illinois is blinding and white. The water reeks of slow decay, of rotting *Lepidodendron* trunks and the blackening stalks of ferns, while the soft ground at the water's edge sinks under the pressure of a fallen log. Across the way, there is another stand of scale trees, but these are not the same single-shafted poles. At the top of each, the trunk branches in two, then in two again, closing the gaps and spreading into a partial canopy. Their trunks are angled drunkenly in the waterlogged soil, but they still rise to a uniform height, so the canopy, some 30 metres above the surface of the water, appears like a Venetian covered market, with finely textured twisted columns and a deep green roof. For an otherwise thick and imprecise tangle, the height of the trees within each stand is notably similar. No saplings intermingle with the brushy juveniles; all the trees across the way are full-grown adults, almost as if planted by a conscientious and geometrically minded landscape gardener. This order is not, of course, a deliberate planting scheme, and neither is it to do with local variations in soil quality or the amount of sunlight – these trees are all the same species, and within each stand, all precisely the same age. Each neighbourhood of trunks is a cohort that has, quite literally, grown up together.[3]

They are densely packed, but this is for a very good reason. Early innovators in the world of plant engineering they may be, with the earliest hardy bark, but *Lepidodendron* are not very woody inside. True wood, the rigid dense material we expect to find in trees, is rare. Only gymnosperms are both common here and made mostly of wood. In scale trees, true wood is only found in very small amounts at the centre of their trunks. Instead, their interior is mostly made of the sort of spongy, light tissue that you might expect to find in a much more herbaceous plant. The bark is strong, and is the only way that scale trees can grow so tall, but their trunks are not as rigid as if they were made from solid timber. This would make the trees rather unstable, but for what is happening beneath the ground.

The roots of scale trees, known as *Stigmaria* because of their wounded, holey texture, grow round one another, tightly interweaving with the roots of their neighbours in the incipiently peaty soil. They form continuous, shallow plates, which make an extensive, firm base to hold all the trees in the ground. They are remarkably dense; the little rootlets that come off the main root axis form a vast surface area across which to absorb water – there are nearly 26,000 rootlets for every square metre of ground. If one tree does in fact collapse, it can easily rip up an immediate neighbour with it, but the strong root system makes a collapse from high winds unlikely – the trees are holding onto one another for stability.[4]

These shallow root plates are transforming the world. The main reasons for the existence of roots may be to anchor plants and to absorb water and nutrients, but they have an impact that goes far beyond the individual. Roots are also transformers of landscapes, quite literally opening the earth to others, so much so that this world of subsoil interaction is called the 'rhizosphere', the root-world. The burrowing of root systems weathers rock, turning it unrelentingly into sand, and traps decaying humus. Without roots, soils do not form, because these fragments are blown or washed away by wind and rain. Without roots securing the compacted earth, that rain merges into wide, flat rivers, coursing down the plantless world in sheets, with crumbling banks and simple, direct routes. The natural meanders of river systems, with their hundreds of ever-shifting channels, floodplains and abandoned oxbows, are a creation of thousands of individuals holding firm against the flow and forcing the river to curve about them. The path of rivers is determined by plants. Roots may delve into the ground, but they, as much as leaves, change the chemistry of the atmosphere. Pushing incessantly through sandstones, rich in silicates of alkaline metals like sodium, calcium and potassium, roots capture and release these minerals into the water with the aid of their associated microbes and fungi. The dissolved metals discharged into these new braided channels would make the river more alkaline, but carbon dioxide, also dissolved in the water, reacts with them, and buffers this change. The continual buffering draws more carbon dioxide out of the air and into the water. The effect of root-weathering of silicates on the atmosphere is so strong that even in the modern day, the promotion of high-weathering plants like bamboo has been suggested as a serious aid in carbon capture. On geological timescales, the change can certainly be vast. Compared with the beginning of the Devonian 110 million years before, the concentration of carbon dioxide in Earth's atmosphere has fallen by about 4,000 parts per million – a number ten times higher than the total amount of carbon dioxide in the atmosphere today. All this has been driven largely by the sinking of roots.[5]

That is not the only change in the weather. It rains more at Mazon Creek than it used to; the rising of the Allegheny Mountains has changed the winds, and brought ever more rain driving down their steepening slopes. Powerful erosive rivers have built up, and the one that flows into the tropical sea at Mazon is a milky tea-brown shade, carrying with it the remains of seed ferns and other upland plants that have been washed from its banks. Soft tides lap twice daily in the bay, and seasonal floods swell through. Mazon Creek is a true swamp – some parts are perpetually flooded, others damply exposed to the air, covered in rotting branches and leaves.[6]

As floods flash through, they churn the landscape. What was sunk becomes exposed, what was land is washed away. Along the water's edge, ecological succession – the sequence of recovery of a community – is perpetually underway. Where the mud is soft and the ground flooded, the roots of scale trees arrive first and stabilize the ground, collecting the muddy silt of the river. Growing straight, they cast little shade and so other species can grow around them. Different species of scale tree are more or less tolerant of the wet; *Lepidodendron* itself will gladly grow with water lapping around its trunk, while others enjoy the water's edge or damp but prefer drained ground. Around the *Lepidodendron* trunks grow tall *Calamites* horsetails. They are anchored by horizontal stems called rhizomes, which usually sit in oxygen-poor water. To overcome this, horsetails pump gas down into their rhizomes to allow them to function efficiently, perhaps up to 70 litres every minute. Following the scale trees and horsetails come the furled fiddleheads of the true ferns, and finally the tree-like seed ferns and the conifer-like cordaites, which will grow only on the well-drained ridges and uplands surrounding the area of Mazon Creek.[7]

For many of the seed ferns and conifers, floods are a rare risk, positioned as they are on drier soil. But even the growing rainfall cannot prevent fires. For those inhabitants of a late Palaeozoic forest, fire is a substantial and particular threat. In the late Carboniferous, wildfire has never been so common, and, with the exception of a peak in the Early Permian, it is as common as it will ever be. Three ingredients make a fire – fuel, oxygen and heat. With the development of the first tall, tree-like plants – *Calamites*, scale trees and the conifer-like cordaites – all of these ingredients are at their greatest abundance during the Carboniferous. Never before has so much organic material been concentrated in plants. Their photosynthesis has also raised the oxygen concentration in the atmosphere. It makes up an astoundingly high 32 per cent of the air, compared with approximately 20 per cent in the modern day. For most of the Carboniferous, the average global temperature has been up to 6 degrees higher than the present day. Even with a recent slide towards icy poles, the heat has not abandoned tropical, equatorial Mazon Creek. It may be wet, and it may be peaty, but when oxygen concentrations rise above about 23 per cent, the wetness of plant matter is less relevant to whether it can catch fire; wood that in the modern day would seem too damp to burn can here be set alight.[8]

And it is the likelihood of fire that is perhaps responsible for the lanky, bare trunks of the scale tree. While there are a few plants that actively require burning in order to germinate, most plants are only able to tolerate fiery environments thanks to certain adaptations, such as rapid growth or the release of seeds only following a fire, when a new one is least likely to strike. Scale trees grow quickly, and as they grow, their lower, spindly leaves fall to the ground. This builds up a continuous leaf litter with a high surface area; anyone who has seen pines burn will see how fast their thin, oily needles catch. This means that, when a fire is sparked, it sweeps quickly over the ground at low temperatures, quickly exhausting its fuel and not being given time to build up enough height to reach the canopy; conifers that live through regular fires have needles that burn far quicker than those that grow elsewhere. A huge gap between the forest floor and the tops of the trees allows space for flames to rise – just not too high.[9]

Set among thousands of insects and centipedes that scuttle and buzz through the feathery carpets of creeping *Mariopteris* seed ferns and the knotted roots of scale trees, are beetles. Mazon Creek is the first place on Earth known to have hosted them. Arthropods are common here, from dragonflies to millipedes, crustaceans to spiders. Round-bodied and left behind by the tide, looking like an upturned colander with limbs, is another, less familiar arthropod, the horseshoe crab *Euproops*. The horseshoe crabs of the modern day are slow-moving, brown, shelly creatures, familiar along the eastern and Caribbean coasts of North America and throughout southern and eastern Asia, where they emerge for their annual mating and egg-laying rituals. *Euproops*, to some eyes, has something of a talent for mimicry unusual in horseshoe crabs. Squint a little, and the spines of this creature look very similar to the leaves of lycopods, and its limbs appear adapted for grasping and pulling over twigs and branches. This horseshoe crab has the appearance of a creature that might thrive on land, but this is coincidental, a result of the way that Mazon Creek will survive the passage of time.[10]

Mazon Creek's life will mostly be preserved elsewhere. A body, even after death, may go on a journey, taken by the waters to be washed into oceans or caves, or broken by scavengers and the elements. Now, at Mazon Creek, those dark-stained floods purge the uplifting mountains of their mud, the land is taken into the sea, and washed in with it are the carcasses and broken plants of the lycopod swamp. In death, land and sea are one, leaving behind a palaeontological palimpsest, seabed overwritten with swamp. The rising sea level has inundated more seaward scale-tree forests, now waterlogged in life position, among which sea scorpions shake off their old carapaces and jellyfish drift. Waterlogged branches of upland conifers and the padded toes of freshwater temnospondyls are deposited in a spoil heap of a bay.[11]

Overall, perhaps the most important substance being laid down in these swamps is being slowly rotted in the underwater muck. Roots, leaves, branches, all turn, slowly, from greenery to peat, and from peat to coal. Laying down coal is precisely what the Carboniferous is famous for, and it is due to one thing only. Death, *en masse*.

There is a juddering in the air among the tallest scale trees, and the 30-metre canopy rustles. A popping like fireworks echoes around their columns, its source seemingly undetectable. For a while, it seems as though this is all that will happen, but then the popping becomes an artillery volley, and the base of a scale tree splinters into interlaced fingers, cracks shooting up one side of the dying tree. The green bark gives a final booming roar as the upright pillar topples, dragging branches with it. With little space to fall, it veers into

a neighbour, already brown with decay, and like dominos both collapse, splashing black peat-water into the air and echoing around
the bay. Their snapped stumps still have jagged swords of bark
pointing proudly at the gap in the canopy, where the sun streams
through stronger. The reason for the lopsided angles and sparse
canopy is becoming clear; the scale trees are falling.

The adult trees that have spread their leaves like umbrellas in the
rain do not have long to last. Each has taken decades to grow, perhaps even up to a century, but the single, critical moment of their
existence is about to arrive. The cones of *Lepidodendron* develop
exactly on the growing tip of the plant, so once the plant becomes
fertile, it can no longer grow larger. Every day is a choice between
continuing to grow or stopping to reproduce, and reproduction is
not something that can be done alone. To maximize their chances
of reproducing successfully, all *Lepidodendron* in a cohort make the
switch together, releasing their spores into the wind in the hope of
landing and producing the next generation. Very few plants do this,
but where they do, they grow faster, reach sexual age more quickly,
and release more seeds. Once the abundant spores of a scale tree are
released, there is simply no point in continuing to grow. The adult
scale trees are now hogging the light that the next generation need,
and serve no further purpose. So, they die as one, and remain standing only as long as the structural integrity of their bark holds, falling
as their lightweight, spongy trunks give way. A creaking and a
crumbling brings a whole generation crashing and splashing down
over a matter of months.[12]

Life is often structured around edges, with diversity highest where
homogenous regions come into contact. A river delta marks just
such a boundary between freshwater and saltwater environments,
with each holding very different physiological challenges. Sometimes, the water is perpetually brackish, with a low level of salt that
acts as an intermediate environment. Where a river runs into a deep
bay, as in the Mazon delta, the division can be maintained surprisingly far out to sea. Saltier water is denser, so that when the river

flows out over the sea, it leaves a plume of open freshwater with firm boundaries above a wedge of salt water, thinning towards land as the seabed rises to meet the estuary. All water is not equal, and water with different temperatures or salinities can remain separate entities even with no physical barrier in place. Usually, this is a horizontal division; in the Arctic, where the Atlantic and Pacific oceans meet, masses of water are stacked on top of one another, mingling only slightly. The longest river in Antarctica, the Onyx, flows inland into Lake Vanda, a lake with three layers of water of different salt concentrations. The differences in salinity are enough to overcome extreme differences in temperature; the bottom layer of Lake Vanda is continually a balmy 23°C, but the uppermost layer is close to freezing. Sometimes, the division is maintained by inertia and can be vertical; where three rivers in modern-day Passau, Bavaria, join, the dark blue Ilz, white Inn and brown Danube continue to flow in the same direction, failing to mingle, and giving rise to a tricolour river for kilometres downstream.[13]

In the salt wedge beneath the Mazon Creek outflow dwells a strange creature, one which utterly confounds understanding. The gift of an experienced naturalist is an ability to identify a species by sight, often having caught only the fleetest of visual clues. To step outside the comfort zone of familiar biology can be deeply disconcerting. An unprepared European birder first faced with North American 'robins', cardinals and mockingbirds will begin entirely adrift in a sea of unfamiliar creatures, such that the sight of a familiar, invasive starling is something of a relief at the recognition. Even so, without having to go as specific as a single common name, there is usually something familiar to grasp onto. You may not know the name of a bluejay, but it has that familiar corvid feeling to it. Somewhere, somehow, the unfamiliar can be fitted into an internal mental classification.

Stepping back in time, for a palaeobiologist, can have the same sort of effect as a journey to a new biome in space. The fossil record is full of near-familiar creatures that can easily be placed in the grand family tree of existence, and can, therefore, be interpreted, their differences noted, perhaps marvelled at, but understood in the context of the wider evolution of the tree of life. Even when hugely diverse extinct groups are discovered, as with the description of Dinosauria, we see similarities in the preserved structures that lead us to recognize that modern birds are a group within the dinosaurs, and this informs our interpretation of their stranger features. But sometimes, as with one particular animal living in the brackish estuary of Mazon Creek, the vagaries of natural selection and the absence of similar creatures in the fossil record create a suite of anatomy so unusual that making any kind of connection is almost impossible. Faced with total novelty, our first instinct is to reach for the metaphor of the supernatural, of the unnatural. Under the waves that top the salt wedge, among the pale, pulsing bells and eerie curtains of *Essexella* jellyfish, swims the elusive creature we call the Tully Monster.[14]

Unlike the fabled monsters of modern cryptozoology – Nessie, the Sasquatch, the chupacabra - the Tully Monster is real, but beyond that, there is very little we understand about it. It is not as if they are rare; these creatures are the size of herring and just as plentiful, found by the hundred. More than thirty times as many body fossils of *Tullimonstrum* have been found as the well-known first bird, *Archaeopteryx*, so, numerically, it should be a simple story. But interpreting their remains is difficult because of what each specimen preserves. They have a segmented torpedo of a body, and at the rear, two rippling tail fins that look a little like the wings of squid. At the front, a long, thin feature, something like the hose of a vacuum cleaner, wiggles, with a tiny, tooth-filled grabbing claw at its end. Adding further confusion, there is a solid bar running from side to side across the top of the creature, horizontal stalks on which are set bulbous organs of some kind, which are generally assumed to be its eyes. All in all, it is unlike anything else that is known in over half a billion years of animal evolution. The closest superficial similarity is with a five-eyed Cambrian oddity called *Opabinia*, a creature not otherwise known in 250 million years – a gap in the fossil record equivalent to the *Rhamphorhynchus* flocks of the European

Jurassic suddenly appearing to mob anglers in the Bodensee, or indeed to a plesiosaur surviving in Loch Ness.

The question for *Tullimonstrum* is not whether it exists, but what it actually is. Over the years, palaeontologists have gazed ever closer at its curious anatomy and concluded variously that it is a kind of worm, maybe a ribbon worm, or related to the annelids, the group that includes earthworms, or nematodes, the group of mostly microscopic worms that exist by the trillion practically everywhere on Earth. Or perhaps it is an arthropod like spiders, crabs or wood-lice, or a mollusc like a snail, or even a vertebrate. Those lumps on the end of the horizontal bar? Are they eyes, or could they be pressure sensors? Are they involved in reproduction, or in stabilizing *Tullimonstrum* as it swims? Nothing provokes debate quite like the hunting of monsters.[15]

Each piece of a *Tullimonstrum* anatomy can be echoed by organisms spread wide in the animal kingdom. In the modern day, the black dragonfish lurks in the deep sea. As an adult, eel-like and with a gaping jaw, it does not seem like a likely candidate to be an analogue of *Tullimonstrum*. But this fish has a larval stage, passing through a time when it exists near the surface as a small and nearly transparent animal, with long eyes extended on stalks, not entirely unlike the bar organs of *Tullimonstrum*. Then again, eyes on stalks are also known in molluscs and arthropods – this is an ecologically useful adaptation that has evolved many times. Even if the melanin within the eye-stalks resembles that of vertebrates and certain smudges resemble the notochord – the basic back support of all vertebrates – the absence of so many other vertebrate traits, including hard tissues besides the 'teeth' in the stethoscope-like grasping proboscis, makes the claim for *Tullimonstrum* as an unusual fish, controversial at best. As with all monsters, the sightings are just that bit too blurred.[16]

Being soft-bodied, it is impressive that it has been preserved. Iron minerals, washed down from the red sandstones inland, react strongly with the carbon dioxide and encase the remains in round nodules, which are then buried. Slowly, these are lithified, turned from river outwash into impenetrable time capsules of hard siderite ironstone. Alongside them, within the peaty swamps, the raw plant material is converted slowly, anaerobically, into coal.[17]

Nobody quite knows why the rate at which organic material is being laid down throughout the equatorial coal belt in the Carboniferous is as high as it is. One idea is that lignin, the main constituent of wood, is a relatively new material, and is simply not yet easily digestible by microbes – they have not evolved the ability to consume it, and so it turns to coal. Others suggest that it is the unique geography of the Carboniferous that led to the laying down of coal, the only time in Earth's history where the tropics have been both extensively wet and dominated by geographic basins. Whether by experimenting with new materials faster than microbes can accommodate, or through happenstance of climate and geography, innovators like the scale trees are radically changing the composition of the atmosphere. The Earth is spiralling towards climate change that will lead to a cooling almost to the point of a global ice age, an increase in seasonality and in aridity, and, ultimately, the wholesale destruction of the very ecosystem that keeps the scale trees alive. This will be their extinction, as the sodden coal swamps of the Carboniferous give way to the drought of the Permian. By removing carbon in such quantities from the atmosphere, they set the stage on which evolution would play out for the next third of a billion years, a stage on which the scale trees were no longer viable. The Carboniferous Rainforest Collapse, only 4 million years after the Mazon Creek lycopods are inundated by local sea level rises, will not just bring down one stand of trees, it will fragment at continental scale the tropical coal forests of the whole of Europe and America. This is one of only two mass-extinction events to seriously affect plants, the other being at the end of the Permian. It is as this dry Permian world takes hold that the earliest amniotes, the first synapsids and sauropsids, which appeared in the Carboniferous, will benefit from their adaptation towards aridity, dispersing through drier channels and becoming cosmopolitan inhabitants of Pangaea.[18]

The irony is that, because of the coal laid down throughout the

Central Pangaean Mountains, these places – Illinois and Kentucky, Wales and the West Midlands in Britain, Westphalia in Germany, will play a critical role in the earliest rapid industrializations of the eighteenth and nineteenth centuries – the driving force behind the re-release of carbon stored beneath the earth for the last 309 million years. Some 90 per cent of all the coal in the earth today was deposited during the Carboniferous. It is the sheer abundance of coal where it was laid down that made it such a cheap, high-energy fuel of choice for industrialization, powering steam engines and becoming part of high-quality steel. The legacy of the scale trees lives on, in a climatic transformation that we undo with every ton of coal burnt. For Mazon Creek itself, the finest of all the Carboniferous fossil beds, there is a further irony. While coal for burning continues to be extracted all around the world, the fossils of Mazon Creek are today effectively impossible to excavate thanks to a very different kind of energy. Transformed into stone under warm swamp water bathed in Carboniferous sunshine, they now lie within a different kind of steaming pool connected to a cleaner, more efficient source of power. The fossil exposures have been flooded to provide a cooling basin for the reactors of the Braidwood Nuclear Power Plant, in Will County, Illinois.

For the inhabitants of the Mazon delta, that is an interminable distance into the future, and, for now, having survived flood, fire and the encroaching seas, life in the lycopod swamps seems indomitable and unchangeable. With the sun glinting through the woody stems of the tree ferns, the surface of the water breaks into an ensorcelling fairytale rainbow, shimmering across the whole spectrum of colours. All that plant matter, dying and sinking into the congealing mire and being broken into peat and coal, releases organic oils, which float to the surface. On a still afternoon like today, the accumulated oil, spread in a layer one molecule thick, is enough to turn the mirror-like water into a psychedelic fantasy land, a swirling soap-bubble palette striped by the shadows of lycopsids and broken only by the minute ripples of fish. It will last until the tide comes in.

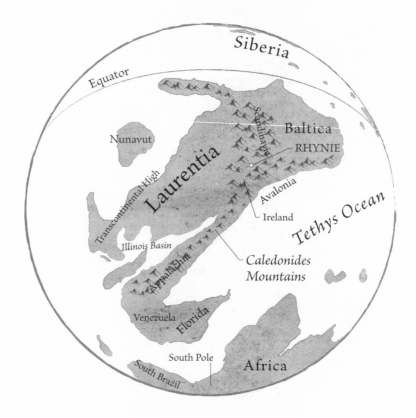

# 12. *Collaboration*

*Rhynie, Scotland, UK*
*Devonian – 407 million years ago*

'Oh now they have gone
To that bonnie highland mountain
For to view the green fields
Likewise its silvery fountain'

– 'The Braes of Balquhidder', traditional Scottish song

'Thin transparent essences, too pure and fine to be called water,
are kept simmering gently in beautiful sinter cups and bowls that
grow ever more beautiful the longer they are used'

– John Muir on Yellowstone, 1898

If anything can be said to unite the Cairngorms of Scotland, the sky-cleaving plains of the Norwegian Hardangervidda, the black hills of Donegal and the Appalachian range of North America, it is the folk fiddle. The primal sound of singing wood, of tree turned noise, earthen and breathing. It is a tradition passed vertically through generations and horizontally across continents, each valley with its own songs, but part of an older, grander culture. The shared history of these mountain ranges, though, is far more than just musical. Individually, the mountains in those places are relatively recently raised, but their roots run deep, pushed down into the mantle by the weight of rock above them. The very foundations on which Appalachia, Ireland, Scotland and Scandinavia are built are part of the same geological event, the same deep-time range. That they are raised places today is a distant echo of their shared, highland past.[1]

*Palaeocharinus rhyniensis*

Mountains and oceans are geologically temporary structures. Mountains build where plates collide, pushed up as others subduct, sinking under them. Mountains shrink as erosion takes their rocks, grain by grain, back to the sea. Oceans form where plates separate at mid-oceanic ridges. Where an oceanic plate subducts under another plate, the ocean gets smaller. In the Late Devonian, the Iapetus, at one time the largest ocean in the world, has dwindled. Through earlier periods, it has shrunk and dragged the continents closer together, and, by now, the gap between them has finally closed. For millions of years, the Iapetus lay in the southern hemisphere between three isolated continents – Baltica (mostly comprising Scandinavia and western Russia), Laurentia (primarily North America and Greenland, but also Scotland and the north and west of Ireland), and Avalonia (including New England, the southern parts of Great Britain and Ireland, and the Low Countries). Laurentia, though, under tectonic force, has been swallowing the seabed, eating up the crust between them, and drawing these landmasses together. By the start of the Silurian, the Iapetus had been reduced to a Mediterranean-sized sea, and has since disappeared entirely, surrounded by crushing continental mass. Baltica is made of denser rock than Laurentia, and so, floating as they are on a magmatic bath, the tendency is for Laurentia to slide on top, forcing the edge of Baltica underneath. It is not a clean process, and the continents crumple, throwing land skywards under its own momentum and down into the mantle, the crust becoming nearly twice as thick as under the average continental plate. The principle is exactly that of a buckling car bonnet in a crash test, where mountains and valleys emerge from the previously flat metal sheet. The landmasses of the Earth are once again converging on one another in their endless cycle of separation and collision. Pangaea, the single global continent until its breakup in the Jurassic, is beginning to come together. The northern half is now complete, and will join with Gondwana in the Carboniferous. The continent that emerges from this three-way pile-up is variously known as Laurussia, the Old Red Continent or Euramerica, and the new peaks are the Caledonides, a range with one end in modern-day Tennessee and the other in Finland. It is now the largest mountain range on the planet.[2]

Once built, as we have already seen in the Triassic, mountain eco-systems tend to be eroded rather than lay down a record of their existence. In the 400 million years between the Devonian and the present day, the Caledonides will be slowly worn down by wind and rain. The landscape of Finland, once mountainous, is now flat Pre-cambrian bedrock, the basement layers of the Caledonides. The only clue to the mountains extending so far east will be an occa-sional lump of resilient younger rock sticking out in the flatland. The Irish Caledonides have been worn down into a rolling glacial landscape, with no traces of their surface features remaining. Only in exceptional circumstances are ecosystems preserved in the moun-tains, and here, in a valley of hot springs, is just such a place. This is Rhynie: in the modern day, hilly farmland for Aberdeenshire cattle, but in the Early Devonian, a colourful and ethereal mountain glen, a landscape of steam, of salts, of life emerging from stone, the home of the earliest ancestors of those fiddle-wood trees.[3]

Compared with the invigorating air of the Carboniferous, the Devonian is oxygen-poor. Plants are rare on land, and Rhynie is a pioneering community, one of the places where the Earth is being greened. Life begets life, and where one species finds a foothold, others will follow, a build-up to the seething swamps to come. There has been land, however inhospitable, for billions of years, and yet it is only in the Devonian that the earliest functioning communities are being established and preserved in great detail. An unplanned exercise in cooperation is taking place, with animal and plant, fun-gus and microbe, competing and collaborating in complex ways. This is an ecosystem discovering itself, and it is here that the funda-mental patterns of life on land are being established.[4]

From the shadowed mountainside, the near-equatorial skies are a cloudless blue. The jagged ridge above is a pale, almost pink, gran-ite grey; but the slopes are black, rough and unforgiving, covered with piles of scree. Across the valley, to the south-east, the crum-bled rockfalls look like a soft dusting against the scattered shine of the igneous surface in the afternoon sun. Here and there, layered, tilted prows, the less resilient stone eroded from around them, project into the air, sharp and pockmarked, the surface crudely excavated by the wind and occasional rain.[5]

Dry stream channels descend the bare slopes into the valley floor, jouking this way and that to dodge tall extrusions of rock. As if choreographed, three quarters of the way down the slope, the gunnels jerk away to the north-east in parallel, their paths guided by the faulted landscape as the mountains themselves trudge up the glen: a line of weakness in the colliding continents, revealed by the paths of raindrops. The rain is infrequent now, although the sandy paths bear miniature channels carved by the last drizzle. Pools formed by the rainwater are generally only ephemeral, but at the foot of rocks that block the stream, they gather and are deep enough to persist between rains, inhabited by the earliest amoebae known from the land. It has not rained here in over a month, and these pools are stagnant and sickly with filaments of algae. Although the sky is empty, the whitened valley contains low, diffuse clouds, and is mottled uncertainly along its length with green. Rising from those verdant patches, scattered pillars are visible, while steaming pools, hot springs, emerge from the pale earth, the water varyingly a striking blue or a palette of rainbows. Beyond them, a floodplain, scattered with the drying remnants of ephemeral lakes, drops towards a bare, tan riverbed through which a shrunken, weaving river takes its northward course. Against the dark Ordovician gabbro, the black volcanic base, Rhynie is a streak of technicolour.[6]

Down in the valley floor, sulphur stings the air, the lofty pink and black walls partially obscured by misty showers from alkaline pools. Faults abound, as continents push against one another. The distance to the Earth's interior is thin here; columns of magma have risen close to the surface and threaten to break through. The cracks of the Earth emerge in this valley of a young and growing mountain chain, as tall as the Himalayas. The great volcanoes to the west, among them the vast Ben Nevis, are spewing lava. Others have already exploded, like the huge 50-square-kilometre supervolcanic

crater that will become Glen Coe, and which collapsed and deto-
nated catastrophically only 13 million years before, in the Silurian.
Ben Nevis's time, too, will come, an eruption that will be heard for
thousands of miles around. The mountain of the modern day is just
the eroded and collapsed heart of a crater whose rim rose hundreds
of metres higher. In Rhynie, an underground lake formed from per-
colated rainwater causes the heat to manifest as a valley of hot
springs a couple of kilometres long. Colourful cauldrons of almost
invisible water spill their contents, conjuring a thin silicon crust over
plants cavalier enough to grow too close. The water gushing over
those tiny sprigs, wet and branched like samphire, is the temperature of a hot bath, about 30°C. Where the spring emerges from the
ground, heated by the molten rock near the surface, it can reach
120°C, kept liquid only by subterranean pressure and rapid cooling
as it emerges.[7]

The springs of Rhynie are extreme environments in many ways.
Much of the water is too hot and too alkaline for most living
things to thrive, and yet they have been colonized. The land, too,
is hostile, but plants have begun to colonize inland – at least forty
different species of plant live in and around the waters of Rhynie.
By collaborating and competing, parasitizing and preying, functional communities have even been established away from the
safety of the water, and the amount of habitable land is increasing. Plants are growing big by entering into deals with fungi, fungi
are growing bigger by co-opting cyanobacteria, and both arthropods and fungi are helping to break down dead organisms, making
soil in which new plants can grow.[8]

The only occupants of the hottest water are microbes that thrive
in these sorts of extreme conditions, so-called alkalithermophiles.
Many of these are sulphur bacteria which, unlike most of the rest of
life, do not obtain their energy from the sun – photosynthesis stops
occurring at temperatures above about 75°C – or from eating those
that do, but by directly breaking down the rock itself. To buffer
themselves against the alkaline conditions, they churn out protein
chains, made of strings of amino acids. These acids, to some extent,

neutralize the alkaline water and allow the normal chemical reactions of life to proceed. In the hotter pools, only these rock-eating cells survive, and the water is entirely clear. Not clear like a fresh river or ocean, still filled with tiny beasts that impart the slightest haze, but clear like distilled alcohol, the only clue to its presence a shimmering as the surface vibrates with bubbles. In the right light, when the sun is at the correct angle, the bare tunnel into the centre of the Earth is lit up as certainly as if it were an empty cave mouth, with only the slightest refraction to break the illusion.[9]

Further from the geological kettle of the underground aquifer, the pools are brightly coloured. The water may still be a sweltering 60 degrees, but cyanobacteria, at least, can survive in these conditions. These are the oldest photosynthesizers in the world and have been eaters of sunlight for 3 billion years. Each traps the energy of light in a special pigment; if a photon – a particle of light – hits the pigment at the right point, the chemical changes shape into a less stable arrangement, and on flipping back generates energy in a form that can be used in other cellular reactions, such as creating sugars and starches. The combined pigments of the millions of cyanobacteria produce stunningly pure colours, each species a subtly different shade. As the changing temperatures from centre to edge are preferred by different species, so the colours change, giving the effect of moving from blues in the centre where the water reflects the sky, through greens to yellows, to oranges and reds. Cyanobacteria are incredibly diverse at Rhynie, from individual cells up to colonial cubes consisting of hundreds of cells.[10]

All around the pools, whether clear or colourful, are layers of white sinter, a silicon-rich sediment left behind as the spilt waters evaporate. The periodic expulsion of this mineral, white and crumbly, like compacted sugar, continually raises the rim of the spring itself, so that the waterhole soon rises centimetre by centimetre on a terraced plateau, a growing stack of dusted pancakes. Where the water level has overflowed the rim, fans of floodwater penetrate the plants below. Among the raised sinter terraces flow the mountain streams, bringing with them the hard-earned dark gabbro sand of the Caledonidean peaks into dark, shallow ponds, cold running water to balance the hot. Away from streams, where stoneworts cling on in the current, the slopes are bare; little as yet can live far from water, and verdure springs only on the valley floor, blanketed in a green forest of stems no bigger than moss, where harvestmen and mites, insects, freshwater myriapods and crustaceans form a miniature ecosystem covering two fifths of the land surface.[11]

The water spilling from the hot pools covers and permeates these low-slung plants, fungi and animals. As it cools, supersaturated silicon precipitates out, finding imperfections to crystallize around, and infuses every aspect of life here. Frozen in place quickly, even subcellular structures act as tiny moulds, now unstably cast in translucent opal. In time, that opal will stabilize as quartz, and, combined with the sandy sediment washed down by the local streams, will form the rock type known as chert, in which entire communities are preserved in three dimensions.[12]

Looming over the other inhabitants of this steamy valley, and exemplifying this tension between cooperation and competition better than any, are pale grey pillars, leathery bollards like smooth cactuses, up to 3 metres in height. *Prototaxites* are skyscrapers in a model village, the largest organisms on the planet. Elsewhere, at about the same time, individuals of this genus have been known to reach up to nearly 9 metres, others have a trunk a metre in diameter. A hundred times larger than the plants on the ground, these are outliers. With only a nubbly, soft skin, they resemble a platoon of half-melted grey snowmen, tall, thin, a single tower without any splits or branches, dominating the landscape. They are entirely unlike anything in the miniature forest beneath, for good reason. *Prototaxites* is not a plant, but, astoundingly, a fungus. Its close relatives today include a bewildering array of fungi including Dutch elm disease, brewer's yeast, penicillium and truffles. Exactly how it got so big is a bit of a mystery – no part of its underground structure is known. One solution that has been proposed is that it, like many of its relatives, is a lichen.[13]

Fungi are the great collaborators of life, forming close associations with species so distantly related to them that we place them in dis-tinct kingdoms. The most intimate association they form is with a photosynthesizing organism, whether a plant or a cyanobacterium, to form a lichen. Excellent at breaking down organic matter, the fungal partner in a lichen can extract huge amounts of mineral nutrient from even the barest surface, sharing it with its photosyn-thesizing partner (known as a photobiont) and protecting it with a tough tissue sheath. In return, with access to light, the photobiont can make energy that will feed the fungus. This powerful combin-ation means that, wherever there is a surface exposed to light and water, a lichen can grow.[14]

Rhynie is the home of two types of lichen, but they are stun-ningly different. *Prototaxites* is the Earth's first really large organism, an initial draft of life on the macroscopic scale. A tangled mesh of hyphae – the exceptionally fine strings of nutrient-absorbing cells that make up most of the structure of a fungus – forms the outer layer. If it is indeed a lichen, this is where it will hold the photobiont in place. Animals bore holes into the sides, so it houses a small eco-system itself, being ecologically more like a smooth, branchless tree. It may have photobionts, but isotopes show that it also rou-tinely consumed other organisms. Its size, perhaps, is a result of exploiting two energy sources, as consumer and collaborator.[15]

Encrusting the many fallen boulders are black, paint-fleck marks, more like the lichens of the modern day. *Winfrenatia* is simple in structure, its flat crust made mostly of undifferentiated fungal hyphae arranged as a mat, anchoring it to the surface. Across the surface of this structure are microscopic pits, within which dwell single cyanobacterial cells, held in place by the fungus like pigs in a pen. The farming comparison is not inappropriate; on the spectrum of mutually beneficial interactions, it's hard to decide exactly what differentiates this relationship from any other domestication. There are even instances of rustling – some fungi only form lichens by kill-ing other lichen-forming fungi and stealing their photobionts before settling down as a lichen themselves. Farming-like relationships between species have evolved a few times in the history of life.

Among animals, leafcutter ants compost leaves in order to grow mushrooms – the fruit of fungi – in special underground chambers; other species of ant guard aphids and milk them for their sugary excretion, or even rear scale bugs for meat. Damselfish take care of gardens of red algae among coral reefs and harvest them for food. Humans rear numerous animals and plants. In each case, the farmer protects the farmed, and takes energy in return. Fungi are certainly in control of the lichen relationship; often, when they extract the energy from their photobiont, they consume the photobiont as well. Are lichens, too, the logical end-product of an ever-closer agricultural relationship? Was the first Aberdeenshire farmer a fungus? If so, it's already diversifying its crop; *Winfrenatia* has not one but two different species of cyanobacterial photobiont, living together in a tight, interdependent three-way relationship.[16]

Every major type of modern fungi is represented in ancestral form in Rhynie, and several of these have interactions with plants. One, a relative of modern bread mould, has grown its fine, hair-like hyphae through the stem wall of a plant called *Aglaophyton*, in what is called a 'mycorrhizal' relationship, with 'myco' referring to the fungus, and 'rhizal' referring to a root. *Aglaophyton* dominates the more established patches of green along this valley. It is a tiny, smooth-stemmed plant, spread across the ground in vertical, forking stalks, each of which ends in an egg-shaped organ for releasing spores. One individual is sprawling, its stalks joined together by little horizontal runners. At intervals, nodules support the recumbent stems like railway sleepers. It is a very loosely structured plant, relying on fine hairs called rhizoids to absorb water. To photosynthesize properly requires a steady and substantial supply of water, and the fungus is a willing trader. It supplies the plant with water and nutrients from the soil, taking a tithe of sugars produced by photosynthesis. In total, mycorrhizae are responsible for helping source the nutrients for about 80 per cent of all modern plant species. That they are present so early in the evolutionary history of plants suggests that this relationship is not just ecologically important but fundamental to the development of life on land.[17]

The takeover of the land has been facilitated not just by interspecies relationships, but by a changing power dynamic between generations played out over geological timescales. Plants have an evolutionary heritage with a very different sexual system from that of animals. In animals, parent and child are physiologically the same; in sexual species, sperm and eggs are produced, with half the adult complement of chromosomes, and these combine to grow a new individual. In asexual species, adults generate eggs with a full complement of chromosomes that grow directly into a new individual. So far, so simple.

In plants, though, offspring do not resemble their parents at all, and this generational complexity has equipped them to conquer the land. The reproduction of green algae, the ancestors of plants, is a two-stage process. First, sperm and ovule fertilize to produce a single-celled generation with twice the number of chromosomes as an adult alga. After shuffling the chromosomes, it separates into two spores, each of which grows into a new full-grown alga, and the cycle begins again.[18]

All plants of the modern day alternate between a generation that produces sperm and ovules (the gametophyte) and a generation that produces spores (the sporophyte), but control has switched. Early land plants developed a spore wall that resisted drying out, a reproductive invention as crucial for life on land as shelled eggs in amniotes. Plants that could make more spores had a better chance of success, and so the sporophyte generation grew ever more important, rising from a single cell to an entirely distinct body from the gametophyte. At Rhynie, we are in the throes of this generational takeover.[19]

Today, the sporophyte in mosses, hornworts and liverworts, damp environment specialists, is still a minor player, living essentially as a parasite on its parent. But it is still important, as the gametophyte must rely on tiny arthropods to transfer its sperm. The main body of a fern is a sporophyte, but you can still find independently-living gametophytes, little heart-shaped mats that eventually reproduce to make a new fern frond. In seed plants, the ancestral gametophyte has withered until it is barely present. Instead, every visible part of a seed plant, from a giant redwood to a daisy, is a spore-producer. Flowering plants have fallen furthest from their ancestors. Pollination moves the male spores – pollen – to the female spores. Within the walls of the female spore, a minute structure develops – all that remains of the giant marine algae – and sperm and eggs are released.[20]

In the Devonian of Rhynie, the sporophyte of *Aglaophyton* is beginning to strike out on its own.* Having evolved from a single-cell developmental stage not too long ago, it has no roots, no leaf-like structures, and is discovering its own anatomy. By associating with fungi, it can access nutrients, bypassing the constraints of its own development, and doing something that no multicellular life has been able to do until now. These plants and fungi are becoming the first groups of organisms to break free from the water, and will become the foundational structures on which the terrestrial ecosystems of the future will be built.

The notion of a single individual is a very animal concept, utterly ignored by other kingdoms of life. The sporophyte does not need to reproduce sexually at all, but, like other plants, can at times clone itself, producing its own cuttings. The presence of mycorrhizal meshes, networks of fungus associated with separate plant entities, further blurs the concept of an individual, as it allows signals and even nutrients to be passed between plants, using the fungal hyphae as a conduit. In a world where your near neighbours are likely to be

---

* All of the Devonian plants of Rhynie exist alternately as a multicellular sporophyte and gametophyte. This point generally presents problems for palaeobotanists, as both can be preserved as body fossils, but they live separately and have radically different shapes. When naming species as found in the fossil record, the only data are features of shape or, occasionally, chemistry. Linking together sporophyte and gametophyte is typically impossible, but at Rhynie, the preservation is so exceptional that even individual sperm cells have been found, along with detailed cell-level structures that demonstrate the shared identity of both generations, with developmental stages linking together the entire life cycle.[21]

your own genetically identical clones, having a fungal partner could allow sharing of resources in difficult times. Collaboration can pay dividends. No species evolves in isolation, but the synergy of plants and fungi has altered the future of life on Earth perhaps more than any other evolutionary innovation.[22]

There are more complex plants still in the silicon pools of Rhynie. *Asteroxylon*, the starwood, resembles a thin green fir-cone, with scale-like structures that photosynthesize like leaves. They are simpler than 'true' leaves, though. The skeletonized structure of a modern leaf has not yet emerged. The internal transport of nutrients and water of modern-day vascular plants is down to the presence of the xylem and phloem, which run from roots all the way to the leaves, with water leaving the plant through their stomata. The earliest plants, though, lack these, and do not even have roots, only rhizoids, hair-like structures that nonetheless absorb water and minerals. The starwood is one of the bigger plants at Rhynie, growing to nearly half a metre in height, and is anchored in the sediment. Its shoots have evolved to resemble roots, an independent origin of roots from the rest of the vascular plants. The root-like shoots of the starwood project down some 20 centimetres into the surface, reaching deeper than the other plants to find new resources. Their tissues have adapted to be able to move plenty of water very quickly in order to photosynthesize and grow big, but in dry periods this is a problem, as they lose more water than they absorb. As a way of balancing the trade-off that all plants face – rapid growth or water efficiency – they have few stomata, widely spaced. For now, *Asteroxylon* prefers growth, with less pressure in general to preserve water. However, they must be picky about when they reproduce. The tropical climate of Rhynie is highly variable. So, all the way up the stem of starwoods are alternating regions of fertility and sterility, another energy-saving solution to an environmental problem.[23]

All this growth eventually comes to an end, and once dead the plants, of no further use to their fungal symbionts, decay. Other fungi, such as ascomycetes, invade through the relaxed stomata to digest the plant from within. The fungi, by extracting the last nutrients from the plant, are developing some of the earliest soils. In time, this will create a softer and better substrate in which plants can grow larger, until they reach the heights of the Carboniferous lycopsid swamps. The rotting vegetation, sinking into the low sward, is eaten not only by static fungi but by little arthropods, the only animal life on land. Nothing with a backbone has yet climbed out of the water; all vertebrates are still fully aquatic, ecologically fish. It will be another 35 million years until one group of Devonian fish, a metre or so in length, with fleshy, lobed fins, emerge onto land, the first four-limbed, or tetrapod, vertebrates, but it will not be far from here. The earliest tetrapod hindlimbs, from the Late Devonian, are from just downhill in Elgin. Only 300 kilometres and 50 million years away in the earliest Carboniferous of what will later become the River Tweed, the diversification of amphibian and reptile will be hothoused, as we vertebrates take our first breathing steps.[24]

Arthropod means 'jointed feet', and refers to the hard outer exoskeleton providing support and articulation to the limbs. Arthropods are the most species-rich phylum of animals of the modern day, and have been so since the initial diversification of animal life around the Cambrian, some 540 million years before the present. In the Early Devonian, they are mostly marine, including crustaceans, sea scorpions, sea spiders and trilobites, but a few have now made their way onto land. Arachnids appeared on land in the early Silurian and were among the first to diversify, rapidly adapting to the dry conditions. By the Devonian, arachnids already include scorpions, mites, harvestmen and the superficially spider-like trigonotarbids.[25]

A decomposing stem of lycopsid is rank with an earthy scent. It swarms with creatures, each only a few millimetres long, six-legged beasts with jointed bodies, long antennae and a short coat of bristles. Springtails, or collembolans, are strictly not insects because of technicalities to do with the position of their mouthparts, but they are the closest cousins of insects. Put a corset on a springtail, tighten up the waist, and you would have a decent impression of an ant.

Feasting on decaying plants, *Rhyniella praecursor*, the 'little Rhynie forebear', crawls around in the undergrowth, but is also small enough to pond-skate out to feed on the surface-floating algae. Despite the low levels of oxygen, *Rhyniella* are minute enough that oxygen can diffuse directly into their bodies.[26]

For a small animal, safety is never assured. From a hiding place within the open-ended starwood stem, the clawed arms of a plated predator emerge and grab an unfortunate *Rhyniella*. Instantly, a firework of small, black springtails scatters widely, demonstrating the reason for their name, a specialized organ called a furcula. In essence, a furcula is a long, rigid stick, held underneath the body at high tension. When the springtail releases that pressure, the stick pushes down into the ground, or even the water's surface, like an upside-down medieval catapult, firing the springtail into the air in a semi-controlled manner. Wherever each springtail lands, it is at least likely to be far away from the animal that startled it.[27]

Pinning *Rhyniella* directly under its body, preventing escape with a cage of eight limbs, is a *Palaeocharinus*, an 'ancient whip spider'. True spiders have not yet arisen, but trigonotarbids are arachnids that are superficially very similar. The distinction is partially cosmetic – they have fewer body segments, both of which are armour-plated; the head sandwiched between two plates, in which are set the eyes and mouth. Extremely hairy limbs are able to sense the vibrations of even the smallest prey approaching the place of ambush. A series of its body plates are perforated underneath with holes for air to enter complex and efficient breathing structures called 'book lungs'; this is an active predator.[28]

Little appears to be happening from outside the trigonotarbid cage, but the fate of the springtail is unpleasant. Without venom or silk to immobilize its prey, the victim must be pierced, crushed and broken. Trigonotarbid mouths are more like a sieve than a hole, so the springtail will be digested outside the predator's body, before being sucked through a series of ever-finer hairs.

In the stagnant freshwater ponds, among the stringy cyanobacterial slime and stonewort algae, life is safer. The pools are ephemeral enough that they do not develop complex internal food webs. Instead, crustaceans that feed on detritus dominate: slender *Lepido-caris*, the scale shrimp, a millimetre-scale, stalk-eyed browser on algae; *Castracollis*, a tadpole-shrimp, long-bodied with those distinctive armoured heads; and the tiny, round, armoured *Ebullitiocaris oviformis* – literally the 'egg-shaped boiled shrimp' in reference to the hot and alkaline conditions.[29]

Despite the simplicity of many of the animal relationships here, the stoneworts, like so many of the other photosynthesizers at Rhynie, have a deep ecological connection to fungi. Stoneworts are a freshwater group of algae closely related to land plants. In the cold-water pools of Rhynie, the most common stonewort is a single, straight axis from which emerge spirals within spirals of side branches. Unlike the teamwork on land between *Aglaophyton* and its mycorrhizae, or within *Prototaxites* and *Winfrenatia*, this relationship is one-sided, and toxic. Water-dwelling fungi attach themselves to stoneworts, embedding themselves into their cell walls or piercing them with tubes. They then proceed to take nutrients, supplying nothing in return. Other fungi, like *Cultoraquaticus*, some of the earliest known fungal parasites, digest the eggs of crustaceans. All four of these fungi are chytrids, a group specializing in parasitism of one kind or another, but especially of algae.[30]

When attacked by parasites, many plants have a response called hypertrophy, still a common symptom of plant disease. In this case, the cell sizes increase by up to ten-fold, attempting to isolate the infection in a single or a few cells. A related response is hyperplasia, where more cells are created to restrict the disease to one part of a tissue, as in galls. Several of the Rhynie stoneworts are infected with parasitic fungi, and show the resulting bulbous swellings along their length.[31]

Parasites are causing problems on land, too. In the stomata of *Aglaophyton*, nematode worms hatch, grow and reproduce, all without ever leaving the plant. *Nothia aphylla* is an early land plant that exists mostly underground, which gives it better access to groundwater than its competitors, most of which are horizontal stems suspended above the sandy soil. This strategy, however, puts it closer to parasitic species, and it has come up with an alternative to hypertrophy to ward off its fungal attackers. If a rhizoid is attacked by a fungus, the cell walls of *Nothia* harden, blocking the hyphae from penetrating any deeper and thereby containing the infection. There is a twist, however. *Nothia* also has mycorrhizal partners, which associate with the plant in many of the same ways as the parasite does, but are not isolated by the immune response. This is an evolutionary exclusivity contract. The fungal symbiont is allowed into the cells of the plant, where the host and symbiont can exchange resources. But any other fungus attempting the same invasion without the required chemical calling card will be surrounded and isolated. The preferred fungus is guaranteed resources denied to other species, while the plant can obtain rare or difficult minerals without being exploited. Organisms are not intrinsic do-gooders, and do not make deals except through generations of haggling via natural selection. It is thought that this relationship came about because, by tolerating fungal activity in some part of the plant, *Nothia* would be better able to identify other fungi as unwelcome strangers. A mutually beneficial relationship is not necessarily forged by peaceful means.[32]

From mutualisms to parasitisms, the conquering of a new environment does not occur in isolation. What began as an inhospitable and unpromising landscape is now teeming with life. For the next 400 million years, this planet will be a plant world, a fungal world, an arthropod world. The big beasts that emerge later, everything that has ever walked or crawled, is dependent on the innovations of communities like Rhynie. Root and hypha grip and sink ever deeper, interlocked as dancers' fingers, into the yielding rock. Together, they will change everything.

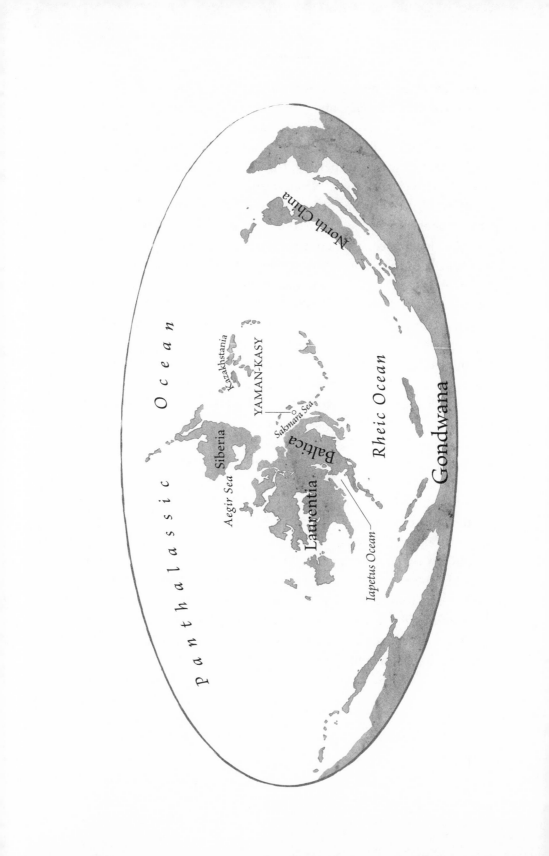

# 13.  *Depths*

*Yaman-Kasy, Russia*
*Silurian – 435 million years ago*

'Я – свет.
И пристально смотрю:
Дыхание в глубинах
возникает.'

'I am made of light.
I peer intensely:
The depths reveal a breath'

– Natalia Molchanova, 'И осознала я небытие' /
'The Depth' (tr. Victor Hilkevich)

'Under every deep a lower deep opens'

– Ralph Waldo Emerson, *Circles*

We on the surface are sun-bound, creatures of light. Inhabiting the thin atmospheric veneer of our planet, we are daily blasted with beams of electromagnetic radiation from our nearest star. It is the energy source that grows all our food, warms the air, evaporates water to bring rain, and sets our own internal biological rhythms. Even the organisms deep within karstic caves depend on the sun they never see. Living in pools formed on a floor of shale, the cave-fishes of Missouri's Ozark Highlands are inhabitants of a single stratigraphic layer. Their ancestors have eschewed the light for so long that even if their rudimentary eyes happened to detect a photon, they would lack the optic nerve with which to alert the brain. Even the cavefish's food chain depends on suspended leaf litter, transported into their caverns through rivers, and the guano of roosting bats, bringing the sun's produce deep below the Earth. To descend into the deep ocean, though, is to truly leave the sun, and everything it means, behind.[1]

*Yamankasia rifeia*

Even in the clearest water, tiny particles float, scattering any light that passes them. Water, too, absorbs the light. The longer the wavelength of light, the sooner it dissipates. Red gives up first, reaching a depth of about 15 metres. Orange, yellow, green – none can penetrate far, a rainbow slowly being consumed. Without green wavelengths, about 100 metres down, at the bottom of what is called the euphotic zone, photosynthesis is no longer possible. Beyond this depth, in what is known as the twilight zone, only deep blue to purple light can pass; everything that lives below is dependent on food falling from above or energy sources other than the sun. A thousand metres beneath the surface, even the last beams of light struggle in vain to pass, and life enters the midnight zone, where it is forever black.

A kilometre of water is a heavy load to bear, with every square metre being pressed down on by a tower of ocean weighing about 10 tons – 100 times the atmospheric pressure. Every 10 metres further down, a sky's weight is added again. No matter whether the ocean floor is at the pole or the equator, no matter when in geological time it is, to live at the bottom of the sea organisms must divorce themselves from the familiar world of the surface. This is not just an experiential difference. The functioning of many parts of an animal's physiology depends on surface conditions. At the bottom of the sea, the constant temperature of around 3°C slows down the critical metabolic pathways of the animal. The crushing weight of the ocean, too, has a profound physiological effect. Proteins often perform their jobs by repeatedly changing shape, and pressures found in the deep sea are enough to squash even those proteins deep within cells into new structures, altering their efficacy unless they evolve more pressure-resistant forms. Having descended from a surface-dweller, a life in the abyss means a transformation even down to the molecular scaffold of your being.[2]

Until 1977, the only deep-sea ecosystems we knew about were the wide expanses of the abyssal plain, the endless, relatively featureless ocean floor that lies between continents, trenches and ridges. These plains are extremely rich in microbes, and host a surprisingly high number of other deep-adapted fish, crustaceans and worms, although food is sparse enough that they are widely scattered. The first challenge to this view came when a camera on a submarine probe, intending to explore the geology and chemistry of oceanic rifts, happened across a dense sprouting of ghostly molluscs and scavenging crabs, the hot vent water shimmering mirage-like in the searchlight. Complex life, though hidden, has existed in the depths of the ocean for just as long as it has in the air. A hydrothermal vent is, at its heart, not so different from the extremophile pools of Rhynie: an ecosystem built not on electromagnetic radiation but on redox chemistry, as microbial alchemists turn swirling potions of dissolved rock into food.[3]

In the Silurian ocean-world, the Ural Ocean, a smallish ocean only about a mile deep – 1,600 metres – straddles the equator. In the low-latitude north, it steeply ascends to meet the shelf of the lifeless island of Siberia. To the east, the young continent of Kazakhstania has risen from the deep. In the south-western corner of the Ural Ocean is a distinct region, the Sakmara Sea, close to the shelf of another continent, Baltica. Earthquakes are common in this particular part of the Sakmara Sea, just off the east coast of Baltica. They reverberate in the water with pitches below human hearing, but the wind and rain at the surface can also send howling and drumming noises to the bottom of the sea, even now, when nothing yet lives that can hear. It is not, strangely, entirely dark. Subtly, almost imperceptibly, the faintest infrared glow permeates the gloom. No living eye can detect it, but it is there, only a slight buzzing of photons. Its source is a haven, an oasis in the deep – the Yaman-Kasy vent. Here, recent geological forces are breathing life into the bottom water, down in the darkened depths. A series of slim barrier islands, the Sakmarians, lie not far offshore, calming the surface between them and the mainland. Their presence, however, is what causes the turbulence on the sea floor. The Sakmarians have been approaching Baltica for millions of years, the plate that bears them subducting under its oceanic neighbour. As this subduction happens, it generates magmatic eddies in the mantle, a complex swirling of liquid rock that causes the plate to split behind the island arc – a long, thin crack that widens, spreading the sea floor, known as a back-arc basin. At Yaman-Kasy, hot mantle magma rushes up and meets the cold seawater, and the Earth's self-destruction is balanced by its self-creation, erupting and solidifying into volcanic rocks – basalts and rhyolites, andesites and serpentinites, and, importantly for life, exhaling an abundance of chemical and heat energy in the form of sulphide-rich fluids. Elsewhere, fossil sites are formed from slowly settling sand, slumping shelves or dunes, a burial of rock that has passed through several stages of existence. At Yaman-Kasy, the rocks that will preserve its inhabitants are freshly made, quenched straight from the forge.[4]

That quenching is what produces the infrared glow. This is earth-light, not sunlight. As the superheated water cools against its surroundings, it emits photons – thermal radiation. The light from vents is strong enough that, in the modern day, a species of bacterium is known to use it for photosynthesis, some 2.5 kilometres below the reaches of the sun. Perhaps one of the Silurian bacteria is doing the same.[5]

There is certainly plenty of ocean floor in which it could happen. Today, 71 per cent of the Earth's surface is salt water, and the average depth of those seas and oceans is 3,700 metres. Even including the tallest mountains and highest plateaus, the average altitude of the Earth's surface today is more than 2 kilometres below sea level. This is nothing compared with the early to mid-Silurian, during which sea levels reached their all-time highs, cycling between 100 and 200 metres higher than the present day. Given the modern continental arrangement, a 150-metre sea-level rise would change the world map utterly. With the Amazon basin largely flooded, Peru would have an eastern seaboard, and the encroaching ocean would turn Beijing, St Louis and Moscow into coastal cities. The surface world is the exception, the continents a scattering of aberrant rises, chunks of rocks embedded in a planet mostly built of low-lying oceanic crust, a crust that cracks, and breathes out its fumes.[6]

At Yaman-Kasy, the full industrial manufacturing architecture of an operational vent chimney is in action. A collection of towers with a mineral sheen rises above crowded stone, their thin heights continually pouring out blackened water several hundred degrees in temperature. Below the smoking cylinders, a teeming mass of life is gathered in a composition straight from an L. S. Lowry painting, a sparsely coloured, urban scene, of thin creatures fed by billowing darkness.[7]

*Yamankasia* is an annelid worm, part of the group that includes common earthworms, whose bodies are divided into ring-shaped segments. They perhaps look like beard worms, specialists of the deep ocean, which are often found around vents, carcasses or other deep-sea oases. Like beard worms, *Yamankasia* lives inside its own chimney, a flexible tube constructed by the worm from a mixture of proteins and polysaccharides like chitin. When feeding, its heads, each covered in hundreds of minute tentacles, rhythmically emerge and retreat like the targets of a fairground mallet game. *Yamankasia* is about the same size as *Riftia*, the giant vent-specialist worm of the modern day, with its tubes being about 4 centimetres in diameter, but it does not obviously share other distinctive features with any particular phylum of worm. It has most likely converged on this lifestyle, happening on the same advantageous partnerships as many other animals have in the deep ocean. They are certainly giants compared with the tiny worms around their base – little *Eoalvinellodes*, whose tubes are barely a couple of millimetres across. The tubes of *Yamankasia* are made of several layers of fibrous organic material, wrinkled longitudinally, and are flexible, although the only currents here in which to bend are the convection currents resulting from the hot vent water rising, cooling, then falling again.[8]

When plants extract energy from sunlight on the surface, it does not happen in structures that are genetically plants. As with the fermenting bacteria in animal herbivores, plants incorporate single-celled organisms called cyanobacteria to do the photosynthesizing for them. These cyanobacteria are so deeply embedded in plant cells that they have, over hundreds of millions of years, lost some of their DNA, and could no longer survive independently. Now, they are known as chloroplasts, little pill-shaped organelles within the cell, working entirely interdependently with the plant to survive. Where the collaborations of symbiotic and mutualistic plants and fungi from the Devonian represent different species living in close proximity, the relationship of bacterial chloroplast and eukaryotic plant is so close and inseparable as to make the whole a single individual. Energy, here, cannot be split by the eukaryote, but by its fellow-travelling bacteria.[9]

In the same way, creatures at Yaman-Kasy and other hydrothermal vents are unable to directly access the energy in the sulphurous vent fluids themselves, but many incorporate bacteria that can. Modern-day beard worms, the largest of the worms that occupy the sulphide-depositing hydrothermal vents in the present, have a specialized organ called a trophosome. In that trophosome, each worm hosts billions of symbiotic sulphur bacteria that they need to extract energy from the vent. Like them, *Yamankasia* has close associations with bacteria living along its tube. The worm protects the symbiont, and the symbiont provides the worm with food. In this case, the interaction is somewhere between that of eukaryotes and their organelles and the closeness of lichen mutualism, further muddying the waters of what, really, an individual is. The halfway-house term here is 'holobiont', the living and inseparable whole that is made of two or more undeniably different organisms. Together, they thrive. Apart, they die. Some modern vent worms, for instance, entirely lack a digestive system; the bacteria make all their food for them. Some vent clams, in a further act of assimilation, become factories for sulphide-binding proteins, helping to enhance the natural ability of the sulphur bacteria. Their internal processes are beginning to be melded, alloyed together.[10]

At Yaman-Kasy, trace elements solidify out of the fluid as ores when the temperature and pressure decrease from rock to ocean.

Differences in chemistry between the hydrothermal fluid and the ocean set up a flow of electrons that, in some vents, hits 700 millivolts, a natural power station. Selenium and tin coat the central conduits, the flues of these unearthly chimneys. Further out, atoms of bismuth, cobalt, molybdenum, arsenic and tellurium, as well as gold, silver and lead, emerge from solution. In the modern day, the ores that form from these elemental oozings are all desirable commodities. Exposed to the air for the first time since their formation, the margins of the Ural Sea have transformed into industrial open-pit mines. Their rocks, containing the fragile tubes of the vestimentiferan worms, are crushed, powdered, dissolved, smelted and shot through by electrical fields to extract the metals within. Yaman-Kasy, the earliest-known hydrothermal vent fauna, continues to produce.[11]

One extremely surprising aspect of the deep-sea ecosystems through time is that, despite their similarity, the species that inhabit them are not closely related to one another. The identity of vent-dwellers has varied considerably through time, and modern vent community members are usually recently descended from species that lived in much shallower water. Given the phenomenal gradient of pressure, temperature and light, it might be expected that adapting to life in the abyss would have been difficult. It seems that this is not the case, and that vent-dwellers have come from all parts of the animal kingdom. Today, no known coral inhabits a hydrothermal vent, but in the Devonian, vent corals appear to have been reasonably common, all independently evolving a second layer to their external hard tissue, the calyx in which the soft polyp lives, presumably as a buffer against the temperature.[12]

The differences in the families that are present in hydrothermal vent faunas tell us that colonizing the dark sea is actually rather commonplace, despite the isolation. Vent fields are often clustered, but still spaced by a few kilometres from one another, the fertile efflux of minerals surrounded by barren sea floor. However, at larger scale, they form lines, associated with the cracks in the crust, a circumstance that gives life a chance to thrive in the depths.

Oceanic currents usually line up well with the direction of cracks in the crust, especially at back-arc basins. This means that larvae can passively drift, perhaps for hundreds of kilometres, before they find a new home. Even distant communities, then, are part of the same connected population, a seascape between which only the newly hatched larvae can disperse and refresh any dwindling populations. Vents act like islands on the surface, together becoming what is known as a metapopulation, a semi-isolated cluster that has limited mixing with the outside world. Each little vent contributes to the genetic diversity of the whole. This is important because vents are temporary, lasting only as long as the heat of the magma remains pressed close to the crack. As tectonic changes occur, that energy source can disappear, and the entire community begin a path towards annihilation. Every time new rifts open, they are as likely to be colonized by species adapting from above as from a chance drifting of a larva from elsewhere. But that colonization will inevitably end. Unlike the constancy of the sun, the deep is a place of long-term impermanence and transience, of novelty and destruction.[13]

Gathered together in a tight crowd are little shellfish, called firediscs. *Pyrodiscus* are brachiopods, the dominant shellfish group of the Palaeozoic ocean from the coast to the deep, before molluscs took that mantle. *Pyrodiscus* shells are something like that of a mussel, tongue-shaped, but holding onto the surface of the rock with a long, tendinous stalk. Any brachiopod in the early Silurian is one of a few lucky species; at the end of the Ordovician, a mass extinction wiped out most of brachiopod diversity. During that extinction event, which was sparked by global cooling, deep-water communities were particularly hard-hit, even those that, at least in theory, had all the characteristics that should make them robust to extinction. Cold chance always has a hand, but the circumstances of the Ordovician mass extinction played a role too. The cooling of the planet would, by itself, seem to have had more of an effect on warm surface waters than on the waters of the deep. But the extent of the cooling was enough to bring changes to deep-ocean circulation, and took dissolved air down onto continental shelves that had otherwise been low in oxygen. When this happened, shallow water species, adapted to higher oxygen conditions, could invade the shelf, and compete with the low-oxygen specialists of the deep water.[14]

Were the waters of Yaman-Kasy to become more oxygenated, the same fate could befall them. The bacteria that form the base of the food chain here are at their best in a low-oxygen environment. If they suffer, the community can rapidly fall apart. Vents are odd places, seemingly able to be everything at once. They are rich in nutrients, stacked city blocks full of animal life among thousands of square kilometres of microbial plain, and yet they are poor in species number. Yaman-Kasy is the oldest, and the most diverse, fossil hydrothermal vent ever discovered, but fewer than ten species are known.[15]

Diversity at vents is usually low, similar to rockpools, with a few dominating and some rare taxa as singletons. They resemble other volatile ecosystems such as fire-dominated forests or tide-washed rockpools, usually with about a third fewer species than comparably productive sites. But they are also constant places; there are no days, no seasons at a vent, no long-term cycles. So growth is fast, and reproduction is continual. Communities can recover easily from small-scale disturbances, but if a major perturbation occurs, they become particularly vulnerable.[16]

They are isolated, with each vent a rise as conspicuous as Mont St Michel. And yet, they are connected within groups – what seems to be important is not the single vent itself, but the whole ridge. Yaman-Kasy is only one vent in a chain, a series of dim beacons running all along the tectonic margins of the Ural Sea. Local or global, now or on the planet's timescale, hydrothermal vents change their traits depending on the scale at which you look.

Although life on the macro scale is impoverished, the microbes around a deep-sea vent are more diverse. The chemistry of the new rock means that many of the reactions that bacteria perform in order to get their food occur much more easily, and this helps to extract organic molecules from the ocean water and fix it into living tissue. Ocean-exposed basalts are host to such bacteria the world over, and give a huge boost to the amount of organic matter in the deep oceans. Up to a billion tons of carbon are fixed every year by transparent films of bacteria coating the deep-sea basalt communities of the world. Around vents, there are even productive communities of bacteria beneath the sea floor itself, taking advantage of the nutritious fluids arriving from beneath. Perhaps most surprisingly, there are hundreds of species of microscopic fungi that make their home only in the deep sea.[17]

In the Sakmara Sea, the magma that makes its way to the crustal surface is a little cooler than average, and particularly rich in silicon, potassium and sodium. These sorts of lavas, which produce the rock rhyolite, are often gassy, and so can form lumps of pumice. The lumps float from the depths to create sulphurous rafts that are often, initially, strong enough for an adult human to walk on. In 2012, a pumice raft from an eruption in a back-arc basin near Tonga produced 400 square kilometres in a single day, eventually dissipating into a thin layer covering 20,000 square kilometres or more. Under the ocean, the lavas solidify into jagged rocks, filled with nooks and crannies, anchor points for the many inhabitants of the Yaman-Kasy vent.[18]

Delicate sea snails, only a couple of millimetres in diameter, shift alongside other small, spiky, white shells. *Thermoconus* – the 'hot-cone' – is a monoplacophoran, a type of mollusc that is generally quiff-topped, as if a limpet had melted in the heat. These, though, are Christmas trees in miniature, stacks of cones, flaring at the base, adding more as they grow. The difference in the fertility of the waters near and far from the vent can be seen starkly in the size of these creatures. Further from the vent, everything is smaller, an effect challenging perspective. At their biggest, closer to the churning water, they can grow fairly large – up to about 6 centimetres tall.[19]

Monoplacophorans are an extremely ancient type of mollusc, the oldest known from the fossil record. With a single, central, rippling foot, they shuffle around in the sediment. Wherever they go, they leave behind the scratchings of their rasping radulae, filing at the rocks to prise off their microscopic food. Monoplacophorans still survive in the present day, but where most fossil monoplacophorans live close to the shore, today they only exist in the deep ocean. The earliest of that group to venture into the deep, though, is *Thermoconus* at Yaman-Kasy. Perhaps, although the fossil record is too sparse to prove it, the monoplacophorans at Yaman-Kasy represent the beginnings of a retreat into a world where no others could survive, an evolutionary hiding place in an inaccessible niche-space, freer of competition.[20]

The deep ocean is an effective hiding place. In 1952, a living monoplacophoran was hauled up off the coast of Mexico from water more than 3,500 metres deep, startling scientists, who had until then understood that the group had gone extinct in the Devonian, 375 million years before. The discovery was hailed as a resurrection; a group that was thought to be dead, but has, cryptically, survived and risen to life once more, is known as a Lazarus taxon. This was not even the first such instance of the deep ocean revealing long-gone secrets. Coelacanths are thick-set, long-lived fish with symmetrical, fleshy tails and equally fleshy fins, part of the group known as lobe-finned fish, more closely related to humans than to haddock. Besides tetrapods, lungfish were long thought to be the only other living group of lobe-fins, but in 1938 a coelacanth, a type of fish thought to have become extinct at the end-Cretaceous mass extinction, appeared in nets in the Indian Ocean, having survived below our ken in the crushing blackness since.[21]

This happens within extinct groups, too. A Devonian fossil site in Germany, preserving another shallower, back-arc basin, the Hunsrück Slates, contains all the classic Devonian fish, but also an anomalocaridid named *Schinderhannes*, a type of predatory arthropod only otherwise known from the Cambrian and Early Ordovician, with 100 million years of missing evolutionary history. The second youngest anomalocaridid yet discovered is a bizarre giant from the deeper Ordovician waters of Fezouata in Morocco, called *Aegirocassis*. This animal reached 2 metres in length and was a bulk filter-feeder like today's baleen whales. It was something quite unlike any other anomalocaridid, implying that there is much that has not been, perhaps cannot be, observed. The deep water is not just a place in which species can escape detection from the eyes of surface-dwellers, but also a place where, without the dusty run-off of the land, a lineage can for a time escape the preserving potential of the fossil record. By the time the Earth captures the image of a hidden lineage once again, it can have changed beyond recognition.[22]

As rocks, the medium in which those images are formed, are made for the first time, the elements mingle into their igneous crystal forms. Each element has many isotopes, forms chemically identical but of different weights, that exist at naturally consistent ratios. Some of these are radioactive, transforming into other elements at predictable rates, a clock that begins to tick as the rock changes from liquid to solid. Where carbon dating works as a short-term clock for life, other elements date deep time, from within the rocks themselves. Zircons, extremely common minerals in igneous rocks, frequently contain uranium but will never contain lead at formation. There are two isotopes of uranium, and both decay to different isotopes of lead, with different half-lives. The amount of lead in a zircon crystal is a direct measure of age. For older rocks, rich in micas and hornblendes, the decay of a radioactive isotope of potassium to argon acts as a timepiece.[23]

Time, for an ocean, lingers. The waters of the global oceans travel round in seemingly eternal cycles from equator to poles, from the deepest ocean to the wave tops, and they do it slowly. The metaphor used is usually that of a giant conveyor belt, but the quickest parts of that belt – such as the Gulf Stream – have a maximum surface speed of about 9 kilometres per hour, a fast walking speed. A droplet of water that travels the full length of this wandering conveyor belt will take a full millennium to do so. The current passing today from Iceland to Greenland and on to Labrador perhaps even now contains some of the self-same water sailed by Leif Erikson and his crew, the first Europeans to cross the Atlantic, returning to those seas for the first time.[24]

Cold, polar water, is denser than warm water, and so it sinks, taking with it oxygen to the deep sea. Water has an unusual property in that its solid form is less dense than its liquid form, which is why ice floats. The density of water is highest at about 4°C, so even though the surface water may fluctuate from warm to cool with season and weather, the water in the nether of the Ural Sea stays the same. Yaman-Kasy is in an area of upwelling, where deep waters begin to rise to the surface, but some effects of the surface still penetrate to the bottom. Big storms, for example, can be efficient at driving sediments from shallow to deep water. In the midnight zone, food falls from above. Hardly manna from heaven, but there is a continuous fall of dead organic material, so-called 'marine snow', the decaying bodies of cyanobacteria and algae, sinking and being buried in the ooze. In the modern day, nearly half of the carbon dioxide captured by life ends up sinking to the bottom of the sea.[25]

In a sense, we are all creatures of the deep. Hydrothermal vents, plumes of superheated water rich in minerals, are bursting with chemical potential and ripe for exploiting, and had an ancient role to play in the origins of life. Primordial oozes, organic soups struck by lightning, raising life in a Frankenstein-like fashion, are the stereotypical portrayal of the emergence of life on a lifeless planet, but these never existed. There is strong evidence, though, that the chemical outlet of deep-ocean vents laid the basis of the internal chemistry of every living thing today.

Three and a half billion years before Yaman-Kasy, so the prevailing science goes, a particular kind of alkaline vent provided the basic environment, in both senses of the word, in which life itself could originate. From deep in the Earth, such vents poured forth hydrogen and methane into mildly acidic ocean water rich in nitrates. In the oxygen-free, alkaline conditions within the vent, bubbles of fatty acids spontaneously form, a structure analogous to cell membranes. These fatty membranes are then in contact with both the vent fluid and seawater, and with a slightly alkaline interior, a protocell. The difference between the acidic seawater and the alkaline vent sets up a flow of hydrogen ions from seawater, through the protocell, to the vent – and wherever there is a flow, work can be generated. Alkaline vents can also naturally produce a molecularly layered mineral called fougèrite, commonly known as 'green rust', which may be the key to some of the mysteries surrounding the origin of life. It acts as a natural catalyst – a facilitator of chemical reactions – that helps to churn out many of the molecules that life is based on, like ammonia, methanol and the fundamental structure of amino acids. Typical fougèrite crystals are small enough to become embedded in protocell membranes, which turns these crystals into natural channels, transporting and concentrating a chemical called pyrophosphate within these membranes.[26]

Today, no matter what the source of its energy, whether from the sun, from minerals, or from digesting other living creatures, every single living thing on Earth converts that energy, first and foremost, into a pyrophosphate compound, ATP, the so-called 'universal energy currency of life'. Across all living things, that conversion only happens when a gradient of hydrogen ions is set up and allowed to flow through our own slightly leaky membranes. To perform any action, from firing nerves to secreting saliva, from contracting a muscle to DNA replication, every cell within every body must first replicate some of the chemistry of the earth bleeding into the sea.[27]

Over the quiet of the deep, the rain drums a persistently unheard rhythm. The dim earthlight, suffusing from the heated water, illuminates the huddled masses gathered around the vents like campfires in midwinter, an insipid infrared glow too weak to be used for energy, but infused with the sulphurous breath of the mantle. Insensitive to the changes above, and unknown by the creatures of the photic zone, the deep-sea faunas continue to do what they always do. Grow, extract, move on and survive. For as long as the fragile surface of the planet continues to fracture and shift around itself, there will be openings in the earth, and an opportunity for those able to thrive in the sunless sea.

Bohemia

Proto-Alpine
Region

North France
& Benelux

Turkey

East Avalonia & Baltica

West Avalonia

Iapetus Ocean

Iberia

South Pole

West African/Amazonian
Ice Sheet

Gondwana

Africa

Florida

Yucatan

Laurentia

South America

Antarctica

Proto-Panthalassic Ocean

SOOM

Sea ice

# 14.  Transformation

*Soom, South Africa*
Ordovician – *444 million years ago*

'broken ice, a sinister chaos'

– Matthew Henson, polar explorer

'With the passing of time, the sea becomes dry land,
and dry land the sea'

– Abu al-Rayhan al-Biruni, 'Chronology of Ancient Nations'
(tr. B. Ghafurov)

Above the blue-grey river of ice, a heavy falling wind, cooled by the frozen uplands and tasting of nothing but snow, roars out onto the ice shelf and dives, crashing, seaward. Such winds are described as katabatic, a blast of cold, dense air pulled from the heights with near hurricane force by the very weight of the Earth. The wind has travelled far from the centre of the retreating Pakhuis ice sheet by the time the glacier finally and ponderously slumps into the bay. Now, the winter wind catches on the irregular surface of floating sea ice and pushes it away from the land, leaving an open, unfrozen zone, a polynya, on the southern margins of Panthalassa. The surface cannot remain uncovered against the frigid air for long, and is in a state of balance. New frazil, a slick and slushy mixture of seawater and spiny ice crystals, bends as the waves sough past, as it continually forms and blows away. With the brash and broken blocks of calved glacier, these ever-forming ices are herded offshore, an unstable geography of fragments.[1]

Further out, the blown ice and glacial calves pack down into an icy crust. The ocean is made solid by this floating landscape of floes

*Orthoconic cephalopod*

and hummocks. As the winds scour through the drumlins of this end of Africa, scraping the worn edges of valleys bare of snow, they become dust-blowers, sweeping up with them the granular remains of rocks rubbed to sand, exposed as the glaciers are retreating. Earth in the sky. The pack-ice is striped with sastrugi, finely grooved ridges and waves of ice, sometimes smooth as crumpled silk, sometimes an angry churned-wave image of the trapped ocean below, surrounded by an orange corona and scattering the winter sun. The sand settles on the ice as it grows through the winter, incorporated into the freezing mixture and held in place, waiting.[2]

Below the glacier, within the polynya, two rivers flow, a river of salt and a river of earth. The headwater of the salt-river is where water is removed, and ice forms. As the surface water freezes, the salt dissolved with those molecules is pushed out, unable to be held as readily in the crystal structure. This makes the surrounding, unfrozen water saltier, and denser. From the surface of the polynya, this brine sinks within the sea, dispersing away from the shore, into the black ocean depths, to gather on the continental margins and away on to join one of the deep-ocean currents. Submarine rivers act just like those on the surface. Where they pass along the sea-floor landscape, they build curving banks, erode canyons, and form briny lakes and waterfalls. The modern Bosporus contains one of these rivers, flowing from the salty Mediterranean for 60 kilometres at the bottom of the Black Sea, but discharging more water than the Mississippi, the Nile and the Rhine put together. In volume, it is one of the ten largest rivers in the world. At Soom, as the salt-river descends from the polynya, out from under the glacier a fresh meltwater river churns out of a submarine ice-cave. It is a dark hole, the outpouring of untravelled streams that flow in the thin zone between rock and ice. The watercourse is squeezed by the weight of the ice, and has picked up mud, so it is expelled as a high-pressure torrent of black sediment. Earth in the water. It gradually rises to the surface, turbid but saltless, giving it a simmering appearance, as if something is lurking beneath. There is little below. Under the sub-zero, hellish mud-cloud, which blooms to tens of metres thick as it meets the warmer ocean water, no light can pass.[3]

Within the flowing ice, sounds ring out. Creaks, long moans and roars reverberate and mark the inevitable shift of ice under its own weight. The glacier carries more than water. Earth in the ice. Erratic rocks and boulders, swept up by the ceaseless flowing sheet hundreds of years earlier, are reaching the end of their entombed journey, and are released into the water, sinking to the floor below. Air in bubbles, little packets of the atmosphere of the past, a frozen archive of earlier centuries and millennia, has been held within the ice as it descended from the top of the Pakhuis ice sheet, some 200 kilometres away. Since then, they have become compressed as the glacier blued and accumulated about them, squashed to perhaps up to twenty times the atmospheric pressure, and are held, waiting. As the glacier melts, its bubbles are finally released, livelily cracking and shattering into existence below the water, a wall of natural seltzer, adding a fat-frying sizzle to the noisescape of creaking glacier and thudding dropstone. As the world warms, millions upon millions of bubbles burst into the water every second, and the volume is only increasing. Even the icebergs sing, and the larger ones – big enough to be floating islands containing internal rivers – sound their basso profundo voices and dark rhythms, vibrating hundreds of kilometres along the coast.[4]

Soom is a stratified world, defined by its layers in life as well as in death. Wind, ice, seawater, fresh earthy plume, and the still bed of this fjord and bay, rendered oxygenless by a lack of mixing. All are stacked on top of or hidden within one another, and released into a single ocean. Summer in Soom brings a colour to the sheet like nothing else. The sun of the late evening reflects off the melting pack-ice and lights the bare mountain slopes on the eastern side of the glacier in deep orange. Low-hanging clouds that covered the polynya in winter disperse, and the wind dies back a little. While icebergs dwindle, they gleam wetly in the sunshine, high-pressure blue where their surface is blown bare of snow and where the air bubbles, kept under pressure for longest, have simply blended into the ice structure. The turbid river continues to throw out silt, but, away from that plume, the clear water in the ever-larger Soom polynya reveals the life of this place. Polynyas are oceanic oases and, as the water warms, the covered parts of the smaller icebergs, visible through the surface water, become a perfect, almost luminously mossy green.[5]

Under the pack-ice, the layers begin to mingle, and the rock rain begins. After hundreds of years of frozen suspension, sea-melted bergs lose their grip on stones, which plummet into the brine, thudding into a bare sea floor below. The settled dust carried by the winter's katabatic wind will sink too, but more sedately; the particles are smaller, and linger longer in suspension. As they hang, they attract the attention of plankton, the microscopic organisms suspended in the upper water. The particles carry with them compounds of phosphorus, and the ocean blooms as it receives it. Microscopic algae that feed on light, until now restricted in their growth by a dearth of minerals, turn the alcohol-clear water bright green in a feeding frenzy. These phytoplankton reproduce rapidly, making use of the minerals while they can. While there is excess food, there is no competition, no struggle for life, just generations of endless growth. The good times cannot, of course, last for ever. As resources dwindle, or as populations grow to the point where they are exhausting the supply faster than it can be renewed, death will follow. The faster the growth, the shorter the boom. Now, they reproduce so quickly, their populations bloom so fast that they clump together, and their dead bodies begin to rain down organic matter into the layers beneath.[6]

Polar ecosystems often show life in the slow lane. Resources are frequently limited, and the cold slows down many biological processes, including growth. When there is a disturbance, a disaster that kills large numbers within an ecosystem, recovery can take an exceptionally long time. They can be subject to the vagaries of scavenging, and generally end up being communities of low diversity and variable composition. This pattern changes where nutrients are blown from the land. Wherever there is local enrichment of nutrients, even in the polar cold, environments can recover entire communities after severe disasters within the space of years, and at Soom, this enrichment comes, year after year after year.[7]

But this pattern is recent, and will not last for ever. The glacial bay at Soom did not exist a few thousand years ago, and it will be deeper water soon enough. It was here that the ice ploughed through from the north more than 100 thousand years ago, and here that it is now retreating. The post-glacial ecosystem is flourishing as the world warms again, after the great freeze of the Ordovician, the first of multicellular life's mass extinctions.

Barely one million years before Soom, the Earth flipped from hot to cold, and sparked an event known as the Hirnantian Glaciation, during which the ecology of the oceans was severely altered, even to the level of microbes. The mass extinction near the end of the Ordovician was the second largest to be endured by complex multicellular life, surpassed in scale only by the Great Dying at the end of the Permian. Before the Hirnantian age, the final division of the Ordovician period, life was balmy in the Ordovician seas. Diversity had skyrocketed through that period, far exceeding what had gone before in the Cambrian. It was in the Ordovician seas that animal reef-building began in earnest, and life took wholeheartedly to free swimming, rather than being confined to communities on and close to the seabed. But over the course of perhaps only 200,000 years, an ice age began, centred on what will become Africa. As part of Gondwana, the supercontinent comprising all of Earth's modern-day southern continents, as well as India, Arabia and parts of southern Europe, Ordovician Africa lies around the South Pole. The site of Soom, in the present-day part of South Africa's Cederberg Wilderness, is in the Ordovician found at about 40° south, a latitude not too far south of where it lies today. Between the present day and the Ordovician, Africa has slipped under the bottom of the Earth; at the time of the Soom glaciers, the South Pole itself is found closer to what will become Senegal. On a globe, Africa looks upside-down. From the pole, Gondwana reaches up an arm that extends past southern Africa through Antarctica to Australia at the equator. The pole itself is not covered in an ice sheet, but two major sites are. One emerges in the south of what is now the Sahara region, with glacial fields drifting north. The other area covers South Africa and parts of central South America, moving towards the tips of the landmasses and emerging into the Peninsula Sea.[8]

During the Ordovician, organisms beginning to adapt to life out of the water became more numerous and more common. As yet, they are mostly microscopic, and merely individual species here and there, not yet the thriving communities that they will form later in the Silurian and Devonian, although some riverways are beginning to be colonized. The early mining efforts of fungi and simple plants have eroded continental surface rocks, releasing phosphorous compounds into watercourses and out, onto the upper ocean, flooding that space with this valuable mineral resource otherwise so rare in seawater. The algal blooms that still occur around Soom were ubiquitous, happening wherever these outwashes occurred, with bigger populations of bigger individuals. This glut increased the marine snowfall to the ocean floor from an occasional flurry to a continuous blizzard. As the algae's carbon-rich bodies settled and were buried, they drew with them carbon dioxide from the atmosphere. At the same time, a coincidental increase in volcanic eruptions as the Caledonides were raised produced a lot more silicate rock. As we have seen, the weathering of silicates causes reactions with carbon dioxide in the air. These fresh silicates helped to reduce the atmospheric concentration of carbon dioxide, too. In the resultant rapid climate change, a full 85 per cent of the species on Earth, almost all marine, went extinct. The glaciation did not last long, but it was long-lived enough to cause devastation. It was the first of the so-called 'Big Five' extinctions, and the only mass-extinction event directly caused by global cooling.[9]

As far as extinction is concerned, the absolute climate is not to blame, nor is the direction of change. It is the rapidity of change that is important. Communities of organisms need time to adapt – if too much change is thrust upon them at once, devastation and loss is the common response. This is true of the end-Cretaceous, when the impact of an extraterrestrial rock caused near-immediate global winter, and of the end-Permian, when skyrocketing greenhouse gases from unprecedented volcanic eruptions sparked global

warming. Within the Hirnantian, as the Earth reverted to a deglaci-
ated state, the warming has caused a second, smaller extinction
pulse. That warming continues at Soom, where the glacier is retreating at speed.[10]

Beneath the water, where the pack-ice still covers the sea but is
thinned by the summer, the ocean is lit with a gentle, diffuse, blue-
green edge. Offshore, the water fades to an invisible darkness
beneath. Rounded bumps and stalactite-like structures texture the
underside of the sheet, and little is swimming past in the cold. Minute
fragments of ice, crystallized out from the supercooled water, hang
motionless, a snowstorm that never falls. Descending deeper, the
sunlight still reaches the shallow shelf, some 50 metres down,
although it is dim. The water is cool and so still and clear that it
seems like air. Hardly a current disturbs life under the ice sheet, but
there is little on the floor to disturb. It is practically bare, almost
devoid of life. The stillness has made the sea floor unbreathable,
deprived of oxygen. The fallen algae are rapidly consumed, not by
grazers or detritivores, but by those ubiquitous sulphur bacteria,
lovers of oxygenless places. In the still water, the waste of their reactions drifts almost by diffusion – hydrogen sulphide clouds that in
water form local pockets of concentrated sulphuric acid.[11]

Where the salt-river drifts by, oxygen is temporarily brought to
patches of the floor, and here little brachiopods, juveniles, less than
half a centimetre long and just beginning to grow, bury themselves
in the sediment, and a few crawling trilobites and soft-bodied,
worm-like lobopods risk a temporary journey to the bottom. In the
water above, fish swim here and there, among them eerie, jawless,
unarmoured, limbless fish, relatives of lampreys. With only dorsal
and anal fins near their tail to guide them, they swim in small
groups, dropping down to the bottom waters to prey on other soft
animals, before fluttering their eel-like tails to gain height and
breathe again. Staying on the black, muddy bottom would only be
lethal, and for more sedentary creatures it is. To have a shell rich in
calcium carbonate in an acidic world is to do nothing more than

become an earnest chemical reaction. During the still times, when the acid clouds concentrate, the carbonate shell materials of the dwellers in the deep will simply dissolve. Because of this, Soom is a place with very low biodiversity compared with similarly deep sites elsewhere. For any creature to live long term in the waters of Soom, it either needs to be able to continually swim, or to find another solution. For the clam-like brachiopods, whose shells are composed mostly of calcium phosphate, this means they have to adopt the life-style of a hitchhiker, attaching themselves to the surfaces of other creatures that can swim in the upper, oxygenated, less corrosive layers of the Soom sea.[12]

Floating near the bubbling blue glacier wall are foot-long conical shells, orthocones, and it is on these that brachiopods grow among the crusty piping of tubeworms. Related to modern nautiluses, orthocone cephalopods demonstrate what would happen if the coil of an ammonite or nautilus were straightened out. Some ortho-cones reach splendid lengths of over 5 metres, but the species common at Soom is more modest in size. Their fleshy arms twitch-ing out of their shells, large eyes inspecting the ocean around them, they propel themselves through the water like jet-skis. The engine of an orthocone is a special, tubular organ at the opening of the shell called a hyponome or siphon, made of a ring of muscle that contracts as a wave. Ordinarily, it swims forward, but the hyponome can give an emergency spritz, sending the hydrodynamic cone float-ing backwards at speed with a powerful jet through long thin streamers of brown seaweed and clouds of scattering shrimp.[13]

A 10-centimetre creature with a large head, two oar-like limbs behind waving graspers and beady, kidney-shaped eyes is a preda-tory sea scorpion, but rather a small one. Sea scorpions are one of the most diverse groups of Palaeozoic animal, although they will peak in the Silurian and Early Devonian, when some will become the largest arthropods ever, more than ten times as long as the spe-cies at Soom. Although they aren't true scorpions, they are not too distantly related, and much of their body structure is very similar, with a thick abdomen and long, thin tail, tapering to a point – although without the venomous sting of a true scorpion. Six pairs of multipurpose appendages – one pair of little grabbers, chelicerae, for grasping food, and five pairs of limbs. Later sea scorpions in these parts will equip their limbs with the abrasive spiny plates with which they feed, sweeping their prey towards mouths that are also composed of appendages.[14]

Using limbs as jaws is a fairly ordinary arthropod approach; the segmented arthropod head and body is a developmental Swiss army knife, with each segment containing a flexible, jointed appendage that can be adapted to a huge variety of functions. Spider fangs are, developmentally, the same structures as insect antennae. What, in developing insects, forms their mouthparts, turns into the first three pairs of legs in a spider. At Soom, the native species of sea scorpion has modified its final limbs into swimming paddles, flattened and sticking out sideways like the oars on a rowing boat. They still retain some of the ancestral anatomy; little gripping claws mark the ends of these paddles, earning the animal the name *Onychopterella*, or 'claw-wing'.[15]

*Onychopterella* is one of the bigger predators at Soom, but there is another mysterious predator somewhere out there in the ocean, which has never actually been seen. It is one of those animals identified indirectly, only known by the traces it leaves behind. The only clues to its existence are droppings, blobs of faecal matter in the mud that contain broken, crunched fragments of shell and teeth. It clearly feeds on some of the swimming crustaceans and on another of Soom's more unusual inhabitants – *Promissum pulchrum*, the creature of the 'beautiful promise'.[16]

*Promissum* is a type of creature called a conodont. Conodonts are chordates – that is, they are related to fish. They are some of the most abundant creatures on the planet, found everywhere from the early Cambrian to the end of the Triassic. The conodont record is so dense, so constantly present throughout time, that the resolution of which species occurred when is exceptionally clear. They are what is known in palaeobiology as an index fossil, meaning that they can be used to date the rocks in which their fossils occur. For over a century, all that was ever found of them was their mysterious teeth. These tooth elements, shaped like spiked tiaras, are very robust, the only hard part of the soft, eel-like creatures. Even before knowing what the full animals were like, they were used as fossil timepieces, their first or last appearances defining the time slices into which geologists divide the world's history. Among these, they mark some of the biggest divisions of all. The beginning and the end of the Permian period are both defined by the first appearances of particular conodont species. Like regnal years of monarchs, our perception of time is shaped by the lives of these creatures.[17]

Lithely sliding through the water, the foot-long *Promissum* hardly looks like the fulfilment of a long-hoped dream, but its remains will be the most detailed soft tissue found from any conodont. Only here and in the Carboniferous Granton Shrimp Beds of Scotland will any musculature, anything other than their teeth be found. *Promissum* swims purposefully but slowly, efficiently slithering in the cold. Its reddish flesh shows that it is a slow-twitcher, its muscles constantly in use, always swimming. Most fish muscles are white, with fast-twitch muscles that do not need to fire constantly but are used for rapid response. In chordates, more muscle use means a higher oxygen requirement, which means that red myoglobin, the muscles' oxygen-carrying protein, is concentrated there. Like haemoglobin in blood, this gives muscles that need to be used frequently a red colour. This is why chicken legs, which need to be able to hold a chicken up all day, are darker than chicken breasts, which are only used rarely for flight. It is also why tuna, which constantly actively swim, have dark flesh. *Promissum* has nothing but slow-twitch fibres with a fairly inefficient V-shaped structure, and a need to be constantly in motion. Soom gives us these details because preservation here is the wrong way around. Hard tissue preserves here very, very badly, while muscles are preserved down to the individual fibre.[18]

The rain of silt and algae that falls in the Soom summer also gives the sea floor an unusual chemistry for fossil formation. If a *Promissum* dies in winter, it will sink to the bottom and become coated in and buried by that persistent subglacial black dirt. The body will decay, the teeth disappear, and nothing will be preserved. But in the summer, when the loess falls too, not all of it can be processed by the zooplankton and organic-eating bacteria, and it settles as a paler layer of rich sediment, enhanced by the organic input from dying plankton, a striped annual deposition called a varve. The preservation of this double-layering at Soom is an almanac, a yearly account of conditions some 440 million years ago, proportionally equivalent to having a daily diary for the earliest humans in western Europe, 1.2 million years before the present.[19]

In the acidic conditions, the cartilaginous skeleton of *Promissum* will still decay, but other elemental forces will take hold. As the proteins in its muscles begin to disassemble, they release a chemical effusion of ammonia and potassium. Reacting with iron minerals, and dissolved within the spaces between individual sand grains, they turn into a rich illite clay. The shape of the muscle fibres determines the ultimate shape of the clay, and muscle is sculpted in mineral, replaced by its own facsimile. The conversion of soft muscles into clay at Soom is unique, and utterly beautiful, a vision of life in a sea that is advancing – as geologists defensively put it, *transgressing* – on the land, chasing the melting ice as the glaciers retreat.

With the warming global climate, the world at Soom is now fairly temperate, despite the ice. The invisible, unnavigable network of meltwater canals pour their water back into the sea and deepen the oceans. Throughout the glacial retreat, sea levels rise, with environments like Soom at the shallow forefront. But this rise happens unevenly around the world; paradoxically, the melt will tend to fill up places furthest from the ice more and faster than the seas close to the ice sheets. This is a declaration of the scale of these periods of enormous glaciations: the ice caps are massive enough objects to quite literally pull the oceans towards them under their own gravitational attraction. Once the ice melts, and the pull is relaxed, the ocean returns to a more even depth.[20]

Perhaps counterintuitively, a rising global sea level does not drown the parts of the world that have been coated in ice for long.

In the short term, seawater may surge onto the continent as the ice retreats, but the Earth's crust is a flexible entity. Water may be thrown about by the wind, contorted by the tide and pulled around easily by gravity, but this all happens in minute, biological time-scales. With the ponderousness of a planet, the Earth floats, rebounds, responds at its own rate. Its crust is exceptionally thin. Under the oceans, it can be as little as 5 kilometres thick, about 0.08 per cent of the distance to the centre of the Earth. Beneath it all is surging liquid, on which the crust floats, just as ice floes float in a polynya. Under our feet is a mirror world, thrust into the mantle. Where peaks exist on Earth, the crust thickens below to create inverted peaks that drive down towards the centre of the Earth. Where our ocean basins drop, the magma rises. The Himalayas have the thickest crust in the modern day, about 70 kilometres thick, but Mount Everest reaches only 9 kilometres above sea level. Mountains are tall because they have deep roots bobbing in the denser mantle. So much of our buoyant land is hidden below the surface. We, too, are walking on icebergs.[21]

When a continent carries an ice sheet, the weight of that sheet distorts the equilibrium, the isostasy that causes the crust to float as it does. The crust is forced down into the mantle, sinking exactly like a laden ship. When the weight eventually melts away, it will rise once more, rebounding upwards over tens of thousands of years, and causing the sea to regress. The regression is not yet happening at Soom, at the forefront of the melt, but it will, in time. Even today, the parts of the Earth that were ice-laden in the Pleistocene are rising, not yet having shaken off the weight of the ice age. Great Britain, for instance, is tilting around a line that can be drawn very approximately from Aberystwyth to York, with the land to the north rising at up to about a centimetre a year, and, as the magma flows into the space below, the land to the south sinking. The process will continue for thousands more years into the future.[22]

The transgression that is happening at Soom is hardly as illicit as the term implies. Indeed, it could almost be seen as a reclaiming of sea floor lost. At the time of the Late Ordovician, before the world iced over, sea levels were exceptionally high. Shallow, epicontinental seas flooded the continents, filled, as they always are, with great diversity. When the ice began to form, that water had to come from somewhere, and the seas fell drastically. The land beneath the epicontinental seas became exposed, and, across hundreds of thousands of square kilometres, life dried out. In the end-Ordovician, this kind of geography is one thing that made the extinction so severe. At the time of the comparable glaciation of Antarctica in the Oligocene, there happened to be very few epicontinental seas, and so there was relatively little to be lost by a fall in sea level.[23]

When environments change, it is often easiest to follow the favourable conditions, those environmental parameters that define a niche. In the sea, this generally means temperature, salinity and, in particular, depth. When the world warms or cools, moving north or south can mean that habitable conditions are also found. In the Late Ordovician, the placement of almost all the world's land south of the equator, centred on the South Pole, makes this near impossible. The coast of Gondwana runs for tens of thousands of kilometres, but much of it is at about the same latitude. If the seas cool, a marine invertebrate cannot escape the encroaching cold by moving north unless it also is able to survive in deeper waters. Once the seas warm again, it cannot survive by moving south unless it moves into shallower water, which may mean dry land and death. Those continents that are not part of Gondwana are no less vulnerable, for they are small, with very limited north–south coastlines, unlike the north–south-oriented continents of the Oligocene. With glacial rapidity, the advancing ice sheets pushed, ground and pressed out of existence the fundamental niches of five in six species.[24]

Glaciers are destroyers, eroders of landforms and biological communities built up over millions of years, with soft rock yielding before them. But they are also constructors like nothing else. The passage of uncountable tons of ice over a landscape marks it for ever, with smooth stripings, broad valleys and rolling drumlin hills. The Pakhuis glaciers charged out of the high ground into valleys and around mountains, and out into the tidewater bays, removing one world and replacing it with another. Glacial landscapes invite a long-term perspective, a planetary pace. From this perspective, ice flows as surely as water; glacial landscapes contain crevasse-rich ice-falls, frozen cascades over precipices and paths of faster flow, rivers of ice within ice. Through Soom, they bulldozed into the quartz sand of the Peninsula Sea, made the ground rear up into enormous undulations, moraines, tills, hummocks. The Earth took a new shape.[25]

From the air, through the layers beneath layers, the mountain sands sink, and cover the nephritic mud of winter with a pale, summer coating. The process that makes the ocean bloom, bringing Soom life, also preserves it in death. The loss of ice throws the world into seeming paradoxes, a reminder of the power of change, and the way that transitions upend everything. Glaciers, icons of indomitability and sluggishness in landscape poetry for centuries, are rapid, noisy beings, stirrers and folders of earth, creators and destroyers of rock formations. Rivers flow where it seems they should not, air within air, ice within ice, and water within water. The states of matter seem to blend into one another, and the transformations of earth to ice to river to plume, of rock to powder to wind to floe, pour vitality out of a continental desert in a seasonal blooming of life. At Soom, even the preservation of this life is an apparent inversion of the ordinary. Soft muscles and gills were preserved in sublime detail, but hard shells and cartilage dissolved into nothing, preserved only as a mould, known by the shape left by their absence. A community whose remains lie only in silt on which it could never have lived. Life turned into clay. The fluid, sighing earth jars and convulses ice-bound Africa a little higher. The weight is being lifted. In time, the rising of the land will outpace the seas.

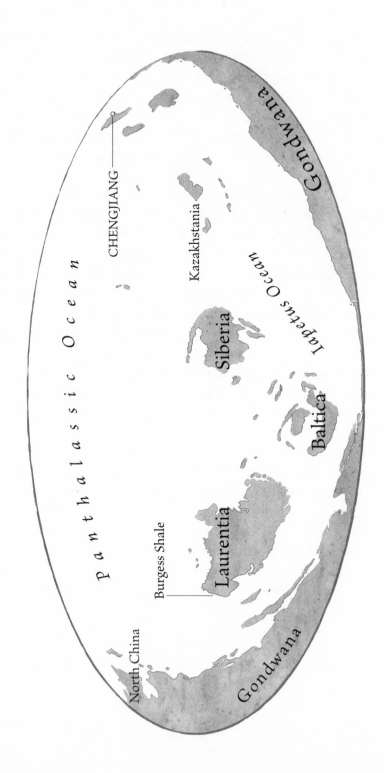

# 15.   Consumers

*Chengjiang, Yunnan, China*
Cambrian – *520 million years ago*

'Consider, once more, the universal cannibalism of the sea; all
whose creatures prey upon each other, carrying on eternal war
since the world began'

– Hermann Melville, *Moby Dick*


'You should save your eyes for sight;
You will need them, mine observer, yet for many another night'

– Sarah Williams (Sadie), 'The Old Astronomer'


The air is stuffy and the sun is baking the earth, although very little
of the planet could be considered earthen. The surface of the
ground is sandpaper rough, an arid crust created by microorgan-
isms inhabiting only the top few millimetres of the land, forming a
poor approximation of soil. The spray of the sea is cooling, but only
in relative terms. With an atmosphere of perhaps over 4,000 parts
per million of carbon dioxide – ten times that of the modern day –
and slightly lower oxygen levels, the air has all the freshness of a
submarine mid-voyage. The latitude is that of Honduras or Yemen,
but the surface of the sea is several degrees warmer than even a
typical day in the Red Sea – well over 35°C. Without shade, with the
sun glaring off the morning sea, the land is desolate and dry, a hot
and dusty desert wind gusting over the bare rock.[1].

To the south of this place, Chengjiang, the Gondwanan landmasses rise into climatically temperate, latitudinally polar mountain
ranges, rare parts of the Earth that are far above the relatively high
seas. In this extreme greenhouse world, the sea level is more than 50

*Omnidens amplus*

metres higher than in the modern day, and much of the continental surface of the planet is below the waves. Chengjiang is part of a flooded continental shelf on the boundary between the desert equatorial part of Gondwana and its rainier south. The northern hemisphere is almost unoccupied by land. Instead, gigantic circular currents, entirely unencumbered by continental barriers, powerfully whirl around the North Pole. A tropical storm belt surges across the northern coasts of the island continents Siberia and Laurentia. Almost on the other side of the world from that turbulent region, Chengjiang is for now clear-skied and scorching. In the stifling humidity along the coastline, there is little reason to stay above the surface of the water, and every reason to delve beneath.[2]

Under the sea, the lifeless calm of the land gives way to frenzy. The sediment is writhing, pockmarked with worm holes. Surging waves above cast shadows as they pass, the seabed pulled gently back and forth. The depths to which waves have an effect on the bed is called the wave base. Beneath the wave base, the floor of the seabed is flat, disturbed only by burrowers. Above the wave base, the ocean floor is rippled, and affected by the surface wind. During storms, the waves become stronger and longer, the wave base deeper, and the limit of the seabed unaffected by waves moves offshore. It is in that liminal zone between the fair-weather and storm wave bases, sometimes still and sometimes rocked by the swell, that the Chengjiang biota, one of the best-known Cambrian ecosystems, thrives in awesome diversity.[3]

Little trilobites called *Eoredlichia* scurry around the floor, hunting for other small arthropods, but there are bigger hunters out there. The crustacean *Odaraia*, a 15-centimetre-long, fat-bodied crustacean with ninety legs and enormous eyes, is scuttling across a rock when it leaps into the water, turns 180 degrees and swims off on its back, stabilized by a three-pronged rudder resembling the tail of an aircraft. *Sidneyia*, an arthropod with an appearance like a flattened lobster, stalks the sea floor mechanically, with long antennae, grasping arms and crushing jaws able to crunch through mollusc and trilobite shells alike. *Fuxianhuia*, a creature resembling a stalk-eyed woodlouse adorned with an earwig's tail, wriggles as it walks.[4]

The holes that litter the seabed are entrances to the burrows of priapulids – 'penis worms', named for their appearance – and their armoured cousins, the palaeoscolecids. *Mafangscolex*, in particular, can commonly be seen poking its spiny head from the sediment, with a twisting, snake-like motion. This worm makes a burrow, nearly twice as long as itself, where it holds its place with hooks attached to its rear end. It excretes a liquid into the sides of the tunnel that sticks the sediment together into a loose cement. The holes do not pass deeply – palaeoscolecids are horizontal burrowers.

On the surface of the sand, sponges dominate, colourful tubes, ropes and mushroom-like growths microscopically filtering the ocean water. Among these are stalked brachiopods, waving their hard, clam-shaped valves to filter-feed, and feather-tentacled *Xianguangia*, an early sea anemone. It is a meadow of animals, where the beautiful but enigmatic stalked creatures, *Dinomischus*, are strewn as wildly as the daisies they resemble. Sometimes, the arrangement is complex; the commonest brachiopod here is one named *Diandongia*, on which smaller creatures can frequently be found growing. Stalked brachiopods, *Longtancunella*, and the anemone-like animal, *Archotuba*, sit tight, jauntily suctioned on to their hosts as the hosts open and close their valves to feed. Arthropods dig, scurry and swim around them, in a cavalcade of forms.[5]

The Cambrian explosion has long been described as a sudden, practically instantaneous origination of all phyla over a period of no more than 20 million years. While this is perhaps simplistic, it is strange that, at both Chengjiang and in the later, more famous biota of the Burgess Shale in faraway Canada, all the modern phyla, all the basic ingredients of modern diversity, are already present. Every animal phylum that exists in the modern day has its origins during the Cambrian or, in some cases, earlier. To be placed together in a phylum, animals must share a basic body plan. The chordate body plan includes a stiffened rod that runs along their upper side – in vertebrates supported by a spine of bone or cartilage – and repeated segments of V-shaped muscles. Cnidarians like jellyfish and corals have single-cell thick layers of tissue housing, among other things, their characteristic hunting cells, each armed with a minuscule toxic harpoon. Arthropods have external plate armour and segmented, jointed limbs.[6] And so on.

Nowadays, creatures in different phyla are very distantly related. At a fundamental level, there are some developmental similarities between, say, the heavily studied *Drosophila* fruit flies and humans, especially at the very basic level of organization. For instance, the same genes define the front-to-back axis of the body in human and fly embryos alike. These genes allow each cell to modify how it will develop, coordinating complexes of genes to determine what organ or tissue ought to be produced. This similarity in regulation persists despite the hundreds of thousands of species more closely related to humans than to flies, and millions more closely related to flies than to humans.[7]

Life is often portrayed as a tree, where a single, ancestral trunk divides ever further into boughs, branches and twigs, phyla, families and species. Travel down from the tips of these twigs, and you find junctions where branches join. One way of measuring how closely related two species are might be to measure the distance from twig-tip to twig-tip: the shorter the distance, the closer the relationships. At Chengjiang, our notions of distance and relatedness within animals begin to blur at a very basic level. Comparing creatures from the modern day with their ancient counterparts produces problems, as time exerts its influence. Nearly 200 species are known from Chengjiang, but the most important from a vertebrate-centric perspective is a deep-bodied animal, only a few centimetres long, shaped like an extended teardrop or fallen leaf, with a fin running lace-like around its tail end. This is the Haikou Fish, *Haikouichthys*, and is one of the earliest candidates to be a true fish, the earliest definite relative of the vertebrates. Although it lacks vertebrae, it has a notochord, the characteristic stiffened back rod of all chordates, on which its relatives, distant in time, will build cartilaginous and bony spines. Besides its tail, it is finless, sinuously sliding through the water. On the sea floor below scurry the charismatic stars of the Palaeozoic – trilobites. Trilobites are so called because their body is made up of three sections or 'lobes', each running from front to back. Perhaps their appeal is down to their sense of fashion – a trilobite will often be adorned with beautiful and unlikely spines, seemingly for no reason other than appearance. Their hard outer skeletons often allowed their shape to be preserved in their entirety and look as if they could crawl away at any minute. Their behaviour is often easily apparent from the posture in which they appear, being found curled up like a woodlouse, or walking behind one another in neat lines to shelter from the current. Perhaps they are most beloved in Dudley, in the West Midlands of the UK. There, quarriers excavating the Silurian limestone regularly turned up 'Dudley bugs', which obtained enough local affection for them to become a symbol of the town. According to folklore, a public lecture about Dudley bugs, held at the quarry in which they were found, attracted 15,000 people in the middle of the nineteenth century. In what is currently known as Utah, Cambrian trilobites were used as jewellery and in medicinal practices by the Pahvant band of the Ute people. In Europe, trilobite pendants 15,000 years old have been discovered, once worn by Pleistocene people.[8]

Long before they are frozen into stone, the legs of the Chengjiang trilobites ripple and their bodies rock in the slight sea swell. Little, wide-eyed *Eoredlichia* is a common genus of trilobite here, sporting a crescent moon for a head and a curtain of tiny spines that protrude from each mottled segment of its thorax. They are just an inch long, but they tower over tiny, scuttling *Yunnanocephalus*, a sixth of their size. Trilobites are archetypal arthropods, segmented evenly in a manner not unlike woodlice, with a limb and a gill protruding from each segment.[9]

Trilobites share a last common ancestor with fruit flies that is, by definition, the earliest arthropod. The Haikou Fish and humans have a last common ancestor that is, by definition, the earliest chordate. From this it might be concluded that *Haikouichthys* and humans, *Eoredlichia* and fruit flies, make pairs that are more closely related to one another than any other combination of the four, and indeed this is how relatedness is usually expressed. But time is important too; the number of mutations that have accumulated in a lineage, although the rates at which they occur vary a little, are roughly proportional to the length of time that lineage has existed.[10]

From the time of the last common ancestor of all four to the time of Chengjiang, the time of *Eoredlichia* and *Haikouichthys*, is a much shorter period than that which separates both humans and fruit flies from this last common ancestor. Indeed, the amount of evolutionary time that separates *Eoredlichia* and *Haikouichthys* is also smaller than that between the two arthropods, *Eoredlichia* and fruit flies, or between the two chordates, *Haikouichthys* and humans.[11]

In this sense, although we share some crucial anatomical features with *Haikouichthys*, from a supporting rod along our backs to an internal protective structure for our sense organs and brain, to V-shaped segmented muscles, the early arthropods and early vertebrates are closer in evolutionary time than kangaroos and humans are today; the phyla are more similar at Chengjiang than they will ever be again.

This begs an important question. If there is less evolutionary time separating *Eoredlichia* and *Haikouichthys* than there is between kangaroos and humans, why are they so anatomically distinct? How can there be such fundamental differences compared with the relative similarity of two mammals? A hundred million years is a long time, but one that can still, remarkably, be overcome. In the modern day, two species of fish – Russian sturgeons and American paddlefish – went on their separate evolutionary ways about 150 million years before the present, and yet they have been documented as producing functional offspring. What is it that makes the separation of the phyla in the Cambrian distinct? The body plans of animals all appear at about the same time. Why then?[12]

Nobody is certain, but two answers have been considered. The first delves into the internal structures of the animals themselves. Perhaps, in the Cambrian and earlier, the development from fertilized egg to embryo to animal was less defined. If so, fundamental changes to tissues and their arrangements would be, on average, less damaging. Once fixed in place, though, fundamentals become very difficult to change. As with the functioning of a computer, fiddling with the code of a single application is relatively simple, and unlikely to damage the overall function of the machine, but editing a line from the operating system is likely to cause problems. Natural selection, then, ends up being a tinkering mechanism, unable – or at least extremely unlikely – to take a sledgehammer to the basic internal structure. In this view, a new phylum cannot arise in the modern day because the anatomy of living beings is simply too complex compared with that of their Cambrian and Precambrian forebears. Evolution today can only be played within the constraints set by the past.[13]

The alternative answer looks outwards, and says that there is nothing intrinsically impossible about a new body plan developing today, were it not for the existence of others. In this view, the world of the Cambrian is fresh for the taking, with the ecosystems simpler, with fewer roles available, fewer possible ways in which to live. The origin of the phyla is described as a 'barrel-filling' model. Establishing the basic roles within an ecosystem is like adding large rocks into a barrel. Now, if a new body plan arose, it would have to compete in an ecological space already occupied by other species that have evolved to fit their niche very well. This is hard, and a natural barrier to novelty. Rather than adding more large rocks, then, evolutionary processes make the ecosystems more integrated, more complex, adding in finer and finer divisions of ecological processes, pebbles and sand falling into the barrel between the gaps left by the larger stones, structures built on other structures.[14]

In Chengjiang, some of the earliest such structures, complex food webs, are being established. This is happening under the incipient domination of the marine world by Bilateria, the group of organisms defined by their mirror symmetry and more complex internal tissue structure. The basic structure of a bilaterian, then, is the shape of a worm. On this model, there are many variations. Some add fins to streamline movement through the water, like the Haikou fish. Others, arthropods, add limbs for crawling along the seabed, and some of these are armoured, modifying the basic worm shape into something more complex. Trilobites, in particular, have difficult armour to crack. Calcified, more like the hi-tech protection of a crab or lobster than the insect-like chitin that most creatures work with, and with turret eyes filled with mineral lenses, they are walking fortresses, hoping to avoid becoming prey, or, perhaps, hoping to catch some of their own. Even in the early days of bilaterally symmetrical animals, it's a worm-eat-worm world.

The world before the Cambrian was one of relatively still ocean floors, with the seabed surface the only habitat occupied by multicellular life. Nothing burrowed far into the mud, and nothing actively swam at high speed through the water above. From a generally stable, filter-feeding world, with animals calmly picking detritus or plankton out of the water as it drifted by, or slowly grazing on microbial communities, some have started to go in search of their food. Animals have become animated; the predator has been born.

Although organisms consuming one another are observed in the fossil record before the Cambrian, it is only in this period that the networks of predator–prey relationships become widespread, complex and clear enough to study. Suddenly, energy does not flow through the ecosystem solely as producer to consumer, then straight into decay. The consumer can itself become the consumed. Animals have adopted all sorts of strategies to avoid this fate, such as armour to prevent or deter attack, eyes and other sense organs to rapidly detect both predator and prey, and efficient movement to catch prey or escape. The innocence of the Precambrian Eden is over, and the arms races have begun. Remarkably, these earliest Cambrian food webs have very similar properties to those of the modern-day, beginning at the bottom.[15]

The structure of a food web can be thought of like a rope

climbing frame. Species act as anchor points, and are connected by their interactions – higher-up species feeding on those lower down. In the Cambrian, some predator–prey relationships are less defined, and the food chain tends to be a little longer, but the principles are essentially the same. When life interacts, it does so under the same rules of energy flow, the same rules that govern probabilities of encounter, the same basic mathematics of the universe. Ever since the dawn of food webs, certain roles have always been present. The ecological structures on which every living community is built have hardly changed in more than half a billion years; there have been only variations on an ancient theme.

At the base of the Cambrian food chain are the dust motes floating in the sunbeams; drifting clouds of phytoplankton, the progenitors of plants, as well as bacteria, make their own food, by photo- and chemosynthesis. Algae die and decompose, and their remains are eaten by deposit-feeders, like the Cambrian herbivore *Wiwaxia*. Along with the odd organic molecule dissolved in the water column, these provide the raw resources from which an eco-system can be built. In any landscape, every living thing draws its atomic make-up from whatever primary producers sit at the bottom of its food web, which in turn have drawn their molecules from the chemistry of whatever surrounds them, be that air, water or rock. All life, ultimately, is built of the minerals in the Earth. In the modern day, of course, our food webs are spread all over the world. A person drinking a mug of tea with a chocolate biscuit in London can be consuming atoms weathered from minerals in several continents, formed across billions of years; ions absorbed by Indian tea plants grown in a patch of Precambrian Gondwanan gneiss soil, thrown up into steep mountain slopes by the Eocene collision of continents; atoms absorbed from redistributed glacial loam by wheat, since ground into flour as if recapitulating the action of Pleistocene glaciers, and Ivorian cacao, grown in fertilizer made from Paleocene phosphate deposits on endlessly recycled rainforest soils, in turn derived from the ancient basement granites, quartzes and schists of the geological heart of West Africa that even at the time of the Chengjiang biota had been lying beneath the ground for perhaps 3 billion years.[16]

A commonly stated statistical fact contends that every breath you take contains atoms once breathed out by Shakespeare, or some variation on that theme. How much more satisfying to think that you continually replenish your atoms with those that, perhaps within the past year, were part of a mountain that was once an ocean floor? There are naturally distant transportations of minerals, in fact – the Amazon basin, for example, depends on the annual influx of blown sand from the Sahara Desert to replenish those minerals lost downriver. For the most part, though, in the natural world, without the luxurious global reach that affluent modern communities afford, most food chains remain stubbornly local. Chengjiang is no exception.[17]

In the slowly undulating Chengjiang current, zooplankton, minute animals that float freely without spending energy, eat the phytoplankton and bacteria. Loofah-shaped sponges filter plankton, while scuttling shrimp-like *Canadaspis* send up whirls of mud as they forage. Penis-worms protrude from burrows, hunting for detritus, or taking the opportunity to scavenge on a carcass. Above, anomalocaridids swoop with jaw-arms raised, snatching in the hunt.[18]

Meandering along the floor is a lobopod, an animal with an earthworm-like body, long and cylindrical, divided into soft rings. But this creature has soft, flexible limbs, seven pairs in all, controlled by hydrostatic pressure and each tipped with a claw. From its back, a long, long spine sticks out, slightly scale-like. It is eerie and faintly alien, and has been named *Hallucigenia*. *Hallucigenia*'s closest relatives that survive to the present day are cryptic, strangely elegant creatures called velvet worms. Looking a little like a limbed slug, but with the dry, squishy texture of a stress toy, they inhabit the soft humus of forest floors. Today, velvet worms catch insects by actively squirting them with a string-shot of sticky slime, but under the sea *Hallucigenia* and its lobopod kin are mostly grazers on sponges.[19]

The early lobopods have a melange of characteristics that makes their position on the tree of life difficult to pin down. The first part of *Hallucigenia*'s gut, the pharynx, is lined with teeth, which looks like a generalized arthropod, while their lower digestive system, in which those sponges are broken down, is very crustacean-like. *Megadictyon* and *Jianshanopodia* are two predatory lobopods whose deeper guts set them apart – they are how they eat. Found between each pair of lobopod limbs, there are eight or nine pairs of blind alleys, set off from the main gut channel, which act as a form of simple gland, used to digest carcasses or carcass fragments that have been consumed, a physiological strategy that may have contributed to the boom in the diversity of arthropods and their kin. But *Hallucigenia* and the other predatory lobopods have another trick up their fleshy sleeves. In order to locate prey and watch out for their own predators, they, along with many of their Cambrian contemporaries, have evolved the ability to do something extraordinary, something new for animals. They are able to detect and make use of electromagnetic radiation. The first eyes are emerging.[20]

For a living being, the world is filled with information, of which only some is useful. Making sense of that information and reacting to it is the basis of all behaviour – an organism that responds to new events in its environment appropriately is going to survive for much longer. The simplest senses are chemical, detecting nearby molecules. This includes the basic chemosensory abilities of bacteria, which seek out gradients in their food concentration, and move in that direction – the equivalent of climbing a hill by feeling the slope of the ground. Animal taste and smell are chemical too, and most species detect salt, acid and other chemicals that might be relevant. But local chemicals are only one form of sense. Magnetic fields, the direction of gravity and temperature all help to work out location, orientation and the appropriate response. These senses have histories that date back billions of years. Light, for most of that time, has been detected, but has only really been useful as a source of energy for cyanobacteria and other photosynthesizing creatures. But with the beginnings of creatures that rapidly move around, of a world where changing growth or slow migration is not sufficient for survival, light has become important not just as a source of energy, but as a source of information.[21]

It is easy to take the ability to see for granted, perhaps because it is such a phenomenally useful ability that it is more or less ubiquitous across multicellular life. Plants detect light waves and grow towards them, but their way of living is slow enough that they do not require specialized organs to focus on that light, merely needing to know that it is there. When animals get more active, their reactions need to be speedier, so many take advantage of the fact that certain wavelengths of the electromagnetic spectrum are reflected off other surfaces. *Hallucigenia*, in particular, has eyes that resemble those of other early arthropods and their kin, but there is immense variety, and many independent originations of sight.[22]

Trilobite eyes are more impressive. As in other arthropods, they are compound, with many individual lenses, each about a tenth of a millimetre in diameter, fixed in place, and pointing in different directions. This gives the trilobite a detailed, if mosaic, picture of the world. Each lens is constructed with a crystal of calcite – a transparent mineral through which light is shaped into a clear image. The lenses in vertebrate eyes require muscles to focus them, moving backward and forward, being bent and distorted so that we can see a detailed image of something at any distance we choose – this is called 'accommodation'. This isn't perfect; at any one time, we can only see accurately objects at one distance. Focusing on your hand in front of your face, you will not see in detail the pictures on the wall beyond. However, the eyes of some trilobites, in existence by the late Cambrian, are bifocal, using a lens made of two materials with different refractive properties. This allows them to simultaneously focus on small objects floating only a few millimetres away and far objects, theoretically at an infinite distance, without any modification, an ability that few other species have ever evolved. Most of the animals at Chengjiang have rather well-developed sight, and ever-more powerful brains to process the information. The selection pressures to develop good sight during the Cambrian must be intense.[23]

That pressure, perhaps, comes from the specialist apex predators of Chengjiang. The ominous name of *Omnidens* – 'all tooth' – is a clue to the predaceous nature of this particular worm. Some of those who have studied it compare it to the sarlacc; the implausibly large and carnivorous sand-dwelling worm in *Star Wars: Return of the Jedi*. Some 150 centimetres long, as wide as a skateboard and flattened to boot, the very real *Omnidens* is an ancient relative of arthropods, probably more distantly related than any living forms. Crawling along the sea floor with twenty-four fleshy limbs, its mouth, a circular mace of up to sixteen witches'-finger spines, is hidden from view. When *Omnidens* is hungry, it can open these guarding spines like a camera lens and bring the real mouthparts outside its body. Up to six spirals of teeth, each six-pointed, surround the entrance to the digestive system.[24]

In Chengjiang, the other top predators are the so-called 'great appendage' arthropods. With their toadstool-shaped stalk eyes, lobster-armour plates that are spread like a dozen pairs of wings, dolphin-like undulation, flared tail and bared appendages, they are alien animals indeed. The great appendage itself is paired: large, spiked fangs in front of the mouth, as flexible as fingers, used for capturing prey. From the way the brains of great appendage arthropods are wired, these fangs appear to be homologous structures – that is, they are versions of the same organ – to the fangs of chelicerates like Rhynie's minute trigonotarbid predators or to the paddling sea scorpions of Soom. *Anomalocaris*, one of the best known of the great appendage arthropods, also has the crustacean-like digestive system of lobopods, making it another mosaic creature. The Chengjiang anomalocaridids are up to 2 metres in length, making almost everything else in this ecosystem look extremely small by comparison.[25]

Modern behaviour shows up even in these early days. Tiny arthropods scutter across the mud, each with a two-part curved shell, a fold in the middle separating the halves. Each half of their shell is a near-full moon like a seed pod of honesty. They have seven pairs of legs that ripple as they walk and they are everywhere, making up three quarters of the community. On each of their legs is a secondary structure, a gill organ, but *Kunmingella douvillei* females show a new evolutionary innovation. Attached to each of the last three pairs of legs are tiny eggs less than a fifth of a millimetre in diameter. Each female can carry about eighty eggs, protected under the armour of the shell. In the entirety of the history of life, *Kunmingella* are among the first animals to brood eggs until they hatch. *Fuxianhuia* is presumed to be another brooder; an adult was found alongside four equally aged juveniles, the oldest example of parental care that extends beyond hatching in the fossil record. All of these creatures have one thing in common. They have a small number of large offspring, one side of a reproductive coin that determines the life history of every creature on the planet.[26]

An organism has only so much energy to spend on reproduction, but, in an evolutionary sense, must do so to avoid going extinct. There is a balance to be struck between expending all one's energy on one single reproductive act and dying in the process and putting all of one's energy towards survival and never reproducing at all. The optimal amount of energy and the time at which reproduction begins varies wildly between species, and is dependent on a few important factors. Death is, as ever, the most pressing one. If the adults of a species have a particularly high mortality rate, it makes evolutionary sense to get reproduction over and done with as young as possible, just in case of an early demise. Where adults have lower mortality than juveniles, it follows that once at adulthood, they will have a longer expected lifespan, and perhaps a greater number of offspring over their entire life. When young can be produced many times, it makes sense to invest heavily in each one, maximizing the chances that those offspring survive past that risky juvenile stage. Other factors, like the way that these death rates vary with population density, food availability or seasonality, add more complexity. Generally, though, a species that takes a long time to mature and has few offspring, like humans, and like *Kunmingella* or *Fuxianhuia*, is one with a high natural infant mortality rate, compensated for by putting a lot of effort into raising each individual. With fewer offspring at any one time, the risks of putting all one's eggs in one basket are balanced by the likelihood of surviving to have more off-spring if the current batch does not survive.[27]

These experiments with quality over quantity as a reproductive method show the long-term stability of the Cambrian fauna. Even though the regular storms might locally damage the sea-floor community, it recovers. Far from being just an explosive period of radical change, the ecology of Chengjiang has been stable enough, predictable enough, that the forces of natural selection have not punished the gamble that having only a few offspring at a time represents.

Along the coast at Chengjiang, there may be no life to mark the passing of the year, no blossom, no leaf-fall, no insect swarms, but the land is still subject to two seasons, wet and dry. Although the diverse community persists through both, no fossils are being preserved during this dry season. It is the crucial paradox at the heart of palaeontology that practically all our information about life comes only from death. The place where fossils are preserved might show the conditions at the time of death, or where a body came to rest. Trace fossils are the closest to direct observations of life, solidified behaviour. But more often than not, there is no body to associate the trace with, and the maker must be inferred. We already know that different settings can alter the probability that a carcass becomes a fossil, but fossilization is affected by time as well as space. In the dry season, the flow of fresh water is slower. The shelf remains salty, and decay is fast. During the wet season, the storm beats hard, the waves crash in, the rivers freshen the local ocean, and the sediment is ripped up into a stormy texture called a tempestite. It is only as the seabed settles and the terrestrial silicate is brought in that carcasses can be buried. In the clayey bottom, rich in iron but very low in carbon, mineral-loving bacteria move in to feed on the remains, reducing the iron, converting the muscles and other soft tissues into pyrite, the touch of a fool's Midas.[28]

The idea of the Cambrian as a frenzied burst after 4 billion years of inactivity is in part an illusion, based on a characteristic of Cambrian animals – their hard parts. Mouthparts, exoskeletons and mineral eyes are all preserved far better in the fossil record than muscle or nerve. These hard parts are thought to be adaptations to the new, predatory world: the Cambrian is the time when multicellular life truly turned on itself for food. The pressure to survive in that world spawned ever-more specialized tools for hunting and for avoiding capture. It is a crucial development in the origin of ecosystems as we know them today, nature red in oral plate and raptorial appendage, but the Cambrian explosion is not the beginning. Before bilaterians burst onto the scene in a great stramash, before the competition and chaos, the rising and falling of a myriad known and unknown organisms, there were other multicellular communities. Attached to the Chengjiang sea floor, a few feather-like animals, *Stromatoveris*, sway gently in the current, late onlookers from another, more peaceful eon. One final site remains to be visited, in the calm before the storm.[29]

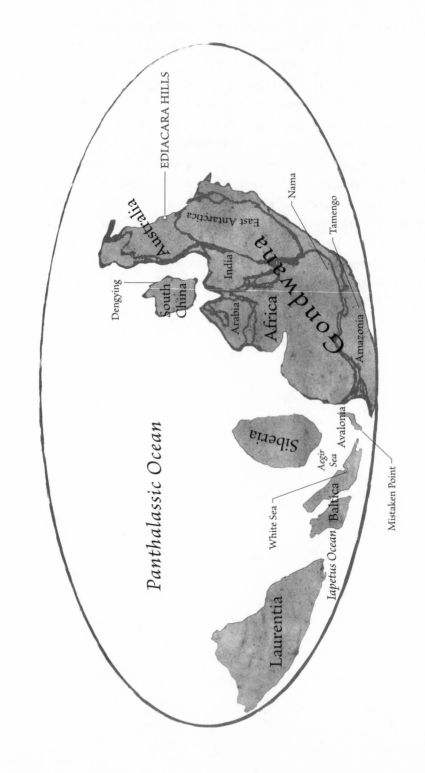

Panthalassic Ocean

EDIACARA HILLS

Australia

Dengying

South China

India

Arabia

Africa

East Antarctica

Gondwana

Nama

Tamengo

Amazonia

Siberia

Aegir Sea

White Sea

Baltica

Avalonia

Iapetus Ocean

Mistaken Point

Laurentia

# 16.   Emergence

*Ediacara Hills, Australia*
Ediacaran – *550 million years ago*

'Nature, in her first hour of creation, did not foresee what her offspring might become. A plant, or an animal?'

– Athénaïs Michelet, *Nature* (tr. W.H.D. Adams)

'Yet portion of that unknown plain
Will Hodge forever be.
His homely Northern breast and brain
Grow up a Southern tree,
And strange-eyed constellations reign
His stars eternally'

– Thomas Hardy, 'Drummer Hodge'

Stand in the centre of Adelaide, the largest city in South Australia, and look north. The road into the midday sun from Adelaide and Port Augusta leads through the Flinders Ranges, one of the most ancient continuous mountain ranges in the world, and into the great deserts of the centre of that wide, sweeping country. Follow it far enough, as it gets smaller, dustier and ever more isolated, through the lands of emu and kangaroo, dry eucalyptus shrubland and the longest dunes on Earth in the Simpson Desert. The land here is older than even the Flinders Ranges, part of the ancient continental centre – the craton – of Australia, and contains minerals laid down as ores billions of years ago. At the mines of Mount Isa, where these metals are extracted at vast scale, the road curves westwards, and eventually it reaches the largest city in the north, Darwin.[1]

The journey through the history of life is at times unfathomably

*Spriggina floundersi*

long. Put into physical terms, imagine that the city of Darwin, perhaps appropriately, marks the time in Earth's history at which all extant life, every living being on the planet, was united as a single species, LUCA, the so-called Last Universal Common Ancestor, and that the centre of Adelaide marks the present day. Travelling along that road, every millimetre through the Flinders Ranges is equivalent to a single year; every kilometre of that 3,500-kilometre journey one million years back into the history of life in Australia. Take but one step, and all colonial influence is gone. A mere 17 metres down the road, and we are back in the Pleistocene, the time of the northern mammoth steppe, when humans shared Australia with cow-sized wombats, giant pythons and *Thylacoleo*, the clambering, cat-like relative of koalas known as 'marsupial lions', with their sharp, bladed secateurs for premolars. A single city block away from our starting point, and human history on the Australian continent is over. By the time we pass the city limits, a marathon's distance from its centre, we are already back in the Eocene, when marsupials inhabit the wide-ranging, lush forests from Australia, across Antarctica, to South America. And there is still a continent of time ahead.[2]

So, we keep walking, and thousands of millennia fall by the roadside, unfolding in reverse. The Australian craton floats around the world, joining and parting with other continents, as species and seas rise and fall, ever changing in reverse, before life abandons the land for the salt water. After a fortnight's hike, 550 kilometres down the road, 550 million years in the past, we find ourselves among the hills of Ediacara, and stop to catch our bearings. Ahead, in the endless outback of early Earth history, are only microbes.[3]

On land, nothing lives, just as it has always been. Seas may evaporate and rain on the lands, but they bring no life to the sandy ground. In the yawning expanse of geological time, mountains have risen in tectonic uplift, have been eroded once more by elemental forces, sand and mud brought down by lifeless rain. In the steep descent towards the ancient Australian coast, a braided river weaves towards the sea, its wide, ever-changing path veering back and forth like rain on a window as the sediment brought down from the hills piles up, building bars and eyots in the flow. Sedimentation, compaction, mineralization, perhaps some metamorphism, uplift and erosion. A tireless cycle of minerals, around and around and around, from sea to shining sea. So it has been since the continents solidified and the oceans formed, 3.9 billion years ago. The seas are certainly shining tonight, under the light of a full and enormous moon in the Ediacaran sky.[4]

To eyes used to the modern constellations, even the sky looks very different. We think of the stars as permanent, fixed in the firmament, but they move with respect to the sun. The Ediacaran is more than two galactic years in the past – that is, the solar system has circled the black hole at the centre of our galaxy more than twice in the intervening time, a total voyage of over 350,000 light-years. Our nearest stellar neighbours are on different trajectories, and we have left them all behind. Even if we had not, many of the stars we are familiar with are yet to be born. We may be in the northern hemisphere, but you won't find Polaris, which first shone out in our Cretaceous. None of the seven stars that make up Orion's distinctive shoulders, feet and belt, is older than the Miocene. Sirius, the brightest night-star of the modern day, has a long history, but even its Triassic birth is further into the Ediacaran future than the Holocene past. Two of the five stars in the W-shaped constellation Cassiopeia are venerable enough to exist somewhere in the galaxy, but, for the most part, the scaffold from which we will paint shapes in the heavens has yet to be constructed.[5]

Even the moon is startling. Ever since a collision between the young, molten Earth and an enormous asteroid placed the moon in the sky, it has slowly been moving away from the Earth, and will continue to do so. Over the minuscule timescale of human history, it has hardly moved at all, but small changes add up over 550 million years, and the Ediacaran moon is 12,000 kilometres closer and 15 per cent brighter than the moon of even the most romantic of poets. Stay a while, and you will discover that the day is shorter, too, a mere twenty-two hours between sunrises, before friction gradually slows the Earth's rotation. This is truly an alien world, more like a watery Mars than the Earth we know today. And yet, in that water, we find complex life.[6]

This is before the beginning of the Phanerozoic, the geological eon where the biological world will begin to resemble our own. The scale of changes since Earth formed has been unfathomable. To get from the origin of life in deep-sea alkaline vents to these seas has taken 3.5 billion years. The first billion or more of those passed with little change, before cyanobacteria discovered the magic of photosynthesis and pumped oxygen into the atmosphere over a 10-million-year period. The added oxygen, a highly reactive gas, caused the dissolved iron in the oceans to oxidize and sink into distinctive red ironstone layers, literally rusting the oceans worldwide. The Earth repeatedly froze in a series of the largest ice ages ever to engulf the planet. In the most extreme cases, ice sheets advanced from the poles and met at the equator, so that most of the planet was covered in snow and ice, the so-called Snowball Earth. The ice was so thick in high latitudes that glaciers flowed on ice sheets that rose hundreds of metres above the surface of the sea, rivers of solid water oblivious to the liquid world beneath. At the equator, glaciers still flowed, although we know that at least some of this tropical water was not permanently covered.[7]

While glaciers coated the world, the only organisms that fed on light beams were cyanobacteria. They thrived better than others in the oxygen-poor ocean, a relative nutrient desert. Only the freezing outflows of meltwater rivers carried enough oxygen into the seas for aerobic life to survive. As the snowball began to thaw, the melting ice wore away the continental surface, and drove millions of tons of phosphate into the ocean, creating a chance for an algal takeover. Suddenly, the advantage that cyanobacteria held – being able to absorb nutrients quickly because of their small size – was no longer relevant. Larger organisms were no longer outcompeted, and cyanobacteria were numerous enough to be preyed upon by larger microbes. Being big is an advantage for a predator, even one on the microscopic cell, and it is at this time that multicellular algae became common. By the beginning of the Ediacaran, the seas were still mostly oxygen-free, but wild fluctuations in their chemistry over the next tens of millions of years eventually flipped the anoxic world into a new, well-mixed ocean, an instability that might itself have driven evolutionary innovation.[8]

Multicellularity had already evolved many times in the past, including possible plant-like organisms and red algae from over a billion years before the present, but during Snowball Earth and its aftermath, self-cooperating organisms emerged to change global ecosystems for ever.[9]

New ecologies became possible and a new way of life opened up. With multicellularity comes division of labour, the specialization of different cells into tissues, each with a particular role. Shape can be controlled and optimized for purpose and reproduction more tightly controlled as groups become individuals. It is in these newly temperate oceans, these shallow, silty seas on the edges of ravaged rocky continents, in the period called the Ediacaran, that life is getting big.[10]

By the time the ecosystem that is preserved in the Ediacara Hills arises, life on the macro-scale has been around for about 20 million years. The earliest known multicellular life comes from a muddy continental seabed on the southern margins of the Iapetus ocean in a landmass, approximately the same size as Madagascar, called Avalonia, named for the island to which the mortally wounded King Arthur was carried after his defeat at Camlann to sleep until he is needed once more. Remnants of Avalonia, though not always of the organisms that inhabited it, are found throughout the lowlands of Frisia and Saxony, through the southern portions of Britain and Ireland, in Newfoundland and Nova Scotia, and in parts of Portugal. It is an ancient place of which only fragments survive, scattered along the broken fringes of continents.[11]

At one of those fringes, the promontory in Newfoundland known, rather poetically, as Mistaken Point, feather-like imprints in submerged volcanic ash reveal traces of the earliest large organisms. Off the shores of Avalonia, the first multicellular life awoke from Cryogenian slumber, and seemingly spread worldwide. By the time the Ediacara Hills have formed, these ecosystems are found from Russia to Australia and are becoming very complex indeed.[12]

The Ediacaran sky may be bright and moonlit, but the water is murky and storm-ridden. Descending into the waves, the surge is strong and cold, and very little can be made out through the dark brown silt. The water clears away from the shore, as the seabed drops down. There are no fish, nothing actively swimming. Almost everything that lives here is attached to the sea floor, escaping the motion above. At the surface, the incoming tide brings with it big rollers, each pocket of water swirling in dizzying vertical circles. Directly below the waterline is turbulence and chaos, but depth replaces this with calm; the circular motion becoming less and less perceptible until the world is dark, blue and still.[13]

In places the ground is coated with a firm, wrinkled layer, hardly distinguishable from the rest of the seabed except in texture, the folds of a rough rhinoceros-hide texture contrasting with the fresh-poured caster-sugar smoothness of ground quartz sand. The rough texture is the microbial mat, coating the interface between earth and water, an ecological structure that has already existed for billions of years. The two simplest domains of life, bacteria and archaea, have fed and reproduced at Ediacara for thousands of years, building layer upon layer as they stabilize the seabed, forming a coherent uppermost layer, like a skin on cold custard. Where the microbes are cyanobacteria, the mats often form distinct clumps called stromatolites that resemble slimy, slowly growing boulders, climbing their way to the light. Elsewhere, as in this delta, the mats are spread flat, a sheet of wrinkled life on the floor of the ocean. Only the top few layers are ever alive at any one time, but the combined generations of millions of microscopic cells nonetheless mean that the mats can reach thicknesses measured in inches.[14]

From these patches of microbial mat, strange, feathery shapes rise up to 30 centimetres into the water. Ridged, rugby-ball-shaped creatures, only a centimetre in diameter, drift between them. Above these hovers an eerie cone, a centimetre-scale flying saucer, spinning as it drifts, before settling once again on the seabed. Close-up, it becomes apparent that this shadowy form is made of eight ridges, spiralling clockwise from the tip of the cone to its base, a coiling helter-skelter, floating hypnotically. Unable to move with great speed in the water, and hardly a natural swimmer, it nonetheless occasionally leaves behind its home on the microbial mat. It is found in the peace and quiet of the calm water below the storm base, and, when swimming, hangs in the water over stranger creatures still. Already, multicellular life is complex, and this is in fact one of the earliest creatures we can with certainty call an animal. *Eoandromeda* – so called because when flattened in fossilization, its eight arms resemble the spiral galaxy Andromeda – is a lantern in the Ediacaran murk, one of the few forms of life here that is remotely recognizable. Exactly what *Eoandromeda* is related to is not precisely known. It has been suggested that its eight-fold symmetry, with those waving structures on each arm, makes it a relative of comb jellies, beautiful animals that today freely swim in the open ocean, glittering with light from iridescent photonic crystals, and using the lure of the light to attract prey. But this similarity may be superficial.[15]

Whenever an extinct creature is placed in the tree of life with confidence, our knowledge about other branches of the tree expands. Knowing the heritage of an organism tells us the sequence of events in evolutionary time, the nature of ambiguous anatomy, and more about life history. For other questions, it is enough to describe what a species does, how it interacts with its neighbours, to judge it not by its heritage but by its actions. *Eoandromeda* is extremely widespread, found worldwide from Australia in the north to China in the south. For all the ambiguity over its anatomy and functional biology, compared with the rest of the biota of the Ediacaran, the suggestion that *Eoandromeda* is a comb jelly is relatively precise, although its accuracy remains uncertain. If it is a comb jelly, it would tell us that other groups, like sponges, the harpoon-feeding cnidarians and bilaterally symmetrical worm-like creatures should all be somewhere around, but identifying them has been another matter.

Ediacaran life has been perplexing scientists since its first specimens were discovered. The Precambrian, so conventional wisdom
went, was devoid of macroscopic fossils. The Cambrian was as far
back as the record went. The first discoveries at the Ediacara Hills
were assumed, therefore, to be early Cambrian. Then, in 1956, in
Charnwood Forest, in Leicestershire, the traces of a peculiar feather-
like fossil were discovered by a fifteen-year-old girl, Tina Negus, in
undoubtedly Precambrian rocks, though nobody initially believed
her. When another student, Roger Mason, showed the site to a local
geology professor, they were described, and named *Charnia*. Later,
they turned up at Mistaken Point, recognized as being the same as
those in Ediacara, as those in Siberia. Some, like *Charnia* itself, still
defy precise classification. In life, *Charnia* is plump and quill-shaped,
a series of fluid-filled vanes off a flexible central shaft, a frond
anchored to the sediment by a fat, buried holdfast. The fronds
reproduce clonally, and thread-like filaments, stolons, link these
holdfasts together in a networking web that allows the organisms to
share nutrients. But although they are unlike anything alive today, it
seems that *Charnia* and many other Ediacaran organisms are nonetheless part of the story of the animal kingdom.[16] The question
that remains is what part they play.

In all, a world of uncertainty is studded into the microbial mat.
There are smoking clusters of anchored creatures, hundreds of
them per square metre, each an upright tower composed of bulges
like knotted rope, as if Gaudi had designed an industrial town. In
between these tenpin arrangements, there are flattish discs, marked
with ridges in a whorl, like oversized fingerprints in the mud.

The milk-white smoke slowly drifting from the cluster of towers
is part of a revolutionary ecological adaptation. Each tower is an
individual of the species called Dorothy's Rope, *Funisia dorothea*,
and although algae have been reproducing sexually already for half
a billion years, these are the earliest known definitively sexual relatives of animals. Sex is so ingrained as something the majority of
animals do that it is easy to forget that it is an ecological strategy.
Without sex, offspring are clones of the parent. Where producing

the most descendants is the only real metric of evolutionary success, this is ideal. However, if all clones are the same, this strategy carries with it other risks. They will be well-adapted to the same environment as their parent, but if the world gets warmer, or more acidic, or food becomes scarcer, they will all fail or succeed equally. Asexual animals do not generally last for long periods of evolutionary time, although there are a few exceptions. In a variable world, sex is a way of scrambling the genetic code with new material, to put all of your (figurative and literal) eggs into different baskets, and to give a higher chance that at least some will survive. Even a little sexual reproduction can overcome the disadvantages of cloning, so Dorothy's Rope does a little bit of both.[17]

The discs lying among the strands of Dorothy's Rope are *Dickinsonia*, grazers of bacteria. Each ridge is a new growth segment, a pair starting at the 12 o'clock mark and slowly migrating in each direction and compressing towards 6 o'clock, like a concertina paper garland. Odd they may be, but they too are closer to animal than to anything else alive, leaving behind them forensic traces of cholesteroids, a molecular signature of animals. Their development suggests that they may even be closer relatives of ours than are sponges. Watch them for long enough, and you will see that they fulfil the 'animated' part of animal life. They move periodically, resting in the microbial mat before relocating when their food has been exhausted.[18]

But there is more to movement than the punctuated resting of *Dickinsonia*. Throughout the shallower water here, where the wave currents are stronger, there are channels in the sand, each a groove with raised edges, just as would be produced by dragging a fingertip through the seabed. Its producer is unknown, but at the end of this trail must be an animal of sorts, sliding on its belly. The most likely candidate is *Ikaria*, a small animal with a defined front and back, the earliest bilaterally symmetric organism, but this has never been found in the exact same fossil beds as the rest of the Ediacara community. Traces with absent makers are found all over the world. In Dengying, in what will become China, there is evidence for something even more remarkable – walking. Tiny holes – burrows, perhaps, a way to seek protection – lead underneath the microbial mat. Linking routes between those holes are minute paired impressions, the footprints of an unknown animal. In this case, there is no groove from a body being dragged; the animal lifted itself up. They are irregular; the organism that made these prints was probably being knocked about by eddying currents. That the prints are paired is astounding; it means that their creator, the mysterious first walker, is bilaterally symmetrical, unlike, for example, sponges or jellyfish.[19] Exactly what made the Dengying tracks is unknown; it is likely never to be found and, regardless of what Sherlock Holmes might lead you to believe, there is only so much detail that can be deduced from a series of footprints.

The decay of fleshy creatures whose lives are preserved only, if at all, in the marks they left on time-hardy stone, bone or shell, is a tantalizing reminder of the incompleteness of the fossil record. So much about this period in Earth's history is indistinct, an inference from a groove here, a pattern in the distribution of alien shapes there. With the silt-slumped casts at Ediacara, we can go beyond mere guesswork, but our perception of this world is incomplete. There are certainly more creatures than as yet meet the eye living in the soft mud of the Ediacaran continental shelves.

The storm is abating, and moving shoreward becomes easier. As the water becomes shallower, the flailing tubes of *Funisia* disappear, replaced by equally rope-like but flat, coiling creatures, up to eighty cm long, that rest passively on the seabed, and by the feathery fronds of the relatives of *Charnia*. *Dickinsonia* still graze on the mat, but the microbial surface covers ever more and more of the sea floor. Already, early on in multicellular life, niches are forming, and species are separating into distinct communities, little patchy habitats within the ecosystem. Such niche structure is extremely new – the earlier biotas of Mistaken Point show a community where the place an individual is found depends more on the location of its parents than on specialization to a particular lifestyle. Nevertheless, species are beginning to divide up resources. The shallower seabed is more affected by the waves, and *Funisia* are less tolerant of the change. Here, the sand is rippled, the cumulative effect of each wave pushing sand up miniature dunes and toppling it over the side. To the creatures above the storm-base, these are natural protective ridges, an undersea equivalent of a windbreak. A group of little volcano-like structures provide even more protection. Around the rim of these cratered cones, there are hard, ruler-straight spines, projecting twice the length of the cone's height. *Coronacollina*, probably related to sponges, is the first organism worldwide to be making hard body parts. Cowering behind is a cluster of *Spriggina*, sheltered from the worst of the turbulent sea. With a head like a crescent moon and a segmented body about 3 centimetres long, it is soft, flattened, worm-like – another early animal.[20]

There is a darkening and a dull jolt as the water thickens and browns like brewing tea. The storm waves have weakened the loose sediment brought down by the braided river. As the storm abates, the forces pushing it back against the coast relax, causing a slump of silt, an underwater landslide that swallows the swimming *Eoandromeda* and buries the microbial mat. The suspended mud settles out of the water very slowly, still being thrown about by the now softening waves. Those species anchored to the bottom will not survive, eventually decaying and leaving behind a perfect impression. All that will remain of the community here will be a Pompeii-style preservation, a perfect cast and mould on the underside of a layer of solidified sand, giving shape but nothing else.[21]

Behind a shallow sandbar, the water is calmer, less affected by waves. There has been a slump of silt here, too, but the mud has settled more quickly, the still water allowing it to simply fall slowly out of suspension, the difference between a bottle continually shaken and one left to sit. All life has been buried, and the sea floor is left barren and homogenous. But then there is movement. Like water down a plughole, a circle of fresh white sand regularly pulses downwards, sucked by an unseen force. A stiff hood of scale-mail armour emerges, its owner shuffling with a rippling muscular foot. Other pulsing circles of sand appear and more armoured creatures follow; a small herd of *Kimberella*. The general impression of each is of a scaly silicone hovercraft: a flexible, rubbery hood reinforced with firmer scales, resting on a muscular bed that bulges under the weight of its armour. From one end, a round head at the end of a stretchable and bendy, almost hydraulic, digger-like arm explores the sand, searching for food.[22]

Before burial, these *Kimberella* were grazing alongside *Dickinsonia* and the fronded *Charniodiscus*, using their heads to rake circles of sediment towards it like a croupier gathering chips. When the underwater landslide happened, they drew themselves under their incipient armour for protection, and each used its slug-like musculature to dig a vertical burrow out of the collapsed fan. Not all have made it. Younger and smaller *Kimberella* can get trapped, unable to muster the force or survive for long enough to make it to the surface. Their futile attempts will be preserved, too; vertical tubes that simply end where others reach the new surface, memorials cast in stone to a doomed last-ditch scramble for survival.[23]

With the exception of accidents like this, nothing truly burrows in the seabed, and the Ediacaran world is made of two-dimensional habitats. Microbes penetrate some small distance into the sediment, but oxygen-loving multicellular organisms are restricted to the surface. So-called 'bioturbation', in which the finely laminated layers are muddled and mixed by living things, introducing some of the seawater's chemistry to the rocks, is essentially a Phanerozoic invention, and one which may have aided in the diversification of animals. The Ediacaran fronds and holdfasts are perhaps sowing the seeds of their own destruction; by existing as protrusions above and below the surface, they are changing the distribution of food resources, and enabling the later evolution of the bilaterian animals that will, ecologically, succeed them. The Ediacarans will persist in places into the Cambrian, including at Chengjiang, but they will be unable to survive in great number in the competitive worm world to come.[24]

If *Eoandromeda* turns out to be a comb jelly, if *Spriggina* is worm-like, *Coronacollina* a sponge, and another free-swimming animal,

*Attenborites*, a cnidarian relative of jellyfish, we have at Ediacara incipient members of many animal groups. Elsewhere, other organisms with animal affinities are emerging through the murk of the turbid sea. In Charnwood, *Auroralumina*, the 'light of dawn', is an animal nearly thirty centimetres tall, each individual consisting of two rigid cups from which tentacles emerge. It seems to be a cnidarian even older than the Ediacara Hills. Like comb jellies, cnidarians are all predators, suggesting a more complex ecosystem than was first imagined at *Charnia*'s discovery.[25]

Around the world, as animals are founding their kingdom, other kingdoms are springing up widely. In Doushantuo in China, and in Tamengo in Brazil, tiny, branching seaweeds sway in the currents, some of the earliest undersea algal gardens. The concept of an entire new kingdom of life rising from nothing is hard to imagine, but this is largely a problem of time. We only define Eumetazoa, the animals, as a 'kingdom' because they began to diverge from one another deep in the history of Earth. Some kingdoms are more independent than others; animals and fungi are closer relatives to one another than either is to plants. If any of the Ediacaran organisms are part of a kingdom that is no longer extant, they are aberrant only from a modern perspective. Within the Ediacaran period itself, the amount of time since these outliers diverged from other multicellular groups may have only been a few tens of millions of years, the same time distance that separates, say, a spider monkey from a ring-tailed lemur.[26]

After travelling so far back in time, it is only by turning and looking back down the road to the present that we begin to classify those that exist in the deep past. With half a billion years and more to play with, any or all of the creatures of the Ediacaran could, in theory, found a kingdom's worth of diversity. The initial divergences occurring in animal life during the Ediacaran will define the body plans of the inhabitants and constituent parts of future ecosystems. From the world of *Dickinsonia*, of *Charnia*, of *Spriggina*, there are many routes we could take. We believe that the Ediacaran biota, for the most part, stand around these divergences, some a junction further along the path than others, but often peering, non-committal. Most paths lead nowhere, others through millions of years and fabulous worlds, only, like almost all others, to fade into the undergrowth. The Ediacaran biota have not fully emerged from their obscure world in part because we are trying to define them the only way we can: on the basis of those few survivors to have found paths to the present over two galactic years, almost an eighth of the age of the solar system.

As with the Paleocene, in which we observed early members of the diversifying placental mammals, the moulds in the rocks of the Ediacara Hills were made by participants in the radiation of multicellular life, of animals and algae, and things that, by the simple fact of having not survived, forfeit a common name. Early after the origin of multicellularity, developmental processes are in flux. As natural selection guides life into specialization through speciation, those processes become fixed; the barrel, perhaps, is being filled. As organisms develop new processes, new functions and new ways of life, new constraints are added. Every junction adds more, life bodging together quick fixes on top of what already exists. Bilaterians are all bilaterally symmetrical. They all have a left and right. Messing with the fundamental mechanisms behind early embryonic division that lead to this symmetry would almost certainly be fatal. That is not to say that these rules cannot be bent; natural selection excels at finding loopholes, but once the basic rules are there, messing with one system too much will cause the rest of the body to fail.

One thing is certain; whether those that we observe are among them or not, there are beings in the Ediacaran seas that are starting out on the long walk – our long walk – to now. The Ediacaran biota are exploring and defining what it means to be an animal.

For the most part, their descendants have left behind their ancestral home. The microbial mat ecosystem is mostly gone, the burrowing of the Cambrian bringing poisonous oxygen to those that did not need it. But the old ways endure in the deep places of the world, places where the rock is too hard to excavate, and where oxygen is low. In these realms, microbial mats stubbornly persist.

Around the White Sea in Russia, they continue to grow where their distant precursors lived half a billion years ago, while stromatolites are still building their green stepping stones off the modern Australian coast, 3.5 billion years after they began.[27]

In the Australian outback, risen out of the water, the casted beasts of Ediacara emerge from the rock, epitaphs on ancestral tombstones. The planet has changed more than can easily be comprehended since they last lay under the night sky. Over their moulds, 10 million years of silt and sand were continuously churned from the land. Their rocks were folded and thrust up into the Flinders Ranges during the Cambrian, and for 540 million years, the ever-eroding peaks in which they have been encased have sailed from north to south, docking with other continents and exchanging inhabitants, the fortunate and changing successors and descendants of these pioneers of multicellularity. Now, emerged into the shade of the eucalyptus, their imprints rest on earth-tilted hills, shallow and indistinct, beneath strange, young westering stars.

# Epilogue

## A Town Called Hope

'The heart may break for lands unseen
For woods wherein its life has been
But not return'

— Violet Jacob, 'The Shadows'

'There's only one way to earn hope, and
that's rolling up our sleeves'

— Diego Arguedas Ortiz, science journalist

In 1978, for the first time in the history of the world, a human, Silvia Morella de Palma, gave birth on the continent of Antarctica. Since then, at least ten children have been born on Antarctica, mostly in the same settlement as the first, a little village called Esperanza – Hope – one of only two permanent civilian settlements at the bottom of the world. At the moment of Emilio Marcos Palma's birth, the slow migration of people to every major landmass of the Earth was completed. Esperanza is an Argentinean community of about 100 people, a collection of squat red-walled houses overshadowed by the snowy black mountains of the West Antarctic Peninsula. It is an active research station, populated almost entirely by families of geologists, ecologists, climate scientists and oceanographers, part of the front line in gathering the data to make predictions about the future of life on our planet.[1]

This is now undoubtedly a human planet. It has not always been, and perhaps will not always be, but, for now, our species has an influence unlike almost any other biological force. The world as it is today is a direct result – not a conclusion or a denouement, but a

285

result – of what has gone before. Much of life in the past happens in a steady state of slow-changing existence, but there are times when everything can become upended. Unavoidable impacts from space, eruptions at a continental scale, global glaciation – the all-pervasive transitions that force life's structures to remodel themselves. Had any of those events happened in another way, or not happened at all, the then-unwritten future could have emerged very differently. It is by looking at the past that palaeobiologists, ecologists and climate scientists can address the uncertainty about the near- and long-term future of our planet, casting backwards to predict possible futures.

Unlike past occasions when a single species or group of species has fundamentally altered the biosphere – the oxygenation of the oceans, the laying down of the coal swamps – our species is in an unusual position of control over the outcome. We know that change is occurring, we know that we are responsible, we know what will happen if it continues, we know that we can stop it, and we know how. The question is whether we will try.

To look into the palaeontological past of the Earth is to see a range of possible outcomes, a truly long-term perspective. On the one hand, life has survived through Snowball Earth and poisoned skies, meteoric impacts and continental-scale volcanos, and the recent world is as diverse and as spectacular as it ever was. Life recovers, and extinction is followed by diversification. That is, in its way, a comfort, but it is not the whole story. Recovery brings radical change, and often startlingly different worlds, into being, while also taking, at a minimum, tens of thousands of years. Recovery cannot replace what has been lost.

As a motto, the community of Esperanza has taken the phrase *'Permanencia, un acto de sacrificio'* – 'Permanence, an act of sacrifice'. As we have seen, in Earth's history, there is no such thing as true permanence. The houses of Esperanza are built on rocks which demonstrate how temporary life can be. They record the shallow seas of the Early Triassic, and the marine setting across the Great

Dying of the end-Permian. They are filled with trace fossils, long-abandoned U-shaped burrows in mudstones, the refilled homes of worms and crustaceans built into sand.[2]

The sea floor of the Hope Bay Formation, a series of rocks formed by collapsed suboceanic fans of silt, was at that time notably oxygen poor. The reason for this, and for similar patterns detected across the world, has been suspected for decades, but has only recently been proven. In 2018, it was determined that the lack of oxygen in the Permo-Triassic ocean was undoubtedly caused by catastrophic global warming on a then unprecedented scale. Volcanic activity in Siberia produced enough greenhouse gas that global temperatures rose starkly, and triggered a mass release of oxygen from the oceans, killing fish and other active marine life worldwide. Bacteria thrived in their absence, releasing as a by-product of their own respiration clouds of hydrogen sulphide, suffusing the atmosphere and poisoning ecosystems on land and sea. Populations crashed, and few survived. The end of the Permian was the time that life – or at least, multicellular life – nearly didn't make it. It stands out as an example to us all of the worst perturbations that an environment can face, in which mere survival relies on pre-existing helpful traits and a dose of luck.[3]

When we compare our world with that of the end-Permian, we can find some worrying similarities. The oxygen loss from the oceans is not restricted to the past. It is happening today. Between 1998 and 2013, the oxygen concentration in the California Current, the major oceanic current running southwards on the west coast of North America, declined by 40 per cent. Globally, since the 1950s, the area of low-oxygen bottom waters has expanded eight-fold, to 32 million square kilometres in 2018 – twice the area of Russia – with more than a gigaton of oxygen lost from the ocean every single year for the last half-century. This is partly because algal blooms are more regularly sparked by runoff of nitrogen from agriculture, but also because the sea is becoming warmer, just as in the end-Permian.[4]

Warmer seas cause a three-pronged problem for aerobic species. The first is pure chemistry; oxygen dissolves less easily in warmer water, so there is less of it around to begin with. Then, there is physics; warm water is less dense than cold and so rises to the top, but if the heat comes from the sun, the surface water warms faster anyway, separating the warm layer from the cold depths. The warm and cold water rarely mix, so any oxygen that does dissolve does not move into the deep ocean. Finally, there is biology. Warmth makes cold-blooded animals metabolize faster, requiring more oxygen, so that any oxygen that has dissolved is used up more rapidly. For active animals, this triple threat spells disaster.[5]

This is not bad news for all concerned – bottom-dwelling animals like crabs and worms can generally survive at lower oxygen concentrations, but another gas poses a different problem. The rate at which carbon dioxide increased at the end of the Permian was high and supplemented by the even more potent greenhouse gas methane. We are easily exceeding those rates of $CO_2$ emission today, and that carbon dioxide is acidifying the oceans.[6]

As carbon dioxide dissolves into seawater – currently at a rate of more than 20 million tons every single day – it produces carbonic acid. This slows corals' ability to produce their carbonate skeletons, with a 30 per cent decline in the rate of new coral production so far. Before the end of the twenty-first century, coral reefs will be dissolving at a higher rate than they are growing. The survivors are likely to be the rounder, blocky corals with lower surface areas rather than the charismatic technicolour filigree trees. As we saw at Soom – admittedly, an extreme case – acidic conditions are a major threat to corals and other shelly creatures like molluscs, but the warmth itself is damaging. The algae that form partnerships with corals become less efficient in warmer water, so they abandon their mutualistic lifestyle, and leave their host to a bleached, helpless fate. There are few true tipping points in the complex Earth system, but coral reefs are one. As the world gets warmer, and as more carbon dioxide enters the oceans, the majority of shallow coral reefs will simply cease to be. As we have seen, though, corals are not the only reef-builders. To everyone's surprise, and in an intriguing mirror of their Jurassic heyday, glass sponge reefs are coming back.[7]

For most of the last 200 million years, glass sponges have cultivated a solitary and beautiful existence in the deep ocean. One species, the Venus' flower basket, *Euplectella aspergillum*, traps a pair of shrimps as cleaners, becoming a crystal cage from which the adults never escape, fed by particles specifically trapped and transferred inside by the sponge. Only the offspring of the shrimp, small enough to fit between the bars that imprison their parents, will leave. Venus' flower baskets live alone, but in the oxygen-depleted waters of British Columbia in Canada, at the top of the California Current, these glass sponges are aggregating, and reefs are now growing again, some already tens of metres tall and several kilometres long. They filter the water gently, so they don't need much oxygen to live, and are mostly made of silicon, which is less affected by acidic waters. If they can face the threats of trawling and oil exploration, the era of the glass sponge reef – and the extraordinary biodiversity it fosters – may be returning, a Lazarus ecosystem in a warming world, a small gain in an ocean of loss.[8]

Out of the water, a consequence of the warming is an evening of global climate. During greenhouse periods of Earth's history such as the Eocene, with its giant forest penguins, the latitudinal temperature gradient from equator to pole has been much lower than it is in the modern day. The records from the time of Seymour Island show that the equator was not substantially warmer than today, even with forested poles. We can see Earth today approaching that more even situation, with the poles warming three times faster than the rest of the planet. This is already beginning to alter the circulation of our atmosphere.[9]

The stability of the atmospheric current system is maintained by the difference in temperature between high and low latitudes. In the northern hemisphere, as polar air moves south and temperate air moves north, they converge into a single current – the jet stream – which is pulled eastwards by the spinning of the Earth. It is hard for a dense pocket of air to merge with one that is less dense, and so, in general, the dense polar air and warmer temperate air do not merge, forming a strong, single flow where the two contact. As the Earth is warming, the temperature difference between high altitude polar and temperate air is reducing, the pockets of air swirl into one another, creating little eddies and gyres, and the flow becomes more turbulent, weakening the coherence of the polar vortex. The boundary between the polar and temperate cells is becoming blurred and unstable, causing the path of the jet stream to swing wildly further north and further south, especially in winter. Over continents, the relative extremes of temperature mean that over North America, for example, the jet stream has a tendency to swing far south during winter, bringing frigid polar air over much of the continent. As a result, North America has regularly been getting regional cold waves in recent years, caused by increases in temperature – and temperature evenness – at a global level. On the 9th of February 2020, a record high temperature for Antarctica in modern times was set at a monitoring station on Seymour Island – 20.75°C – and the average temperature has been steadily rising, year on year, for decades.[10]

This should not be surprising. We can predict what global climates should look like by comparing our atmosphere with those of the past. The atmosphere today has a composition similar to that of the Oligocene, that transitional phase between greenhouse and icehouse. The Intergovernmental Panel on Climate Change – or IPCC – projects that within the lifetimes of children already born, we will reach – under currently implemented plans – levels of carbon dioxide in our atmosphere not seen since the Eocene. If we reach that atmospheric composition, we will also eventually reach Eocene temperatures. The uncertainty is not in the final temperature, only in the time it takes for the atmosphere to adjust, because the feedback systems of the planet's environment ensure that there is a lag between reaching atmospheric stability and a final temperature plateau. The only way to ensure that we don't reach these concentrations, and therefore these temperatures, is to reduce carbon emissions at a greater rate than is currently planned.[11]

Most of those carbon emissions come from fossil fuels: oil from the bodies of marine plankton, and coal from the lycopod swamps. So far, 3 trillion tons of carbon in fossil fuel deposits have been discovered, of which about only half a trillion tons have been burnt, and yet we are already feeling the effects. The fossil record shows us the conditions that led to their burial, and the extensive tropical swamps of the Carboniferous are not about to return today. Simply, the world is not in a state to lay down carbon reserves naturally in the quantity required to buffer climate change. Plants are still the biggest carbon sink in the modern day, and increased $CO_2$ levels will stimulate photosynthesis a small amount, but we don't have the forested ecosystems and extensive swamps required to form enough coal to counter our burning.[12]

With warming also comes increased decomposition, and the rerelease of the carbon stored as peat since the paludification of the mammoth steppe. Across large parts of Canada and Russia, the vast peat deposits are within permafrost – ground that is perpetually frozen. Frozen peatlands in the northern hemisphere hold 1.1 trillion tons of carbon, approximately half of all the organic material in the world's soils, and more than twice the total amount of carbon released by humans from fossil fuels since 1850. But that carbon is unstably stored. Now, along the northern coast of Alaska's North Slope, at the edge of the Beaufort Sea, the permafrost is thawing, and the land is being eroded. Chunks of upended turf, still held together by the ice in its soil, are found along the coast, toppled into the worryingly liquid Arctic Ocean.[13]

As the permafrost thaws, the peat soils relax, shrinking as the ice melts and settling. As it softens, slumping clay soil tilts its trees, their trunks leaning lopsidedly in all directions. Known as 'drunken forests', entire patches of woodland can be felled at once without a chainsaw in sight. Once unfrozen, the organic material in the soil begins to decompose, and release greenhouse gases, a process that can take a long time. If all the carbon held in the permafrost were to be released as carbon dioxide and methane, the warming effect would be utterly without precedent. But this is not going to happen all at once; very local factors mean that some parts of the permafrost, little warm, wet hollows that heat up faster, or south-facing slopes, will melt quicker. Permafrost can refreeze, and the decomposition takes decades. As it was in the Permian, Siberia, high in the north of the world, is an ominous presence, but this time it is less of a time-bomb waiting to suddenly go off, and more a source of continual pressure. Its slow rate of emission can be slowed further, and even stopped. Current policy, current behaviour, is to allow the permafrost to melt, but we can change that policy, and by doing so solve the problem. We know, from the fossil record and modern climate modelling, what the consequences will be if we do not.[14]

The permafrost is not the only holdover from the Last Glacial Maximum. Ice is still locked up not only in polar ice sheets and the glaciers that marched out of the poles, but also in those glaciers at high altitude. Although the polar sheets have vastly declined since the Last Glacial Maximum, the Himalayan glaciers are still there, having consistently existed for tens of thousands of years during these glacial and interglacial periods. But, as warming comes to the high mountains, these glaciers are melting too, changing the distribution of water – the fundamental chemical on which all life depends – in South and Central Asia.

Many major rivers of India, in particular the Indus, the Ganges and the Brahmaputra, are dependent on mountain glaciers and the annual snow-melt for their seasonal flow. In total, more than a third of the Brahmaputra's flow comes from meltwater. In the short term, the increased snow-melt is causing more frequent flash floods and substantial erosion of the catchment area. This increase is only fed by the raising of the snow line in the mountains, and cannot continue indefinitely. The flow of the Brahmaputra is already highly variable, and in the medium term, later in the twenty-first century, as the glaciers melt into nothing, the dry season will turn to predictable drought. We have seen, in the Miocene, how an entire sea can evaporate in a thousand years – the glaciers of the Himalayas contain far less water than the Mediterranean. For the 700 million people that live along the banks of the rivers fed by Himalayan ice, this is probably now an inevitable catastrophe, with 90 per cent of glacial volume in the Hindu Kush projected to disappear. For 10 per cent of the world's human population, water will, at some point, fail to arrive. The people of Bangladesh, the inhabitants of the vast Ganga-Brahmaputra delta, where two great rivers meet the sea, face this as one part of a triple threat. Increased heat at the equator generates more evaporation of the sea surface, with early and intense monsoons already occurring. Warming water physically expands, and this is topped up by the melting glaciers and ice sheets of the Antarctic and Greenlandic mountains, raising sea levels. Bangladesh, mostly less than 10 metres above current sea levels, is likely to be inundated. Land, river and sky are all under threat for a nation of a quarter of a billion people. In total, worldwide, about one billion people live fewer than 10 metres above current high tide lines.[15]

The population of humans has expanded at a mind-boggling rate. There are now more than 7 billion of us on the planet, and we are the dominating force on all but a very few ecosystems. One reason for that is lower child-mortality rates, an undeniably good thing, but one common concern that people bring up is the issue of overpopulation. All being equal, more humans would, of course, consume more resources, but all is not equal. As someone who has bought this book, you are likely to live a lifestyle of relatively high consumption. The average carbon emissions per person world-wide was 4.8 tons of carbon dioxide in 2018, but wealthy countries dominate that pattern. Americans averaged 15.7 tons, Australians 16.5 tons and Qataris 37.1 tons. By contrast, the only countries in Africa to have above-average per capita emissions were South Africa and Libya, with the majority emitting less than 0.5 tonnes per person.[16]

Overpopulation, as an issue, is one that is solving itself. Fertility rates around the world have been declining for decades, and global population is predicted to peak during this century alongside greater urbanization and education of women. The real and pressing problem is what that population consumes. The 2018 IPCC report found that net carbon dioxide emissions would have to fall by 45 per cent globally to limit global warming to 1.5°C. If the average American rate of carbon emissions could be reduced to that of, for example, the EU average – hardly a drop in living standards – that alone would reduce global carbon emissions by 7.6 per cent. By comparison, stopping all international flights completely would represent a drop of 1.5 per cent. But emissions are not the whole story, and wealthy countries are responsible for higher rates of consumption of our other resources too.[17]

Alongside $CO_2$, plastic has become the public face of our environmental impact. We see pictures of huge rotating gyres of plastic waste swirling in the oceans, and hear reports of fragments being increasingly found in the stomachs of ocean-dwelling animals. The effects go beyond biology. Loss of cultural heritage for seafaring people, collapse of fisheries as the accumulated effect of plastic on fish populations takes hold, and a measurable impact for mental health from the despoiling of beaches from washed-up litter, all add less immediately visible costs. Leaving aside the massive biological and social loss, the plastic damage in the ocean has been estimated as causing annual global economic costs of up to 2.5 trillion US dollars.[18]

The all-pervasive nature of plastic is most radically shown in the way that microbes are evolving. The fossil record shows us time and time again that, whenever a new niche opens up, whenever there is a new resource to exploit, something evolves to exploit it. Nature is nothing if not inventive, and the proliferation of plastic products through the latter part of the twentieth century has resulted in a new, largely unexploited resource. In 2011, a fungus, *Pestalotiopsis microspora*, from the Ecuadorian rainforest, was discovered to have some ability to digest polyurethane. In 2016, the mud near a plastic recycling plant in Sakai, Japan, was found to contain a bacterium, *Ideonella sakaiensis*, which has evolved to digest polyethylene terephthalate, breaking it down into two products that do not harm the environment. This is the first known life form of many known to be entirely plastivorous, able to safely decompose an entire plastic bottle in about the same amount of time as a hot compost heap degrades plant matter, with obvious potential in the world of recycling. Not since the oxygen efflux over a billion years ago has there been such a fundamental change to the type of resource available for biochemistry to act upon, and the smallest, most quickly reproducing organisms are keeping up with the change.[19]

As was true for the creatures of the mammoth steppe, another way to keep up with perpetual change is simply to migrate. The penguins of Brown Bluff, south of Esperanza, demonstrate this kind of climate-driven migration. They are mostly Adelies, living on the peninsula itself, but they also populate islands throughout the Ross Sea. Living in enormous colonies, their guano has seeped into the soil, and, deposited as it is year after year, the layers of waste hold a record of the length of time that penguins have lived in the same site. Antarctica has become warmer, more habitable since the last ice age, and digging down through centuries of Adelie excrement has shown us that the island colonies have been continually occupied for nearly 3,000 years. On Brown Bluff, where the ice sheet was piled for longer, the penguin colony has only been around for 400 years. Species can change their ranges; warming temperatures made Brown Bluff a sensible place to raise chicks, and so a new colony was formed.[20]

Penguins, though, are relatively good at moving from place to place, and when changing currents turn their marine paradise into a barren expanse of water, they can adapt and move with the times. Other species simply cannot move quickly enough to escape climate change. Long-lived plants cannot easily follow the weather, for example, as each has a limit to its environmental tolerance. I remember a small-leaved lime tree (*Tilia cordata*) that grew on the hill by my home, deep in rural Perthshire. It fruited every year, but, unlike the rowans, the birches and the pines, it was the only one of its kind that I ever saw. Native to much of the UK but adapted to a warmer climate than it had found itself in, its fruit were unlikely to have been fertile. A vagrant seed, dispersed northwards by some chance event, had planted and grown but was stuck beyond the edge of its reproductive range. As climates change, the balance of power shifts, with the optimum set of conditions migrating, and the ranges of species following behind. Between 1970 and 2019, the Great Plains ecosystem of North America moved north by an average distance of 365 miles – that is, on average, a metre every forty-five minutes. In a wide and flat continent, there is space to move, but if you are on a small island, at the higher latitude coasts or on a mountain, adapted to cool high altitudes, there will eventually be nowhere left to run. Long-distance dispersal is rare in the natural world, and, being pushed to the end of their range, many species are effectively walking the plank.[21]

We are also introducing new ecosystems. Perhaps the modern equivalent to the post-extinction ecosystems, denuded of trees and of large animals, with low productivity, is that of the city. Many species simply cannot survive in these new worlds, and those that can must adapt, even down to their most basic behaviours. Even the cacophony of a jungle is practically silent compared with a city and, for species that signal their presence to mates or potential rivals through sound, this noise is highly disruptive. The volume of songbird calls in cities is higher in pitch, faster and shorter than that of rural members of the same species. Only those that squeak at those high frequencies can be heard above the low rumble of machinery. Scent-based signals are affected by climate change too. In higher temperatures, marks left by male lizards to attract mates are more volatile, disappearing sooner, and so mating opportunities are missed. As sea ice fragments, scent trails left by polar bear footpads are also gone, affecting everything from breeding behaviour to territoriality. The survival of a species is about more than the environmental tolerance of an individual's physiology. It relies also on the resilience of its behaviours. There is no corner of the Earth where we have not touched the way of life of its inhabitants in some way.[22]

By sheer number, we are incredibly common, and since 2 November 2000, there has even been a continuous human presence outside the planet's atmosphere. Humans make up, by mass, 36 per cent of all mammals. A further 60 per cent of the mass of all mammals are domesticated animals – cattle, swine, sheep, horses, cats and dogs. Only 4 per cent of mammal mass on this planet is wild. For birds, it is even more stark. Sixty per cent of birds on Earth are from a single species – domestic chickens. Taken as a whole, the mass of human-produced material is, as of 2020, about equal to the mass of living material on Earth. If we were to sample the planet today in the same way that we sample the fossil record, we would look at the distribution of bones, and conclude that something very strange was going on that so much of the vertebrate biomass was made up of so few species. We would be talking in terms of catastrophic environmental damage, of mass extinction. Indeed, the biomass of wildlife has declined at a horrifying rate. The world into which Emilio Marcos Palma was born in 1978 was home to 2.5 times as many wild vertebrates as the one in 2018. In a geological snap of the fingers, we have lost more than half of the living individual vertebrates on the planet.

Since the last ice age, the largest species have either been wiped out from every continent, or are well on the way to extinction. The planet is beginning to resemble a post-extinction world, with the human ecosystem the refuge of the disaster taxa. Those adapted to our world – versatile animals able to live off refuse like rats, European foxes, raccoons, herring gulls or the Australian white ibis, or those who we have teamed up with or breed for our own purposes – chickens, cattle and dogs in particular – are thriving. Many plants and less mobile animals have benefited from human-mediated long-distance dispersal, accidental or deliberate. Shipping lanes have replaced uncommon raft events in bringing physically unconnected continents closer to one another in a dispersal sense. By taking organisms from their habitat, we often remove them from their competitors, allowing them to thrive and outcompete other ecologically important native organisms.[23]

Where so many species are disappearing at the rate of mass

extinctions, it is easy to look at what we have done and despair. But we must not become despondent. Human-induced change is, in itself, not new and, to a large extent, can be considered natural. We are part of the biological realm, inhabiting the tree of life. There is solid evidence that humans, like so many species that have come before us, have always been natural ecosystem engineers. Humans have been creating pastures for nearly 8,000 years. The burning of forests and grasslands to introduce cattle at about the same time changed the way in which parts of Eurasia reflect sunlight, affecting the absorption of heat, and altering the monsoon pattern of India and South East Asia. Humans have deliberately moved species around with them since well into the Pleistocene; there is evidence from the Solomon Islands that the common cuscus, a tree-dwelling possum that is an important game species, was introduced to those islands from New Guinea more than 20,000 years ago, apparently alongside the human trade in obsidian.[24]

We are such effective ecosystem engineers that the idea of a pristine Earth, unaffected by human biology and culture is impossible. Such an Eden simply does not and, since the emergence of humans, never has existed. While the damage that is being done to global ecosystems is unprecedented in our species' lifetime, conservation programmes need to decide what degree of human impact is desirable and achievable for any ecosystem. Pre-industrial? Pre-colonial? Pre-human? These are difficult questions. Returning current ecosystems to an entirely wild state often disproportionately negatively affects indigenous and poor communities who depend on them, adding a complex social context to environmental decision-making. The Bangladeshi philosopher Nabil Ahmed, in his article *Entangled Earth*, says of his country: 'it is not possible to differentiate between land and river, human populations, sedimentation, gas, grains and forests, politics and markets'. All coalesce into a single entity, and all bear the legacy of interaction between political and natural actors. That nation, he argues, was born directly from the 1970 Bhola cyclone, and drew its independence from a political reaction to a natural and humanitarian disaster.[25]

Just as in the sweltering Permian of Pangaea, when superstorms raged across a global ocean, we are seeing an increase in tropical storm events the world over. The number of Atlantic hurricanes per season has been increasing steadily since records began in the early twentieth century, in 2020 reaching thirty named storms, three times the long-term average. In 2018, there was even an unprecedented hurricane-strength storm within the Mediterranean. This is happening because warmer water increases the rate of air rising around the tropical latitudes, which means that hurricanes can become stronger more quickly, giving them a better chance to build up to severe strength by the time they reach land, with severe consequences for the nations in their paths.[26]

It is impossible to ignore the social implications of climate change. From the race among the rich nations of the Arctic circle and beyond to exploit the resources in the seabed beneath the melting ice, to the continuing international wrangling over dams being built in East Africa to gain control of a diminishing water supply, changes to the environment have been affecting political decisions for decades already. That one is a race for wealth and the other a battle for a fundamental resource is indicative of the degree to which the costs of climate change are borne most by those that have contributed to it the least and reaped the lowest rewards. Today, we can see the changes that are about to unfold. The geological history of our planet paints, in broad but unmistakeable brushstrokes, a picture of possible futures. We are undergoing a humanitarian and natural disaster that encompasses the whole planet, but it is something we can manage.[27]

That the worlds of yesterday are strange and beautiful is a lesson in the adaptability of life. There is, however, a second lesson that the rocks teach – that of our own world's impermanence. I started this book with a reference to Shelley's well-known poem 'Ozymandias'. Less well known is that this poem was written in a sonnet competition against his friend Horace Smith, a comradely game among literary tourists inspired by the same artefact. Where Shelley looks to the past, mocking the hubris of those in power, Smith sombrely looks to the future in a more explicit sense. After using the first eight lines in considering the city, indicated by the pedestal but long since gone, he reflects on the temporariness of the city he knows best, and says:

> We wonder – and some Hunter may express
> Wonder like ours, when thro' the wilderness
> Where London stood, holding the Wolf in chace
> He meets some fragment huge, and stops to guess
> What powerful but unrecorded race
> Once dwelt in that annihilated place.

Those landscapes we take for granted are not integral parts of the world; life will continue without them, without us. Eventually, the carbon dioxide we emit will be absorbed, once again, into the deep ocean, and the cycles of life and of mineral will continue. We, like every other inhabitant of our planet, have evolved alongside the current cohort of species, interacting with them in complex ways. We are part of the global ecosystem and always have been, and it is folly to think that we ourselves will not be affected by the changes we are imposing on the world.

As a species, we are well placed to survive the mass-extinction event we are currently generating. With our technology, from clothing to dykes, air conditioning to desalinators, we have consistently modified our own environments to survive where we otherwise would not. But the ecosystems that have been built since the last mass extinction 66 million years ago are under severe stress. By destroying communities and changing the chemistry of the world, we are tugging at the strands of the spider's web once more, and several strands have already broken. If enough snap, the consequences for the way we interact with the world could become a biological and social catastrophe unlike any ever faced. This might seem at first glance to be overwhelming, paralysing. But the very fact that we can reflect on the state of our environment at all, that we have the analytical ability to look at the past and find analogues for the present, is the reason that we can be positive.

We know what can happen during environmentally turbulent periods like the one in which we live. In mapping the past, we can predict the future, and find the routes that avert disaster. Where some disastrous outcomes are inevitable, we can plan for them, minimize the damage and mitigate them. Since at least the 1970s, infrastructure has been built with the effects of climate change in mind. The Thames Barrier, London's primary flood defence, was specifically designed with the expectation that sea levels would rise 90 centimetres by about 2100, with capacity up to 2.7 metres. We also know that international collaboration works; the Montreal Protocol of 1987, which was signed by 197 governments, phased out the production and use of chlorofluorocarbons, which were responsible for the thinning of the ozone layer. The 'hole' in the ozone layer is well on its way to recovery thanks to such measures. These measures are paid for by a fund through which the countries with the greatest per capita contribution to the problem aid economically developing nations with compliance.[28]

During the course of writing this book, two things happened that showed the importance of a more focused view into the past and the future. In early 2019, amid only a little fanfare, a plaque was placed on the site of Okjökull, the first of Iceland's glaciers to lose its status as a river of ice by melting enough that it now fails to move under its own weight. The plaque brands itself in Icelandic and English 'A letter to the future', and after explaining Okjökull's demotion to an ice lake, says 'This monument is to acknowledge that we know what is happening and what needs to be done. Only you know if we did it.'[29] Look, it is saying, on our works.

The second event, the pandemic spreading of the SARS-CoV-2 coronavirus, forced humanity to confront radical change in a more immediate way. Within the space of a month, a third of the world's population went into enforced or voluntary lockdown, fundamentally changing many aspects of their lives to counter an existential threat. The effect of these changes was immediate. Los Angeles, a city synonymous with traffic jams, reported clean air conditions not seen in generations. Venice, for so long clogged by tourist boats, had clearer water than ever before. Carbon emissions dropped, although only by about 8 per cent, and oil became less than worthless as stores remained full and deliveries piled up. Several outlets reported these instances as examples of 'the Earth healing itself', carrying with it the subtext that humanity was the true virus. Such misanthropy is not necessary. Humans can indeed live exploitatively, but there are better lessons here. We can alter our behaviour and respond to a crisis, and the changes we make can have immediately beneficial effects. The suffering of those in other nations affects us all, and it is only by working together, and pooling resources and support where it is needed, that the damage caused by such international crises can be minimized. By listening to experts, by taking threats seriously, and by putting well-being first, some countries managed the pandemic far more effectively than others. The coordinated international action to develop functioning vaccines in record time is testament to our capacity to respond quickly and effectively to a deadly threat. The lack of international cooperation in distributing those vaccines, and the resultant subsequent waves of infection and death, demonstrate the naivety of an insular, defensive response to a global crisis.

In the face of environmental change, it is complacency that is deadly. The 'business as usual' approach, in which no changes are made to rates of ecosystem destruction or greenhouse gas emission, will generate climates no hominid has ever faced before. However, people who speak of unavoidable doom are equally unhelpful. In conservation, success and failure is not a binary choice. When newspapers report that we have five years, or ten years, to stop climate change, these are not all-or-nothing deadlines. Making changes on time does not mean that everything will be as it was, and failing to do so does not mean annihilation. The ecosystems that existed in the first half of the twentieth century and before are permanently changed, but damages continue to accrue. The sooner and more strongly we act, the less comprehensive the damage will be. Whether we choose to act collectively to counter the causes and effects of the changing climate is up to us. The spire may have fallen, but the cathedral yet stands, and we must choose whether to douse the flames.

Only by altering our habits, and by endeavouring to live less exploitatively, can we prevent the changes to the environment from becoming an unparalleled catastrophe, another Great Dying. The planet cannot provide the resources required to support a life as profligate as that now enjoyed in economically developed nations, let alone enough extra for other species to feed and mate, and live their own lives. The only reliable way of keeping the wild worlds of today from becoming another forgotten set of ecosystems, another gallery in the museums of a future epoch, is to reduce consumption and stop relying on climate-changing sources of energy. Inevitably, these solutions meet with resistance. People are understandably worried that it might lower our quality of life in the short term, and involve some personal and societal effort. Yet within decades, without action at the level of the community, of the nation, of the globe, we will certainly suffer even more. For our long-term well-being as a species and as individuals, we must enter into a more mutualistic relationship with our global environments. Only then can we preserve not just their infinite variety, but also our place within them. Change, eventually, is inevitable, but we can let the planet take its own time, as we allow the shifting sands of geological time to lead us gently into the worlds of tomorrow. Sacrifice, an act of permanence. Then, we too will live in hope.

# Notes

## Introduction: The House of Millions of Years

1   Bell, E. A. & others. *PNAS* 2015; 112:14518–21; Chambers, J. E. *Earth and Planetary Science Letters* 2004; 223:241–52; El Albani, A. & others. *Nature* 2010; 466:100–104; Miller, H. *My Schools and Schoolmasters*. Edinburgh, UK: George A. Morton; 1905.

2   Leblanc, C. *Museum International* 2005; 57:79–86; Parr, J. *Keats-Shelley Journal* 1957; 6:31–5.

3   Ullmann, M. 'The Temples of Millions of Years at Western Thebes'. In: Wilkinson, R. H. and Weeks, K. R., eds. *The Oxford Handbook of the Valley of the Kings*. Oxford, UK: Oxford University Press; 2016. Pp. 417–32.

4   Dunne, J. A. & others. *PLoS Biol.* 2008; 6:693–708; Gingerich, P. D. *Paleobiology* 1981; 7:443–55; Gu, J. J. & others. *PNAS* 2012; 109:3868–73; Pardo-Pérez, J. M. & others. *J. Zool.* 2018; 304:21–33; Rayfield, E. J. *Annual Review of Earth and Planetary Sciences* 2007; 35:541–76; Smithwick, F. M. & others. *Curr. Biol.* 2017; 27:3337.

5   Black, M. *The Scientific Monthly* 1945; 61:165–72; Cunningham, J. A. & others. *Trends in Ecology and Evolution* 2014; 29:347–57.

6   Frey, R. W. *The Study of Trace Fossils: A Synthesis of Principles, Problems, and Procedures in Ichnology*. Berlin: Springer-Verlag; 1975; Halliday, T. J. D. & others. *Acta Palaeontologica Polonica* 2013; 60:291–312; Nichols, G. *Sedimentology and Stratigraphy*. Oxford, UK: Blackwell; 2009.

7   Herendeen, P. S. & others. *Nature Plants* 2017; 3:17015; Prasad, V. & others. *Nature Communications* 2011; 2:480; Strömberg, C. A. E. *Annual Review of Earth and Planetary Science* 2011; 39:517–44.

8   Breen, S. P. W. & others. *Frontiers in Environmental Science* 2018; 6:1–8; Ceballos, G. & others. *PNAS* 2017; 114:E6089–96; Elmendorf, S. C. & others. *Ecology Letters* 2012; 15:164–75.

9   Ezaki, Y. *Paleontological Research* 2009; 13:23–38.

10  Hutterer, R. and Peters, G. *Bonn Zool. Bull.* 2010; 59:3–27.

11  Ashe, T. *Memoirs of Mammoth*. Liverpool, UK: G. F. Harris; 1806; O'Connor, R. *The Earth on Show: Fossils and the Poetics of Popular Science, 1802–1856*. Chicago: University of Chicago Press; 2013; Peale, R. *An historical disquisition on the mammoth: or, great American incognitum, an extinct, immense, carnivorous*

*animal, whose fossil remains have been found in North America.* C. Mercier and Co.; 1803.

# 1. *Thaw – Pleistocene*

1  Berger, A. & others. *Applied Animal Behaviour Science* 1999; 64:1–17; Bernáldez-Sánchez, E. and García-Viñas, E. *Anthropozoologica* 2019; 54:1–12; Beyer, R. M. & others. *Scientific Data* 2020; 7:236; Burke, A. and Cinq-Mars, J. *Arctic* 1998; 51:105–15; Chen, J. & others. *J. Equine Science* 2008; 19:1–7; Feh, C. 'Relationships and communication in socially natural horse herds: social organization of horses and other equids'. In: MacDonnell, S. and Mills, D., eds. Dorothy Russell Havemeyer Foundation Workshop. Holar, Iceland 2002; Forsten, A. J. *Mammalogy* 1986; 67:422–3; Gaglioti, B. V. & others. *Quat. Sci. Rev.* 2018; 182:175–90; Guthrie, R. D. and Stoker, S. *Arctic* 1990; 43:267–74; Janis, C. *Evolution* 1976; 30:757–74; Mann, D. H. & others. *Quat. Sci. Rev.* 2013; 70:91–108; Turner Jr, J. W. and Kirkpatrick, J. F. *J. Equine Veterinary Science* 1986;6: 250–58; Ukraintseva, V. V. *The Selerikan horse. Mammoths and the Environment*. Cambridge, UK: Cambridge University Press; 2013. Pp. 87–105.

2  Burke, A. and Castanet, J. *J. Archaeological Science* 1995; 22:479–93; Carter, L. D. *Science* 1981; 211:381–3; Gaglioti, B. V. & others. *Quat. Sci. Rev.* 2018; 182:175–90; Packer, C. & others. *PLoS One* 2011; 6:e22285; Sander, P. M. and Andrássy, P. *Palaeontographica Abteilung A* 2006; 277:143–59; Wathan, J. and McComb, K. *Curr. Biol.* 2014; 24:R677-R679; Yamaguchi, N. & others. *J. Zool.* 2004; 263:329–42.

3  Bar-Oz, G. and Lev-Yadun, S. *PNAS* 2012; 109:E1212; Barnett, R. & others. *Molecular Ecology* 2009; 18:1668–77; Chernova, O. F. & others. *Quat. Sci. Rev.* 2016; 142:61–73; Chimento, N. R. and Agnolin, F. L. *Comptes Rendus Palevol* 2017; 16:850–64; de Manuel, M. & others. *PNAS* 2020; 117:10927–34; Nagel, D. & others. *Scripta Geologica* 2003:227–40; Stuart, A. J. and Lister, A. M. *Quat. Sci. Rev.* 2011; 30:2329–40; Turner, A. *Annales Zoologici Fennici* 1984:1–8; Yamaguchi, N. & others. *J. Zool.* 2004; 263:329–42.

4  Guthrie, R. D. *Frozen Fauna of the Mammoth Steppe: The Story of Blue Babe.* Chicago, USA: The University of Chicago Press; 1990; Kitchener, A. C. & others. 'Felid form and function'. In: Macdonald, D. W. and Loveridge, A. J., eds. *Biology and Conservation of Wild Felids.* Oxford, UK: Oxford University Press; 2010. Pp. 83–106; Rothschild, B. M. and Diedrich, C. G. *International J. Paleopathology* 2012; 2:187–98.

5  Sissons, J. B. *Scottish J. Geology* 1974; 10:311–37.

6   Gazin, C. L. *Smithsonian Miscellaneous Collections* 1955; 128:1–96; Jass, C. N. and Allan, T. E. *Can. J. Earth Sci.* 2016; 53:485–93; Merriam, J. C. *University of California Publications of the Geological Society* 1913; 7:305–23; Upham, N. S. & others. *PLoS Biol.* 2019; 17.

7   Bennett, M. R. & others. Science 2021; 373:1528–1531. Goebel, T. & others. *Science* 2008; 319:1497–1502; Kooyman, B. & others. *American Antiquity* 2012; 77:115–24; Seersholm, F. V. & others. *Nature Communications* 2020; 11:2770; Vachula, R. S. & others. *Quat. Sci. Rev.* 2019; 205:35–44; Waters, M. R. & others. *PNAS* 2015; 112:4263–7.

8   Krane, S. & others. *Naturwissenschaften* 2003; 90:60–62; Madani, G. and Nekaris, K. A. I. *J. Venomous Animals and Toxins Including Tropical Diseases* 2014; 20; Nekaris, K. A. I. and Starr, C. R. *Endangered Species Research* 2015; 28:87–95; Nekaris, K. A. I. & others. *J. Venomous Animals and Toxins Including Tropical Diseases* 2013; 19; Still, J. *Spolia Zeylanica* 1905; 3:155; Wuster, W. and Thorpe, R. S. *Herpetologica* 1992; 48:69–85; Zareyan, S. & others. *Proc. R. Soc. B* 2019; 286:20191425.

9   Begon, M. & others. *Ecology: From Individuals to Ecosystems*. Oxford, UK: Blackwell Publishing; 2006.

10  Alexander, R. M. *J. Zoology* 1993; 231:391–401; Ellis, A. D. 'Biological basis of behaviour in relation to nutrition and feed intake in horses'. In: Ellis, A. D. and others, eds. *The impact of nutrition on the health and welfare of horses*. Netherlands: Wageningen Academic Publishers; 2010. Pp. 53–74; Kuitems, M. & others. *Arch. and Anth. Sci.* 2015; 7:289–95; van Geel, B. & others. *Quat. Sci. Rev.* 2011; 30:2289–303.

11  Beyer, R. M. & others. *Scientific Data* 2020; 7:236; Hopkins, D. M. 'Aspects of the Paleogeography of Beringia during the Late Pleistocene'. In: Hopkins, D. M. and others, eds. *Paleoecology of Beringia*: Academic Press; 1982. Pp. 3–28; Paterson, W. S. *Reviews of Geophysics and Space Physics* 1972; 10:885; Tinkler, K.J. & others. *Quaternary Research* 1994; 42:20–29.

12  Ager, T. A. *Quaternary Research* 2003; 60:19–32; Anderson, L. L. & others. *PNAS* 2006; 103:12447–50; Brubaker, L. B. & others. *J. Biogeog.* 2005; 32:833–48; Fairbanks, R. G. *Nature* 1989; 342:637–42; Holder, K. & others. *Evolution* 1999; 53:1936–50; Quinn, T. W. *Molecular Ecology* 1992; 1:105–17; Shaw A. J. & others. *J. Biogeog.* 2015; 42:364–76; *Paleodrainage map of Beringia*: Yukon Geological Survey; 2019; Zazula, G. D. & others. *Nature* 2003; 423:603.

13  Guthrie, R. D. *Quat. Sci. Rev.* 2001; 20:549–74; *Paleodrainage map of Beringia*: Yukon Geological Survey; 2019.

14  Batima, P. & others. 'Vulnerability of Mongolia's pastoralists to climate extremes and changes'. In: Leary, N. and others, eds. *Climate Change and Vulnerability*. London: Earthscan; 2008. Pp. 67–87; Clark, J. K. and Crabtree, S. A. *Land* 2015; 4:157–81; Fancy, S. G. & others. *Can. J. Zool.* 1989; 67:644–50; Mann, D. H. & others. *PNAS* 2015; 112:14301–6.

15 Clark, J. & others. *J. Archaeological Science* 2014; 52:12–23; Lent, P. C. *Biological Conservation* 1971; 3:255–63; Sommer, R. S. & others. *J. Biogeog.* 2014; 41:298–306.

16 Guthrie, R. D. and Stoker, S. *Arctic* 1990; 43:267–74.

17 Kuzmina, S. A. & others. *Invertebrate Zoology.* 2019; 16:89–125; Mann, D. H. & others. *Quat. Sci. Rev.* 2013; 70:91–108.

18 Begon, M. & others. *Ecology*: From Individuals to Ecosystems. Oxford, UK: Blackwell Publishing; 2006; Beyer, R. M. & others. *Scientific Data* 2020; 7:236; Kazakov, K. 2020. *Pogoda i klimat.* <http://www.pogodaiklimat.ru>.

19 Churcher, C. S. & others. *Can. J. Earth Sci.* 1993; 30:1007–13; Emslie, S. D. and Czaplewski, N. *J. Nat. Hist. Mus. LA County Contributions in Science* 1985; 371:1–12; Figueirido, B. & others. *J. Zool.* 2009; 277:70–80; Figueirido, B. & others. *J. Vert. Paleo.* 2010; 30:262–75; Kurtén, B. *Acta Zoologica Fennica* 1967; 117:1–60; Sorkin, B. *J. Vert. Paleo.* 2004; 24:116A.

20 Chernova, O. F. & others. *Proc. Zool. Inst. Russ. Acad. Sci.* 2015; 319:441–60; Harington, C. R. *Neotoma* 1991; 29:1–3; Matheus, P. E. *Quaternary Research* 1995; 44:447–53.

21 Grayson, J. H. *Folklore* 2015; 126:253–65; Hallowell, A. I. *American Anthropologist* 1926; 28:1–175; Huld, M. E. *Int. J. American Linguistics* 1983; 49:186–95.

22 Mann, D. H. & others. *Quat. Sci. Rev.* 2013; 70:91–108; Zimov, S. A. & others. 'The past and future of the mammoth steppe ecosystem'. In: Louys, J., ed. *Paleontology in Ecology and Conservation.* Berlin: Springer Verlag; 2012. Pp. 193–225.

23 Guthrie, R. D. *Quat. Sci. Rev.* 2001; 20:549–74.

24 Chytrý, M. & others. *Boreas* 2019; 48:36–56; Guthrie, R. D. *Quat. Sci. Rev.* 2001; 20:549–74; Kane, D. L. & others. *Northern Research Basins Water Balance* 2004; 290:224–36; Mann, D. H. & others. *Quat. Sci. Rev.* 2013; 70:91–108.

25 Pečnerová, P. & others. *Evolution Letters* 2017; 1:292–303; Rogers, R. L. and Slatkin, M. *PLoS Genetics* 2017; 13:e1006601; Vartanyan, S. L. & others. *Nature* 1993; 362:337–40.

26 Currey, D. R. *Ecology* 1965; 46:564–66; Gunn, R. G. *Art of the Ancestors: spatial and temporal patterning in the ceiling rock art of Nawarla Gabarnmang, Arnhem Land, Australia.* Archaeopress Archaeology; 2019; Paillet, P. *Bulletin de la Société préhistorique française* 1995; 92:37–48; Valladas, H. & others. *Radiocarbon* 2013; 55:1422–31.

27 Martínez-Meyer, E. and Peterson, A. T. *J. Biogeog.* 2006; 33:1779–89.

## 2. Origins – Pliocene

1 Kassagam, J. K. *What is this bird saying? – A study of names and cultural beliefs about birds amongst the Marakwet peoples of Kenya.* Kenya: Binary Computer Services; 1997.

2   Field, D. J. *J. Hum. Evol.* 2020; 140:102384; Hollmann, J. C. *South African Archaeological Society Goodwin Series* 2005; 9:21–33; Owen, E. *Welsh Folk-lore.* Woodall, Minshall, & Co.; 1887; Pellegrino, I. & others. *Bird Study* 2017; 64:344–52; Rowley, D. B. and Currie, B. S. *Nature* 2006; 439:677–81; Ruddiman, W. F. & others. *Proc. Ocean Drilling Program, Scientific Results* 1989; 108:463–84.

3   Chorowicz, J. *J. African Earth Sciences* 2005; 43:379–410; Feibel, C. S. *Evol. Anthro.* 2011; 20:206–16; Furman, T. & others. *J. Petrology* 2004; 45:1069–88; Mohr, P. A. *J. Geophysical Research* 1970; 75:7340–52.

4   Feibel, C. S. *Evol. Anthro.* 2011; 20:206–16; Furman, T. & others. *J. Petrology* 2006; 47:1221–44; Hernández Fernández, M. and Vrba, E. S. *J. Hum. Evol.* 2006; 50:595–626; Kolding, J. *Environmental Biology of Fishes* 1993; 37:25–46; Olaka, L. A. & others. *J. Paleolimnology* 2010; 44:629–44; Van Bocxlaer, B. *J. Hum. Evol.* 2020; 140:102341; Yuretich, R. F. & others. *Geochimica Et Cosmochimica Acta* 1983; 47:1099–1109.

5   Alexeev, V. P. *The origin of the human race.* Moscow: Progress Publishers; 1986; Brown, F. & others. *Nature* 1985; 316:788–92; Leakey, M. G. & others. *Nature* 2001; 410:433–40; Lordkipanidze, D. & others. *Science* 2013; 342:326–31; Ward, C. & others. *Evolutionary Anthropology* 1999; 7:197–205.

6   Aldrovandi, U. *Ornithologiae.* Bologna: Francesco de Franceschi; 1599; Hedenström, A. & others. *Curr. Biol.* 2016; 26:3066–70; Henningsson, P. & others. *J. Avian Biol.* 2010; 41:94–8; Hutson, A. M. *J. Zool.* 1981; 194:305–16; Liechti, F. & others. *Nature Communications* 2013; 4; Manthi, F. K. *The Pliocene micromammalian fauna from Kanapoi, northwestern Kenya, and its contribution to understanding the environment of* Australopithecus anamensis. Cape Town: University of Cape Town; 2006; Mayr, G. *J. Ornithology* 2015; 156:441–50; McCracken, G. F. & others. *Royal Society Open Science* 2016; 3:160398; Zuki, A. B. Z. & others. *Pertanika J. Tropical Agricultural Science* 2012; 35:613–22.

7   Delfino, M. *J. Hum. Evol.* 2020; 140:102353; Field, D. J. *J. Hum. Evol.* 2020; 140:102384; Kyle, K. and du Preez, L. H. *Afr. Zool.* 2020; 55:1–5; Manthi, F. K. and Winkler, A. J. *J. Hum. Evol.* 2020; 140:102338; Werdelin, L and Manthi, F. K. *J. African Earth Sciences* 2012; 64:1–8.

8   Geraads, D. & others. *J. Vert. Paleo.* 2011; 31:447–53; Lewis, M. E. *Comptes Rendus Palevol* 2008; 7:607–27; Stewart, K. M. and Rufolo S. J. *J. Hum. Evol.* 2020; 140:102452; Van Bocxlaer, B. *J. Systematic Palaeontology* 2011; 9:523–50; Van Bocxlaer, B. *J. Hum. Evol.* 2020; 140:102341; Werdelin, L and Lewis, M. E. *J. Hum. Evol.* 2020; 140:102334; Werdelin, L. and Manthi, F. K. *J. African Earth Sciences* 2012; 64:1–8.

9   Stewart, K. *Nat. Hist. Mus. LA County Contributions in Science* 2003; 498:21–38; Stewart, K. M. and Rufolo, S. J. *J. Hum. Evol.* 2020; 140:102452.

10  Field, D. J. *J. Hum. Evol.* 2020; 140:102384; Owry, O. T. *Ornithological Monographs* 1967; 6:60–63; Rijke, A. M. and Jesser, W. A. *Condor* 2011; 113:245–54.

11  Field, D. J. *J. Hum. Evol.* 2020; 140:102384; Kozhinova, A. https://ispan.waw.pl/ireteslaw/handle/20.500.12528/1832017; Louchart, A. & others. *Acta Palaeontologica Polonica* 2005; 50:549–63; Meijer, H. J. M. and Due, R. A. *Zoo. J. Linn. Soc.* 2010; 160:707–24; Ogada, D. L. & others. *Conservation Biology* 2012; 26:453–60; Pomeroy, D. E. *Ibis* 1975; 117:69–81; Szyjewski, A. *Religia Słowian.* Warsaw: Wydawnictwo WAM; 2010; Warren-Chadd, R. and Taylor, M. *Birds: Myth, lore & legend.* London: Bloomsbury; 2016. P. 304.

12  Basu, C. & others. *Biology Letters* 2016; 12:20150940; Brochu, C. A. *J. Hum. Evol.* 2020; 140:102410; Geraads, D. & others. *J. African Earth Sciences* 2013; 85:53–61; Geraads, D. and Bobe, R. *J. Hum. Evol.* 2020; 140:102383; Harris, J. M. *Annals of the South African Museum* 1976; 69:325–53; Nanda, A. C. *J. Palaeont. Soc. India* 2013; 58:75–86.

13  Harris, J. M. *Annals of the South African Museum* 1976; 69:325–53; Solounias, N. *J. Mamm.* 1988; 69:845–8; Spinage, C. A. *J. Zool.* 1993; 230:1–5.

14  Sengani, F. and Mulenga, F. *Applied Sciences* 2020; 10:8824; Wynn, J. G. *J. Hum. Evol.* 2000; 39:411–32.

15  Cerling, T. E. & others. *PNAS* 2015; 112:11467–72; Wagner, H. H. & others. *Landscape Ecology* 2000; 15:219–27.

16  Farquhar, G. D. and Sharkey, T. D. *Annual Reviews* 1982; 33:317–45; Waggoner, P. E. and Simmonds, N. W. *Plant Physiology* 1966; 41:1268.

17  Pearcy, R. W. and Ehleringer, J. *Plant, Cell, and Environment* 1984; 7:1–13; Spreitzer, R. J. and Salvucci, M. E. *Ann. Rev. Plant Biol.* 2002; 53:449–75; Westhoff, P. and Gowik, U. *Plant Physiology* 2010; 154:598–601.

18  Caswell, H. & others. *American Naturalist* 1973; 107:465–80; Cerling, T. E. & others. *PNAS* 2015; 112:11467–72; Pearcy, R. W and Ehleringer, J. *Plant, Cell, and Environment* 1984; 7:1–13.

19  Cerling, T. E. & others. *PNAS* 2015; 112:11467–72; Field, D. J. *J. Hum. Evol.* 2020; 140:102384; Franz-Odendaal, T. A and Solounias, N. *Geodiversitas* 2004; 26:675–85; Geraads, D. & others. *J. African Earth Sciences* 2013; 85:53–61; Harris J. M. *Annals of the South African Museum* 1976; 69:325–53; Uno, K. T. & others. *PNAS* 2011; 108:6509–14; Wynn, J. G. *J. Hum. Evol.* 2000; 39:411–32.

20  Cerling, T. E. & others. *PNAS* 2015; 112:11467–72; Sanders, W. J. *J. Hum. Evol.* 2020; 140:102547; Valeix, M. & others. *Biological Conservation* 2011; 144:902–12.

21  Žliobaitė, I. *Data Mining and Knowledge Discovery* 2019; 33:773–803.

22  Gunnell, G. F and Manthi, F. K. *J. Hum. Evol.* 2020; 140:102440; Wynn, J. G. *J. Hum. Evol.* 2000; 39:411–32.

23  Dávid-Barrett, T. and Dunbar, R. I. M. *J. Hum. Evol.* 2016; 94:72–82; Head, J. J. and Müller, J. *J. Hum. Evol.* 2020; 140:102451; Stave, J. & others. *Biodiversity and Conservation* 2007; 16:1471–89; Ungar, P. S. & others. *Phil. Trans. R. Soc. B* 2010; 365:3345–54; Ward, C. & others. *Evolutionary Anthropology* 1999; 7:197–205;

Ward, C. V. & others. *J. Hum. Evol.* 2001; 41:255–368; Ward, C. V. & others. *J. Hum. Evol.* 2013; 65:501–24.

24 Stave, J. & others. *Biodiversity and Conservation* 2007; 16:1471–89.

25 Almécija, S. & others. *Nature Communications* 2013; 4; Brunet, M. & others. *Nature* 2002; 418:145–51; Haile-Selassie, Y. & others. *American J. Physical Anthropology* 2010; 141:406–17; Parins-Fukuchi, C. & others. *Paleobiology* 2019; 45:378–93; Pickford, M. and Senut, B. *Comptes Rendus A* 2001; 332:145–52; Sarmiento, E. E. and Meldrum, D. J. *J. Comparative Human Biology* 2011; 62:75–108; Ward, C. V. & others. *Phil. Trans. R. Soc. B* 2010; 365:3333–44; Wolpoff, M. H. & others. *Nature* 2002; 419:581–2.

26 Rose, D. 'The Ship of Theseus Puzzle'. In: Lombrozo, T. and others, eds. *Oxford Studies in Experimental Philosophy.* Volume 3. Oxford, UK: Oxford University Press; 2020. Pp. 158–74.

27 Wagner, P. J. and Erwin, D. H. *Phylogenetic Patterns as Tests of Speciation Models.* New York: Columbia University Press; 1995. Pp. 87–122.

28 Kimbel, W. H. & others. *J. Hum. Evol.* 2006; 51:134–52.

29 Lewis, J. E. and Harmand, S. *Phil. Trans. R. Soc. B* 2016; 371:20150233; McHenry, H. M. *American J. Physical Anthropology* 1992; 87:407–31; Reno, P. L. & others. *PNAS* 2003; 100:9404–9; Ward, C. V. & others. *Phil. Trans. R. Soc. B* 2010; 365:3333–44.

30 Geraads, D. & others. *J. African Earth Sciences* 2013; 85:53–61; Sanders, W. J. *J. Hum. Evol.* 2020; 140:102547.

31 Faith, J. T. & others. *Quaternary Research* 2020; 96:88–104; Fortelius, M. & others. *Phil. Trans. R. Soc. B* 2016; 371:20150232; Werdelin, L. and Lewis, M. E. *PLoS One* 2013; 8:e57944.

32 Bobe, R. and Carvalho, S. *J. Hum. Evol.* 2019; 126:91–105; Harmand, S. & others. *Nature* 2015; 521:310; Department of Agriculture, Turkana County Government, Kenya. https://www.turkana.go.ke/index.php/ministry-of-pastoral-economies-fisheries/department-of-agriculture. Accessed 07/08/2020.

33 Olff, H. & others. *Nature* 2002; 415:901-904; Ripple, W. J. & others. *Science Advances* 2015; 1:e1400103.

## *3. Deluge – Miocene*

1 Audra, P. & others. *Geodinamica Acta* 2004; 17:389–400; Fauquette, S. & others. *Palaeo3* 2006; 238:281–301; Mao, K. S. & others. *New Phytologist* 2010; 188:254–72; Young, R. A. 'Pre-Colorado River drainage in western Grand Canyon: Potential influence on Miocene stratigraphy in Grand Wash Trough'. In: Reheis,

M. C. and others, eds. *Late Cenozoic Drainage History of the Southwestern Great Basin and Lower Colorado River Region: Geologic and Biotic Perspectives*: The Geological Society of America; 2008. Pp 319–33.

2 Cita, M.B. 'The Messinian Salinity Crisis in the Mediterranean'. In: Briegel, U. and Xiao, W., eds. *Paradoxes in Geology*: Elsevier; 2001. Pp. 353–60.

3 Hou, Z. G. and Li, S. Q. *Biological Reviews* 2018; 93:874–96.

4 Hsü, K. J. 'The desiccated deep basin model for the Messinian events'. In: Drooger, C.W, ed. *Messinian Events in the Mediterranean*. Amsterdam: Noord-Halland Publ. Co.; 1973. Pp. 60–67; Madof, A. S. & others. *Geology* 2019; 47:171–74; Popov, S. V. & others. *Palaeo3* 2006; 238:91–106; Wang, F. X. and Polcher, J. *Sci. Reports* 2019; 9:8024.

5 Barber, P. M. *Marine Geology* 1981; 44:253–72; Cita, M. B. 'The Messinian Salinity Crisis in the Mediterranean'. In: Briegel, U. and Xiao, W., eds. *Paradoxes in Geology*: Elsevier; 2001. Pp. 353–60; El Fadli, K. I. & others. *Bull. Am. Meteorological Soc.* 2013; 94:199–204; Haq, B. U. & others. *Global and Planetary Change* 2020; 184:103052; Kontakiotis, G. & others. *Palaeo3* 2019; 534; Murphy, L. N. & others. *Palaeo3* 2009; 279:41–59; Natalicchio, M. & others. *Organic Geochemistry* 2017; 113:242–53.

6 Anzidei, M. & others. 'Coastal structure, sea-level changes and vertical motion of the land in the Mediterranean'. In: Martini, I. P. and Wanless, H. R., eds. *Sedimentary Coastal Zones from High to Low Latitudes: Similarities and Differences*. Volume 388. London: Geological Society of London Special Publications; 2014; Dobson, M. and Wright, A. *J. Biogeog.* 2000; 27:417–24; Meulenkamp, J. E. & others. *Tectonophysics* 1994; 234:53–72.

7 Fauquette, S. & others. *Palaeo3* 2006; 238:281–301; Freudenthal, M. and Martín-Suárez, E. *Comptes Rendus Palevol* 2010; 9:95–100.

8 Kleyheeg, E. and van Leeuwen, C. H. A. *Aquatic Botany* 2015; 127:1–5; Meijer, H. J. M. *Comptes Rendus Palevol* 2014; 13:19–26; Pavia, M. & others. *Royal Society Open Science* 2017; 4:160722.

9 Mas, G. & others. *Geology* 2018; 46:527–30; van der Geer, A. & others. *Gargano. Evolution of Island Mammals: Adaptation and Extinction of Placental Mammals on Islands*, 1st edition: Blackwell Publishing Ltd; 2010. pp 62–79; Willemsen, G. F. *Scripta Geologica* 1983; 72:1–9.

10 Kotrschal, K. & others. 'Making the best of a bad situation: homosociality in male greylag geese'. In: Sommer, V. and Vasey, P. L., eds. *Homosexual Behaviour in Animals: An Evolutionary Perspective*. Cambridge, UK: Cambridge University Press; 2006. pp 45–76; Meijer, H. J. M. *Comptes Rendus Palevol* 2014; 13:19–26; Pavia, M. & others. *Royal Society Open Science* 2017; 4:160722.

11 Alcover, J. A and McMinn, M. *Bioscience* 1994; 44:12–18; Ballmann, P. *Scripta Geologica* 1973; 17:1–75; Brathwaite, D. H. *Notornis* 1992; 39:239–47; Wehi, P. M. & others. *Human Ecology* 2018; 46:461–70.

12  Guthrie, R. D. *J. Mamm.* 1971; 52:209–212; Mazza, P. P. A. and Rustioni, M. *Zoo. J. Linn. Soc.* 2011; 163:1304–333.

13  Bazely, D. R. *Trends in Ecology & Evolution* 1989; 4:155–56; Wang, Y. & others. *Science* 2019; 364:1153.

14  Mazza, P. P. A. *Geobios* 2013; 46:33–42; Patton, T. H. and Taylor, B. E. *Bull. Am. Mus. Nat. Hist.* 1971; 145:119–218.

15  Jaksić, F. M. and Braker, H. E. *Can. J. Zool.* 1983; 61:2230–2241; Leinders, J. J. M. *Scripta Geologica* 1983; 70:1–68; Mazza, P. & others. *Palaeontographica Abteilung A* 2016; 307:105–147.

16  Freudenthal, M. *Scripta Geologica* 1971; 3:1–10.

17  Van Hinsbergen, D. J. J. & others. *Gondwana Research* 2020; 81:79–229.

18  Angelone, C. and Čermák, S. *Palaeontologische Zeitschrift* 2015; 89:1023–38; Ballmann, P. *Scripta Geologica* 1973; 17:1–75; Delfino, M. & others. *Zoo. J. Linn. Soc.* 2007; 149:293–307; Mazza, P. *Bull. Palaeont. Soc.* Italy 1987; 26:233–43; Moncunill-Solé, B. & others. *Geobios* 2018; 51:359–66.

19  Benton, M. J. & others. *Palaeo3* 2010; 293:438–54; Itescu, Y. & others. *Global Ecology and Biogeography* 2014; 23:689—700; Lomolino, M. V. *J. Biogeog.* 2005; 32:1683–99; Marra, A. C. *Quaternary International* 2005; 129:5–14; Meiri, S. & others. *Proc. R. Soc. B* 2008; 275:141–48; Mitchell, K. J. & others. *Science* 2014; 344:898–900; Nopcsa, F. *Verhandlungen der zoologische-botanischen Gesellschaft.* Volume 54. Vienna 1914. Pp. 12–14; van Valen, L. M. *Evolutionary Theory* 1973; 1:31–49; Worthy, T. H. & others. *Biology Letters* 2019; 15:20190467.

20  Alcover, J. A. & others. *Biol. J. Linn. Soc.* 1999; 66:57–74; Bover, P. & others. *Geological Magazine* 2010; 147:871–85; Köhler, M. & others. *PNAS* 2009; 106:20354–58; Kurakina, I. O. & others. *Chemistry of Natural Compounds* 1969; 5:337–39; Quintana, J. & others. *J. Vert. Paleo.* 2011; 31:231–40; Welker, F. & others. *Quaternary Research* 2014; 81:106–16; Winkler, D. E. & others. *Mammalian Biology* 2013; 78:430–37.

21  Caro, T. *Phil. Trans. R. Soc. B* 2009; 364:537–48; Freudenthal, M. *Scripta Geol.* 1972; 14:1–19; Nowak, R. M. *Walker's Mammals of the World I.* 5th ed. Baltimore, Maryland: Johns Hopkins University Press; 1991. Pp. 1–162; Wilson, D. E. and Reeder, D. M. *Mammal Species of the World. A Taxonomic and Geographic Reference.* Baltimore, Maryland, USA: Johns Hopkins University Press; 2005.

22  Abril, J. M. and Periáñez, R. *Marine Geology* 2016; 382:242–56; Balanyá, J. C. & others. *Tectonics* 2007; 26:TC2005; Garcia-Castellanos, D. & others. *Nature* 2009; 462:778-U. &96; Pliny the Elder. *Natural History.* Volume II855.

23  Garcia-Castellanos, D. & others. *Nature* 2009; 462:778–U96; Micallef, A. & others. *Sci. Reports* 2018; 8:1078.

24  Marra, A. C. *Quaternary International* 2005; 129:5–14; Northcote, E.M. *Ibis* 1982; 124:148–58.

25 Ermakhanov, Z. K. & others. *Lakes & Reservoirs* 2012; 17:3–9; Hammer, U. T. *Saline Lake Ecosystems of the World.* Springer Netherlands; 1986; Lehmann, P. N. *American Historical Review* 2016; 121:70–100; O'Hara S. L. & others. *Lancet* 2000; 355:627–8; Rögl, F. and Steininger, F. F. 'Neogene Paratethys, Mediterranean and Indopacific Seaways'. In: Brenchley, P., ed. *Fossils and Climate.* London: Wiley and Sons; 1984. Pp. 171–200; Walthan, T. and Sholji, I. *Geology Today* 2002; 17:218–24; Yechieli, Y. *Ground Water* 2000; 38:615–623; Yoshida, M. *Geology* 2016; 44:755–8.

26 Billi, A. & others. *Geosphere* 2007; 3:1–15.

27 Black, T. *Ecology of an island mouse, Apodemus sylvaticus hirtensis:* University of Edinburgh; 2013; Bover, P. & others. *Holocene* 2016; 26:1887–91; Kidjo, N. & others. *Bioacoustics* 2008; 18:159–81; Vigne, J. D. *Mammal Review* 1992; 22:87–96; Vigne, J. D. 'Preliminary results on the exploitation of animal resources in Corsica during the Preneolithic'. In: Balmuth, M. S. and Tykot, R. H., eds. *Sardinian and Aegean Chronology.* Oxford, UK: Oxbow Books; 1998. Pp. 57–62.

## 4. Homeland – Oligocene

1 Diester-Haass, L. and Zahn, R. *Geology* 1996; 24:163–6; Flynn, J. J. & others. *Palaeo3* 2003; 195:229–59; Kedves, M. *Acta Bot. Acad. Sci. Hung.* 1971; 17:371–8; Kohn, M. J. & others. *Palaeo3* 2015; 435:24–37; Liu, Z. & others. *Science* 2009; 323:1187–90; Prasad, V. & others. *Science* 2005; 310:1177–80; Sarmiento, G. *Boletín Geológico Ingeominas* 1992; 32; Strömberg, C. A. E. *Annual Review of Earth and Planetary Sciences,* Vol. 39 2011; 39:517–44.

2 Croft, D. A. & others. *Arquivos do Museu Nacional* 2008; 66:191–211; Folguera, A. and Ramos, V. A. *J. South American Earth Sciences* 2011; 32:531–46; Lockley, M. & others. *Cretaceous Research* 2002; 23:383–400.

3 Houston, J. and Hartley, A. J. *Int. J. Climatol.* 2003; 23:1453–64; Mattison, L. and Phillips, I. D. *Scottish Geographical Journal* 2016; 132:21–41; Nanzyo, M. & others. 'Physical characteristics of volcanic ash soils'. In: Shoji, S. and others, eds. *Volcanic Ash Soils, Genesis, Properties, and Utilization.* Tokyo: Elsevier; 1993. Pp. 189–207; Williams, M. A. J. 'Cenozoic climate changes in deserts: a synthesis'. In: Abrahams, A. D. and Parsons, A. J., eds. *Geomorphology of Desert Environments.* London: Chapman and Hall; 1994. Pp. 644–70.

4 Hernández-Hernández, T. & others. *New Phytologist* 2014; 202:1382–97.

5 Croft, D. A. & others. *Fieldiana* 2003; 1527:1–38; Hester, A. J. & others. *Forestry* 2000; 73:381–91; McKenna, M. C. & others. *Am. Mus. Nov.* 2006; 3536:1–18; Milchunas, D. G. & others. *American Naturalist* 1988; 132:87–106; Scanlon, T. M. &

others. *Advances in Water Resources* 2005; 28:291–302; Simpson, G. G. *South American Mammals.* In: Fittkau, J. J., editor. *Biogeography and Ecology in South America.* The Hague: Dr. W. Junk N.V; 1969. Pp. 879–909.

6　De Muizon, C. & others. *J. Vert. Paleo.* 2003; 23:886–94; De Muizon, C. & others. *J. Vert. Paleo.* 2004; 24:398–410; Delsuc, F. & others. *Curr. Biol.* 2019; 29:2031; McKenna, M. C. & others. *Am. Mus.* Nov. 2006; 3536:1–18; Patiño, S. & others. *Hist. Biol.* 2019, DOI: 10.1080/08912963.2019.1664504; Urbani, B. and Bosque, C. *Mammalian Biology* 2007; 72:321–29.

7　Croft D. A. & others. *Annual Review of Earth and Planetary Sciences* 2020; 48:259–90; Hautier L. & others. *J. Mamm. Evol.* 2018; 25:507–23.

8　Barry, R. E. and Shoshani, J. *Mammalian Species* 2000; 645:1-7; Croft, D. A. *Evolutionary Ecology* Research 2006; 8:1193-1214; Croft, D. A. *Horned Armadillos and Rafting Monkeys: The Fascinating Fossil Mammals of South America. Bloomington and Indianapolis:* Indiana University Press; 2016; Flynn, J. J. & others. *Palaeo3* 2003; 195:229–59.

9　Croft D. A. & others. *Annual Review of Earth and Planetary Sciences* 2020; 48:259–90; Winemiller, K. O. & others. *Ecology Letters* 2015; 18:737–51.

10　Rose, K.D. & others. 'Xenarthra and Pholidota'. In: Rose, K. D. and Archibald, J. D., eds. *The Rise of Placental Mammals: Origins and Relationships of the Major Extant Clades.* Baltimore, USA: Johns Hopkins University Press; 2005. Pp. 106–26.

11　Costa, E. & others. *Palaeo3* 2011; 301:97–107; Köhler, M. and Moyà-Solà, S. *PNAS* 1999; 96:14664–7.

12　Guerrero, E. L. & others. *Rodriguésia* 2018; 69.

13　Bond, M. & others. *Nature* 2015; 520:538; Martin, T. *Paleobiology* 1994; 20:5–13.

14　Capobianco, A. and Friedman M. *Biological Reviews* 2019; 94:662–99; Chakrabarty, P. & others. *PLoS One* 2012; 7:e44083; Martin, C. H. and Turner, B. J. *Proc. R. Soc. B* 2018; 285:20172436; Pyron, R. A. *Syst. Biol.* 2014; 63:779–97; Richetti, P. C. & others. *Tectonophysics* 2018; 747:79–98.

15　Bertrand, O. C. & others. *Am. Mus.* Nov. 2012; 3750:1–36.

16　Linder, H. P. & others. *Biological Reviews* 2018; 93:1125–44.

17　Cully, A. C. & others. *Conservation Biology* 2003; 17:990–98; Hooftman, D. A. P. & others. *Basic and Applied Ecology* 2006; 7:507–19; Pereyra, P. J. *Conservation Biology* 2020; 34:373–7; Preston, C. D. & others. *Bot. J. Linn. Soc.* 2004; 145:257–94; Thomas, C. D. and Palmer, G. *PNAS* 2015; 112:4387–92; van de Wiel, C. C. M. & others. *Plant Genetic Resources* 2010; 8:171–81; Wildlife and Countryside Act. Parliament of the United Kingdom 1981.

18　Ameghino, F. *Anales del Museo Nacional* (Buenos Aires) 1907; 9:107–242; Benton, M. J. *Palaeontology* 2015; 58:1003–29; Gaudry, A. *Bulletin de la Société Géologique de France* 1891; 19:1024–35; Podgorny, I. *Science in Context* 2005; 18:249–83; Vilhena, D. A. and Smith, A. B. *PLoS One* 2013; 8:e74470.

19  Hochadel, O. *Studies in Ethnicity and Nationalism* 2015; 15:389–410; McPherson, A. *State Geosymbols: Geological Symbols of the 50 United States*. Bloomington: AuthorHouse; 2011; Rowland, S. M. 'Thomas Jefferson, extinction, and the evolving view of Earth history in the late eighteenth and early nineteenth centuries'. In: Rosenberg, G. D., ed. *The Revolution in Geology from the Renaissance to the Enlightenment*: Geological Society of America Memoir 2009; 203: Pp. 225–46.

20  McKenna, M. C. & others. *Am. Mus.* Nov. 2006; 3536:1–18; Waitt, R. B. *Bulletin of Volcanology* 1989; 52:138–57.

21  Flynn, J. J. & others. *Palaeo3* 2003; 195:229–59; Travouillon, K. J. and Legendre, S. *Palaeo3* 2009; 272:69–84.

22  Barton, H. & others. *J. Archaeological Science* 2018; 99:99–111; Lucas, P. W. & others. *Annales Zoologici Fennici* 2014; 51:143-52; Massey, F. P. & others. *Oecologia* 2007; 152:677-683; Massey, F. P. & others. *Basic and Applied Ecology* 2009; 10:622–30; Rudall, P. J. & others. *Botanical Review* 2014; 80:59–71; Veits, M. & others. *Ecology Letters* 2019; 22:1483–92.

23  McHorse, B. K. & others. *Integrative and Comparative Biol.* 2019; 59:638–55; Mihlbachler, M. C. & others. *Science* 2011; 331:1178–81; Saarinen, J. *The Palaeontology of Browsing and Grazing*. In: Gordon, I. J. and Prins H. H. T., eds. *The Ecology of Browsing and Grazing II*. Cham: Springer Nature Switzerland; 2019. Pp. 5–59; Tapaltsyan, V. & others. *Cell Reports* 2015; 11:673–80.

24  Bacon, C. D. & others. *PNAS* 2015; 112:6110–15; Woodburne, M. O. *J. Mamm. Evol.* 2010; 17:245–64.

25  Barnosky, A. D. and Lindsey, E. L. *Quaternary International* 2010; 217:10–29; Barnosky, A. D. & others. *PNAS* 2016; 113:856–61; Frank, H. T. & others. *Revista Brasileira de Paleontologia* 2015; 18:273–84; MacPhee, R. D. E. & others. *Am. Mus. Nov.* 1999; 3261:1–20; McKenna, M. C. and Bell, S. K. *Classification of Mammals Above the Species Level*. New York Columbia University Press; 1997; Vizcaíno, S. F. & others. *Acta Palaeontologica Polonica* 2001; 46:289–301.

26  MacPhee, R. & others. *Society of Vertebrate Palaeontology 74th Annual Meeting*. Berlin, Germany 2014; Welker, F. & others. *Nature* 2015; 522:81–4.

27  Bai, B. & others. *Communications Biology* 2018; 1; Osborn, H. F. *Bull. Am. Mus. Nat. Hist.* 1898; 10:159–65; Rose, K. D. & others. *Nature Communications* 2014; 5.

# 5. Cycles – Eocene

1  Bowman, V. C. & others. *Palaeo3* 2014; 408:26–47; Case, J. A. *Geological Society of America Memoirs* 1988; 169:523–30; Doktor, M. & others. *Acta Palaeontologica*

*Polonica* 1996; 55:127–46; Marenssi, S. A. & others. *Sedimentary Geology* 2002; 150:301–21; Poole, I. & others. *Annals of Botany* 2001; 88:33–54; Poole, I. & others. *Palaeo3* 2005; 222:95–121; Pujana, R. R. & others. *Review of Palaeobotany and Palynology* 2014; 200:122–37; Seddon, P. J. and Davis, L. S. Condor 1989; 91:653-659; Tatur, A. and Keck, A. *Proceedings of the NIPR Symposium on Polar Biology* 1990; 3:133–50; Zinsmeister, W. B. and Camacho, H. H. 'Late Eocene (to possibly earliest Oligocene) molluscan fauna of the La Meseta Formation of Seymour Island, Antarctic Peninsula'. In: Craddock, C., ed. *Antarctic Geoscience*. Madison, Wisconsin: University of Wisconsin Press; 1982. Pp. 299–304.

2  Buffo, J. & others. *USDA Forest Service Research Paper* 1972; 142:1–74.

3  Wyatt, B. M. & others. J. *Astrophysics and Astronomy* 2018; 39:0026.

4  Fricke, HC. & others. *Earth and Planetary Science Letters* 1998; 160:193–208; Frieling, J. & others. *Paleoceanography and Paleoclimatology* 2019; 34:546–66; Gehler, A. & others. *PNAS* 2016; 113:7739–44; Gingerich, P. D. *Paleoceanography and Paleoclimatology* 2019; 34:329–35; Higgins, J. A. and Schrag D. P. *Earth and Planetary Science Letters* 2006; 245:523–37; Storey, M. & others. *Science* 2007; 316:587–9; Zachos, J. C. & others. *Science* 2003; 302:1551–4.

5  D'Ambrosia, A. R. & others. *Science Advances* 2017; 3:e1601430; Hooker, J. J. and Collinson, M. E. *Austrian J. Earth Sciences* 2012; 105:17–28; Porter, W. P. and Kearney, M. *PNAS* 2009; 106:19666–72; Shukla, A. & others. *Palaeo3* 2014; 412:187–98; Sluijs, A. & others. *Nature* 2006; 441:610–13; Zachos, J. C. & others. *Science* 2005; 308:1611–15.

6  Bijl, P. K. & others. *PNAS* 2013; 110:9645–50; Dutton, A. L. & others. *Paleoceanography* 2002; 17:6-1-6-13.

7  Slack, K. E. & others. *Mol. Biol. Evo.* 2006; 23:1144–55; Tambussi, C. P. & others. *Geobios* 2005; 38:667–75.

8  Acosta Hospitaleche, C. *Comptes Rendus Palevol* 2014; 13:555–60; Davis, S. N. & others. *PeerJ* 2020; 8; Jadwiszczak, P. *Polish Polar Research* 2006; 27:3–62; Levins, R. *Evolution in Changing Environments: Some Theoretical Explorations*. Princeton, New Jersey: Princeton University Press; 1968.

9  Acosta Hospitaleche, C. & others. *Lethaia* 2020; 53:409–20; Dzik, J. and Gaździcki, A. *Palaeo3* 2001; 172:297–312; Jadwiszczak, P. and Gaździcki, A. *Antarctic Science* 2014; 26:279–80; Reguero, M. A. & others. *Rev. Peru. Biol.* 2012; 19:275–84; Schwarzhans, W. & others. *J. Systematic Palaeontology* 2017; 15:147–70.

10  Reguero, M. A. & others. *Rev. Peru. Biol.* 2012; 19:275–84; Scher, H. D. & others. *Science* 2006; 312:428–30.

11  Randall, D. *An Introduction to the Global Circulation of the Atmosphere*. Princeton: Princeton University Press; 2015.

12  Acosta Hospitaleche, C. and Reguero, M. J. *South American Earth Sciences* 2020; 99; Bourdon, E. *Naturwissenschaften* 2005; 92:586–91; Ivany, L. C. & others.

*Bull. Geol. Soc. Am.* 2008; 120:659–78; Jadwiszczak P. & others. *Antarctic Science* 2008; 20:413–14; Ksepka, D. T. *PNAS* 2014; 111:10624–9; Louchart, A. & others. *PLoS One* 2013; 8:e80372; Phillips, G. C. *Survival Value of the White Coloration of Gulls and Other Sea Birds*: Oxford University, UK; 1962.

13 Ksepka, D. T. *PNAS* 2014; 111:10624–9; Mackley, E. K. & others. *Marine Ecology Progress Series* 2010; 406:291–303.

14 Reguero, M. A. & others. *Rev. Peru. Biol.* 2012; 19:275–84; Wueringer, B. E. & others. *PLoS One* 2012; 7:e41605; Wueringer, B. E. & others. *Curr. Biol.* 2012; 22:R150–R151.

15 Buono, M. R. & others. *Ameghiniana* 2016; 53:296–315; Gingerich, P. D. & others. *Science* 1983; 220:403–6; Nummela, S. & others. *J. Vert. Paleo.* 2006; 26:746–59.

16 Ekdale, E. G. and Racicot, R. A. *J. Anatomy* 2015; 226:22–39; Park, T. & others. *Proc. R. Soc. B* 2017; 284:20171836.

17 Bond, M. & others. *Am. Mus. Nov.* 2011; 3718:1–16; Mörs, T. & others. *Sci. Reports* 2020; 10:5051.

18 Reguero, M. A. & others. *Palaeo3* 2002; 179:189–210; Reguero M. A. & others. *Global and Planetary Change* 2014; 123:400–413.

19 Gelfo, J. N. *Ameghiniana* 2016; 53:316–32; Gelfo, J. N. & others. *Antarctic Science* 2017; 29:445–55.

20 Amico, G. and Aizen, M. A. *Nature* 2000; 408:929–30; Goin, F. J. & others. *Revista de la Asociación Geológica Argentina* 2007; 62:597–603; Goin, F. J. & others. *J. Mamm. Evol.* 2020; 27:17–36; Muñoz-Pedreros, A. & others. *Gayana* 2005; 69:225–33; Springer, M. S. & others. *Proc. R. Soc. B* 1998; 265:2381–6.

21 Tambussi, C. P. & others. *Polish Polar Research* 1994; 15:15-20; Torres, C. R. and Clarke, J. A. *Proc. R. Soc. B* 2018; 285:20181540.

22 Alvarenga, H. M. F. & others. *Pap. Avulsos Zool.* 2003; 43:55–91; Bertelli, S. & others. *J. Vert. Paleo.* 2007; 27:409–19; Mazzetta, G. V. & others. *J. Vert. Paleo.* 2009; 29:822–30; Tambussi, C. and Acosta Hospitaleche, C. *Revista de la Asociación Geológica Argentina* 2007; 62:604–17; Worthy, T. H. & others. *Royal Society Open Science* 2017; 4:170975.

23 Degrange, F. J. & others. *International Congress on Vertebrate Morphology* 2016. Volume 299. Washington, DC, USA, 29 Jun–03 Jul 2016. P. 224.

24 Arendt, J. *Chronobiology International* 2012; 29:379–94; Geiser, F. *Clinical and Experimental Pharmacology and Physiology* 1998; 25:736–9; Grenvald, J. C. & others. *Polar Biology* 2016; 39:1879–95; Peri, P. L. & others. *Forest Ecology and Management* 2008; 255:2502–11; Williams, C. T. & others. *Physiology* 2015; 30:86–96.

25 Goin, F. J. & others. *Geological Society of London Special Publications* 2006; 258:135–44; Krause, D. W. & others. *Nature* 2014; 515:512; Krause D. W. & others. *Nature* 2020; 581:421–7; Monks, A. and Kelly D. *Austral Ecology* 2006; 31:366–75.

26  Case, J. A. *Geological Society of London Special Publications* 2006; 258:177–86.

27  Goldner, A. & others. *Nature* 2014; 511:574; Ivany, L. C. & others. *Geology* 2006; 34:377–80; Kennedy, A. T. & others. *Phil. Trans. R. Soc. A* 2015; 373:20150092; Zachos, J. C. and Kump, L. R. *Global and Planetary Change* 2005; 47:51–66.

28  Burckle, L. H and Pokras, E. M. *Antarctic Science* 1991; 3:389–403; Holderegger, R. & others. *Arctic Antarctic and Alpine Research* 2003; 35:214–17; Peat, H. J. & others. *J. Biogeog.* 2007; 34:132–46; Veblen, T. T. & others. *The Ecology and Biogeography of* Nothofagus *forests*. New Haven and London: Yale University Press; 1996; Zitterbart, D. P. & others. *Antarctic Science* 2014; 26:563–64.

29  Bonadonna, F. & others. *Proc. R. Soc. B* 2005; 272:489–95.

## 6. *Rebirth – Paleocene*

1   Alvarez, L. W. & others. *Science* 1980; 208:1095–108; Arthur, M. A. & others. *Cretaceous Research* 1987; 8:43–54; Byrnes, J. S. & others. *Science Advances* 2018; 4:eaao2994; Chiarenza, A. A. & others. *PNAS* 2020; 117:17084–93; Collins, G. S. & others. *Nature Communications* 2020; 11:1480; DePalma, R. A. & others. *PNAS* 2019; 116:8190–99; Goto, K. & others. 'Deep sea tsunami deposits in the Proto-Caribbean Sea at the Cretaceous/Tertiary Boundary'. In: Shiki, T. and others, eds. *Tsunamites*: Elsevier; 2008. Pp. 251–75; Jablonski, D. and Chaloner, W. G. *Trans. R. Soc. B* 1994; 344:11–16; Kaiho, K. & others. *Sci. Reports* 2016; 6:28427; Morgan J. & others. *Nature* 1997; 390:472–6; Sanford J. C. & others. *J. Geophysical Research-Solid Earth* 2016; 121:1240–61; Tyrrell, T. & others. *PNAS* 2015; 112:6556–61; Vajda, V. and McLoughlin S. *Science* 2004; 303:1489; Vajda, V. & others. *Science* 2001; 294:1700–1702; Vellekoop J. & others. *PNAS* 2014; 111:7537–41; Witts, J. D. & others. *Cretaceous Research* 2018; 91:147–67.

2   Alvarez, L. W. & others. *Science* 1980; 208:1095–108; Field, D. J. & others. *Curr. Biol.* 2018; 28:1825; Harrell, T. L. and Martin, J. E. *Netherlands J. Geosciences* 2015; 94:23–37; Henderson, M. D. and Petterson, J. E. *J. Vert. Paleo.* 2006; 26:192–5; Kaiho, K. and Oshima, N. *Sci. Reports* 2017; 7:14855; Robinson, L. N. and Honey, J. G. *PALAIOS* 1987; 2:87–90; Schimper, W. D. *Traité de paléontologie végétale*. Paris: Ballière; 1874; Swisher III, C. C. & others. *Can. J. Earth Sci.* 1993; 30:1981–96; Weishampel, D. B. & others. 'Dinosaur Distribution'. In: Weishampel, D. B. and others, eds. *The Dinosauria*. 2nd ed.: University of California Press; 2004. Pp. 517–606; Wilf, P. and Johnson, K. R. *Paleobiology* 2004; 30:347–68; Wilson, G. P. 2014; 503:365–92.

3   Smith, S. M. & others. *Bull. Geol. Soc. Am.* 2018; 130:2000–2014; Wells, H. G. *A Short History of the World*. New York: The MacMillan and Company; 1922.

4 Berry, K. *Rocky Mountain Geology* 2017; 52:1–16; Diemer, J. A. and Belt E. S. *Sedimentary Geology* 1991; 75:85–108; Fastovsky, D. E. *PALAIOS* 1987; 2:282–95; Fastovsky. D. E. and Bercovici, A. *Cretaceous Research* 2016; 57:368–90; Robertson, D. S. & others. *J. Geophysical Research* 2013; 118:329–36; Russell, D. A. & others. *Geological Society of America Special Paper* 361; 2002. Pp. 169–76; Slattery, J. S. & others. *Wyoming Geological Association Guidebook 2015*; 2015:22–60.

5 Correa, A. M. S. and Baker, A. C. *Global Change Biology* 2011; 17:68–75; Harries, P. J. & others. *Biotic Recovery from Mass Extinction Events 1996*:41–60; Jolley, D. W. & others. *J. Geol. Soc.* 2013; 170:477–82; Lehtonen, S. & others. *Sci. Reports* 2017; 7:4831; Vajda, V. and Bercovici, A. *Global and Planetary Change* 2014; 122:29–49; Walker, K. R. and Alberstadt, L. P. *Paleobiology* 1975; 1:238–57.

6 Johnson, K. R. *Geological Society of America Special Papers* 361; 2002. Pp. 329–91.

7 Arakaki, M. & others. *PNAS* 2011; 108:8379-8384; Ivey, C. T. and DeSilva, N. *Biotropica* 2001; 33:188–91; Malhado, A. C. M. & others. 2012; 44:728–37.

8 Bush, R. T. and McInerney, F. A. *Geochimica Et Cosmochimica Acta* 2013; 117:161–79; Lichtfouse, E. & others. *Organic Geochemistry* 1994; 22:349–51; Tipple, B. J. & others. *PNAS* 2013; 110:2659–64.

9 Simpson, G. G. *J. Mamm.* 1933; 14:97–107; Wilson, G. P. & others. *Nature* 2012; 483:457–60.

10 Ameghino, F. *Revista Argentina de Historia Natural* 1891; 1:289–328; Bonaparte, J. F. & others. *Evolutionary Monographs* 1990; 14:1–61; Fox, R. C. & others. *Nature* 1992; 358:233–35; Rich, T. H. & others. *Alcheringa* 2016; 40:475–501; Wible, J. R. and Rougier, G. W. *Annals of Carnegie Museum* 2017; 84:183–252.

11 Behrensmeyer, A. K. & others. *Paleobiology* 2000; 26:103–47; Grossnickle, D. M. & others. *Trends in Ecology & Evolution* 2019; 34:936–49; Trueman, C. N. . *Palaeontology* 2013; 56:475–86.

12 Friedman, M. *Proc. R. Soc. B* 2010; 277:1675–83; Grossnickle, D. M. and Newham, E. *Proc. R. Soc. B* 2016; 283:20160256; Wilson, G. P. & others. *Nature Communications* 2016; 7:13734.

13 Dos Reis, M. & others. *Biology Letters* 2014; 10:20131003; Goswami, A & others. *PNAS* 2011; 108:16333–338; Halliday, T. J. D. & others. *Proc. R. Soc. B* 2016; 283:20153026; O'Leary M. A. & others. *Science* 2013; 339:662–7; Prasad, G. V. R. and Goswami, A. *12th Symposium on Mesozoic Terrestrial Ecosystems* 2015. Pp. 75–7; Wible, J. R. & others. *Bull. Am. Mus. Nat. Hist.* 2009; 327:1–123.

14 Halliday, T. J. D. & others. *Biological Reviews* 2017; 92:521–50; Halliday, T. J. D. & others. *Proc. R. Soc. B* 2016; 283:20153026.

15 Lindqvist, C. and Rajora, O. P. *Paleogenomics: Genome-Scale Analysis of Ancient DNA*. Cham, Switzerland: Springer Nature; 2019.

16 Archibald, J. D. 'Archaic ungulates ("Condylarthra")'. In: Janis, C. M. and others, eds. *Evolution of Tertiary Mammals of North America. Terrestrial Carnivores,*

*Ungulates, and Ungulate-like Mammals*. Cambridge, UK: Cambridge University Press; 1998. Pp. 292–331; De Bast, E. and Smith, T. *J. Vert. Paleo.* 2013; 33:964–76.

17 Emerling, C. A. & others. *Science Advances* 2018; 4:eaar6478.

18 Barbosa-Filho, J. M. & others. 'Alkaloids of the Menispermaceae'. In: Cordell, G. A., ed. *The Alkaloids: Chemistry and Biology*. Volume 54: Elsevier; 2000. Pp. 1–190; Clemens, W. A. *PaleoBios* 2017; 34:1–26; Field, D. J. & others. *Curr. Biol.* 2018; 28:1825; Johnson, K. R. *Geological Society of America Special Papers* 361; 2002. Pp. 329–91; Parris, D. C. and Hope, S. *Proceedings of the 5th Symposium of the Society of Avian Paleontology and Evolution* 2002:113–24.

19 Anderson, A. O. and Allred, D. M. *The Great Basin Naturalist* 1964; 24:93–101; Botha-Brink, J. & others. *Sci. Reports* 2016; 6:24053; Robertson, D. S. & others. *Bull. Geol. Soc. Am.* 2004; 116:760–68.

20 Holroyd, P. A. & others. *Geological Society of America Special Paper* 503; 2014. Pp. 299–312; Milner, A. C. *Geological Society Special Publications* 140; 1998. Pp. 247–57; O'Connor, P. M. & others. *Nature* 2010; 466:748–51; Turner, A. H. and Sertich, J. J. W. *J. Vert. Paleo.* 2010; 30:177–236; Young, M. T. & others. *Zoo. J. Linn. Soc.* 2010; 158:801–59.

21 Bryant, L. J. *Non-dinosaurian lower vertebrates across the Cretaceous-Tertiary Boundary in Northeastern Montana*. Berkeley: University of California Press; 1989; Katsura, Y. *Paleoenvironment and taphonomy of the fauna of the Tullock Formation (early Paleocene), McGuire Creek area, McCone County, Montana*. Bozeman: Montana State University; 1992; Keller, G. & others. *Palaeo3* 2002; 178:257–97; Puértolas-Pascual, E. & others. *Cretaceous Research* 2016; 57:565–90; Wilson, G. P. & others. *Geological Society of America Special Paper* 503; 2014. Pp. 271–97.

22 Johnson, K. R. *Geological Society of America Special Papers* 361; 2002. Pp. 329–91; Lofgren, D. L. *The Bug Creek problem and the Cretaceous-Tertiary transition at McGuire Creek, Montana*. Berkeley, California: University of California Press; 1995; Shelley, S. L. & others. *PLoS One* 2018; 13:e0200132; Wilson, M. V. H. *Quaestiones Entomologicae* 1978; 14:13–34.

23 Donovan, M. P. & others. *PLoS One* 2014; 9:e103542.

24 Labandeira, C. C. & others. *Geological Society of America Special Paper* 361; 2002. Pp. 297–327.

25 Crossley-Holland, K. *The Penguin Book of Norse Myths: Gods of the Vikings*. London: Penguin Books Ltd; 1993.; van Valen, L. M. *Evolutionary Theory* 1978; 4:45–80.

26 Carroll, R. L. *Vertebrate Paleontology and Evolution*. New York, USA: W.H. Freeman and Company; 1988; Hostetter, C. F. *Mythlore* 1991; 3:5–10; van Valen, L. M. *Evolutionary Theory* 1978; 4:45–80.

27 Cooke, R. S. C. & others. *Nature Communications* 2019; 10.

28 Halliday, T. J. D. and Goswami, A. *Biol. J. Linn. Soc.* 2016; 118:152–68; Halliday, T. J. D. & others. *Biological Reviews* 2017; 92:521–50; Puechmaille, S. J. &

others. *Nature* Communications 2011; 2; Smith, F. A. & others. *Science* 2010; 330:1216–19.

29  Coxall, H. K. & others. *Geology* 2006; 34:297–300; Dashzeveg, D. and Russell, D. E. *Geobios* 1992; 25:647–50; Storer, J. E. *Can. J. Earth Sci.* 1993; 30:1613–17.

30  Koenen, E. J. M. & others. *Syst. Biol.* 2020; 70:508–26; Lowery, C. M. & others. *Nature* 2018; 558:288; Lyson, T. R. & others. *Science* 2019; 366:977–83.

## 7. Signals – Cretaceous

1  Hone, D. W. E. and Henderson, D. M. *Palaeo3* 2014; 394:89–98; Henderson, D. M. *J. Vert. Paleo.* 2010; 30:768–85; Lü, J. *Memoir of the Fukui Prefecture Dinosaur Museum* 2003; 2:153–60; Lü, J. & others. *Acta Geologica Sinica* 2005; 79:766–9; Martill, D. M. & others. *Cretaceous Research* 2006; 27:603–10; Modesto, S. P. and Anderson, J. S. *Syst. Biol.* 2004; 53:815–21.

2  Chen, P. J. & others. *Science in China Series D* 2005; 48:298–312; Fricke, H. C. & others. *Nature* 2011; 480:513–15; Wang, X. R. & others. *Acta Geologica Sinica* 2007; 81:911–16.

3  Falkingham, P. L. & others. *PLoS One* 2014; 9:e93247; Mallison, H. 'Rearing Giants: Kinetic-dynamic modeling of sauropod bipedal and tripedal poses'. In: Klein, N. and others, eds. *Biology of the Sauropod Dinosaurs*. Indianapolis: Indiana University Press; 2011. Pp. 237–50; Taylor M. P. & others. *Acta Palaeontologica Polonica* 2009; 54:213–20.

4  Cerda, I. A. and Powell, J. E. *Acta Palaeontologica Polonica* 2010; 55:389–98; Gallina, P. A. & others. *Sci. Reports* 2019; 9:1392; Gill, F. L. & others. *Palaeontology* 2018; 61:647–58; Twyman, H. & others. *Proc. R. Soc. B* 2016; 283:20161208; Wedel, M. J. *Paleobiology* 2003; 29:243–55; Wedel, M. J. *J. Exp. Zool. A* 2009; 311A:611–28.

5  Chen, P. J. & others. *Science in China Series D* 2005; 48:298–312; Xing, L. D. & others. *Lethaia* 2012; 45:500–506.

6  Gu, J. J. & others. *PNAS* 2012; 109:3868–73; Heads, S. W. and Leuzinger, L. *Zookeys* 2011; 77:17–30; Li, J. J. & others. *Mitochondrial DNA Part A* 2019; 30:385–96; Moyle, R. G. & others. *Nature Communications* 2016; 7:12709; Wang, B. & others. *J. Systematic Palaeontology* 2014; 12:565-574; Wang, H. & others. *Cretaceous Research* 2018; 89:148–53.

7  Frederiksen, N. O. *Geoscience and Man* 1972; 4:17–28; Hethke, M. & others. *International J. Earth Sciences* 2013; 102:351–78; Labandeira, C. C. *Annals of the Missouri Botanical Garden* 2010; 97:469–513; Wu, S. Q. *Palaeoworld* 1999; 11:7–57; Yang, Y. & others. *American J. Botany* 2005; 92:231–41.

8  Dilcher, D. L. & others. *PNAS* 2007; 104:9370–74; Eriksson, O. & others. *International J. Plant Sci.* 2000; 161:319–29; Friis, E.M. & others. *Nature*; 410:357–60; Gomez, B. & others. *PNAS* 2015; 112:10985–8; Ji, Q. & others. *Acta Geologica Sinica* 2004; 78:883–96.

9  Chinsamy, A. & others. *Nature Communications* 2013; 4; Hou, L. H. & others. *Chinese Science Bulletin* 1995; 40:1545-1551; Ji, S. & others. *Acta Geologica Sinica* 2007; 81:8–15; Xing, L. D. & others. *J. Palaeogeography* 2018; 7:13.

10  Chen, P. J. & others. *Science in China Series D* 2005; 48:298–312; Hedrick, A. V. *Proc. R. Soc. B* 2000; 267:671–5; Igaune, K. & others. *J. Avian Biology* 2008; 39:229–32; Yuan, W. & others. *Naturwissenschaften* 2000; 87:417–20.

11  Chen, P. J. & others. *Science in China Series D* 2005; 48:298–312; Clarke, J. A. & others. *Nature* 2016; 538:502–5; Habib, M. B. *Zitteliana* 2008;B28:159–66; Kojima, T. & others. *PLoS One* 2019; 14:e0223447; Senter, P. *Hist. Biol.* 2008; 20:255–87; Vinther, J. & others. *Curr. Biol.* 2016; 26:2456–62; Woodruff, D. C. & others. *Hist. Biol.* 2020: DOI: 10.1080/08912963.2020.1731806; Xu, X. & others. *Nature* 2012; 484:92–5.

12  Bestwick, J. & others. *Biological Reviews* 2018; 93:2021–48; Lü, J. C. & others. *Acta Geologica Sinica* 2012; 86:287–93; Pan, H. Z. and Zhu, X. G. *Cretaceous Research* 2007; 28:215–24; Tong, H. Y. & others. *Am. Mus.* Nov. 2004; 3438:1–20; Zhou, Z. H. & others. *Can. J. Earth Sci.* 2005; 42:1331–8.

13  Gao, T. P. & others. *J. Systematic Palaeontology* 2019; 17:379–91; Li, L. F. & others. *Systematic Entomology* 2018; 43:810-842; Zhang, J. F. *Cretaceous Research* 2012; 36:1–5.

14  Schuler, W. and Hesse, E. *Behavioral Ecology and Sociobiology* 1985; 16:249–55.

15  Lautenschlager, S. *Proc. R. Soc. B* 2014; 281:20140497; Xu, X. & others. *PNAS* 2009; 106:832–4.

16  McNamara, M. E. & others. *Nature Communications* 2018; 9:2072.

17  Nel, A. and Delfosse, E. *Acta Palaeontologica Polonica* 2011; 56:429–32; Shang, L. J. & others. *European J. Entomology* 2011; 108:677–85; Wang, M. M. & others. *PLoS One* 2014; 9:e91290; Wang, Y. J. & others. *PNAS* 2010; 107:16212–215.

18  De Bona, S. & others. *Proc. R. Soc. B* 2015; 282:20150202; Dong, R. *Acta Zootaxonomica Sinica* 2003; 28:105–9.

19  Pérez-de la Fuente, R. & others. *Palaeontology* 2019; 62:547–59; Wang, B. & others. *Science Advances* 2016; 2:e1501918.

20  Hu, Y. M. & others. *Nature* 1997; 390:137–42; Hurum, J. H. & others. *Acta Palaeontologica Polonica* 2006; 51:1–11; Smithwick, F. M. & others. *Curr. Biol.* 2017; 27:3337; Wong, E. S. W. & others. *PLoS One* 2013; 8:e79092.

21  Li, J. L. & others. *Chinese Science Bulletin* 2001; 46:782–6; Xu, X. and Norell, M. A. *Nature* 2004; 431:838–41; Hu, Y. M. & others. *Nature* 2005; 433:149–52.

22 Angielczyk, K. D. and Schmitz, L. *Proc. R. Soc. B* 2014; 281:20141642; Cerda, I. A. and Powell, J. E. *Acta Palaeontologica Polonica* 2010; 55:389–98; Schmitz, L. and Motani, R. *Science* 2011; 332:705–7.

23 Arrese, C. A. & others. *Curr. Biol.* 2002; 12:657–60; Hunt, D. M. & others. *Vision Research* 1998; 38:3299–306; Onishi, A. & others. *Nature* 1999; 402:139–40.

24 Evans, S. E. and Wang, Y. *J. Systematic Palaeontology* 2010; 8:81–95.

25 Evans, S. E. & others. *Senckenbergiana Lethaea* 2007; 87:109–18; Hechenleitner, E. M. & others. *Palaeontology* 2016; 59:433–46; Norell, M. A. & others. *Nature* 2020; 583:406–10; Rogers, K. C. & others. *Science* 2016; 352:450–53; Sander, P. M. & others. *Palaeontographica Abteilung A* 2008; 284:69–107; Vila, B. & others. *Lethaia* 2010; 43:197–208; Wilson, J. A. & others. *PLoS Biol.* 2010; 8:e1000322.

26 Amiot, R. & others. *Palaeontology* 2017; 60:633–47; Ji, Q. & others. *Nature* 1998; 393:753–61; Moreno, J. and Osorno, J. L. *Ecology Letters* 2003; 6:803–6; Wiemann, J. & others. *PeerJ* 2017; 5; Wiemann, J. & others. *Nature* 2018; 563:555; Yang, T.R. & others. *Acta Palaeontologica Polonica* 2019; 64:581–596.

27 Yang, Y. and Ferguson, D. K. *Perspectives in Plant Ecology Evolution and Systematics* 2015; 17:331–46.

28 Jiang, B. Y. & others. *Sedimentary Geology* 2012; 257:31–44.

29 Zhang, X. L. and Sha, J. G. *Cretaceous Research* 2012; 36:96–105.

30 Wu, C. E. *Journey to the West* (tr. Jenner, W. J. F.). Beijing: Collinson Fair; 1955.

## 8. Foundation – Jurassic

1 Bennett, S. C. *J. Paleontology* 1995; 69:569–80; Frey, E. and Tischlinger, H. *PLoS One* 2012; 7:e31945; Frey, E. & others. *Geological Society of London Special Publications* 2003; 217:233–66; Hone, D. W. E. and Henderson, D. M. *Palaeo3* 2014; 394:89–98; Upchurch, P. & others. *Hist. Biol.* 2015; 27:696–716; Wellnhofer, P. *Palaeontographica A* 1975; 149:1–30; Witton, M. P. *Geological Society of London Special Publication* 2018; 455:7–23.

2 Arkhangelsky, M S. & others. *Paleontological Journal* 2018; 52:49–57; Lanyon, J. M. and Burgess E. A. *Reproductive Sciences in Animal Conservation* 2014; 753:241–74; Vallarino, O. and Weldon, P. J. *Zoo Biology* 1996; 15:309–14.

3 Davies, J. & others. *Nature Communications* 2017; 8; Foffa, D. & others. *J. Anatomy* 2014; 225:209–19; Foffa, D. & others. *Nature Ecology & Evolution* 2018; 2: 1548–555; Jones, M. E. H. and Cree, A. *Curr. Biol.* 2012; 22:R986–R987; Schweigert, G. & others. *Zitteliana* 2005; B26:87-95; Stubbs, T. L. and Benton, M. J. *Paleobiology* 2016; 42:547–73; Thorne, P. M. & others. *PNAS* 2011; 108:8339–44; Young, M. T. & others. *PLoS One* 2012; 7:e44985.

4 Collini, C. A. *Acta Theodoro-Palatinae* Mannheim 1784; 5 Physicum:58–103; O'Connor, R. *The Earth on Show: Fossils and the Poetics of Popular Science 1802–1856.* Chicago: University of Chicago Press; 2013; Ruxton, G. D. and Johnsen, S. *Proc. R. Soc. B* 2016; 283:20161463; Torrens, H. *British Journal for the History of Science* 1995; 28:257–84.

5 Danise, S. and Holland, S. M. *Palaeontology* 2017; 60:213–32; Scotese, C. R. *Palaeo3* 1991; 87:493–501; Sellwood, B. W. and Valdes, P. J. *Proceedings of the Geologists' Association* 2008; 119:5–17; Vörös, A. and Escarguel, G. *Lethaia* 2020; 53:72–90.

6 Gill, G. A. & others. *Sedimentary Geology* 2004; 166:311–34; Hosseinpour, M. & others. *International Geology Review* 2016; 58:1616–45; Korte, C. & others. *Nature Communications* 2015; 6; Maffione, M. and van Hinsbergen, D. J. J. *Tectonics* 2018; 37:858–87; Scotese, C. R. *Palaeo3* 1991; 87:493–501.

7 Armstrong, H. A. & others. *Paleoceanography* 2016; 31:1041–53; Korte, C. & others. *Nature Communications* 2015; 6.

8 Morton, N. *Episodes* 2012; 35:328–32.

9 Ereskovsky, A. V. and Dondua, A. K. *Zoologischer Anzeiger* 2006; 245:65–76; Lavrov, A. I. and Kosevich, I. A. *Russ. J. Dev. Biol.* 2014; 45:205–23; Leinfelder, R. R. 'Jurassic Reef Ecosystems'. In: Stanley, G. D., ed. *The History and Sedimentology of Ancient Reef Systems.* Boston, MA, USA: Springer; 2001; Ludeman, D. A. & others. *BMC Evol. Biol.* 2014; 14; Reitner, J. and Mehl, D. *Geol. Palaeont.* 1995. Mitt. Innsbruck: Helfried Mostler Festschrift; 335–47.

10 Leys, S. P. *Integrative and Comparative Biology* 2003; 43:19-27; Leys S. P. & others. *Advances in Marine Biology,* Vol. 52 2007; 52:1–145; Müller, W. E. G. & others. *Chemistry of Materials* 2008; 20:4703–11.

11 Colombié, C. & others. *Global and Planetary Change* 2018; 170:126–45; Leinfelder, R. R. 'Jurassic Reef Ecosystems'. In: Stanley, G. D., ed. *The History and Sedimentology of Ancient Reef Systems.* Boston, MA, USA: Springer; 2001.

12 Tompkins-MacDonald, G. J. and Leys. S. P. *Marine Biology* 2008; 154:973–84; Vogel, S. *PNAS* 1977; 74:2069–71; Yahel, G. & others. *Limnology and Oceanography* 2007; 52:428–40.

13 Krautter, M. & others. *Facies* 2001; 44:265-282; Pisera, A. *Palaeontologia Polonica* 1997; 57:3–216.

14 Brunetti, M. & others. *J. Palaeogeography-English* 2015; 4:371–83; Krautter, M. & others. *Facies* 2001; 44:265–82; Leinfelder, R. R. 'Jurassic Reef Ecosystems'. In: Stanley, G. D., ed. *The History and Sedimentology of Ancient Reef Systems.* Boston, MA, USA: Springer; 2001.

15 Dommergues, J. L. & others. *Paleobiology* 2002; 28:423–34; Landois, H. *Jahresb. Des Westfälischen Provinzial-Vereins für Wissenschaft und Kunst* 1895; 23:99–108.

16 Inoue, S. and Kondo, S. *Sci. Reports* 2016; 6:33489; Lukeneder, A. and Lukeneder, S. *Acta Palaeontologica Polonica* 2014; 59:663–80; Stahl, W. and Jordan,

R. *Earth and Planetary Science Letters* 1969; 6:173; Ward, P. *Paleobiology* 1979; 5:415–22.

17  Kastens, K. A. and Cita, M. B. *Bull. Geol. Soc. Am.* 1981; 92:845–57; Schweigert, G. & others. *Zitteliana* 2005; B26:87–95; Solé, M. & others. *Biology Open* 2018; 7:bio033860; Zhang, Y. & others. *Integrated Zoology* 2015; 10:141–51.

18  Allain, R. *J. Vert. Paleo.* 2005; 25:850–58; Mazin, J. M. & others. *Geobios* 2016; 49:211–28; Meyer, C. A. and Thüring, B. *Comptes Rendus Palevol* 2003; 2:103–17; Moreau, J. D. & others. *Bulletin de la Société Géologique de France* 2016; 187:121–7; Owen R. *Rep. Brit. Ass. Adv. Sci* 1842; 11:32–7; Wellnhofer, P. *Palaeontographica A* 1975; 149:1–30; Witton, M. P. *Zitteliana* 2008; 28:143–59.

19  Elliott, G. F. *Geology Today* 1986; Jan-Feb: 20–23; Schweigert, G. and Dietl, G. *Jb. Mitt. Oberrhein Geol. Ver. NF* 2003; 85:473–83; Schweigert, G. & others. *Zitteliana* 2005; B26:87–95; Uhl, D. & others. *Palaeobiodiversity and Palaeoenvironments* 2012; 92:329–41.

20  Mazin, J. M. and Pouech, P. *Geobios* 2020; 58:39–53; Unwin, D. M. *Geological Society of London Special Publications* 2003; 217:139–90.

21  Bennett, S. C. . *Neues Jahrbuch für Geologie und Palaontologie-Abhandlungen* 2013; 267:23–41.

22  Bennett, S. C. *J. Paleontology* 1995; 69:569–80; Bennett, S. C. *J. Vert. Paleo.* 1996; 16:432–44; Bennett, S. C. *J. Paleontology* 2018; 92:254–71; Black, R. 'A Flock of Flaplings'. *Laelaps: Scientific American*; 2017; Lü, J. C. & others. *Science* 2011; 331:321–4; Prondvai, E. & others. *PLoS One* 2012; 7:e31392; Unwin, D. and Deeming, C. *Proc. R. Soc. B* 2019; 286:20190409.

23  Frey, E. and Tischlinger, H. *PLoS One* 2012; 7 e31945; Hoffmann, R. & others. *Sci. Reports* 2020; 10:1230.

24  Briggs, D. E. G. & others. *Proc. R. Soc. B* 2005; 272:627–32; Klug, C. & others. *Lethaia* 2010; 43:445–56; Mazin, J. M. and Pouech, P. *Geobios* 2020; 58:39–53; Mazin, J. M. & others. *Proc. R. Soc. B* 2009; 276:3881–6.

25  Hoffmann, R. & others. *J. Geol. Soc.* 2020; 177:82–102; Knaust, D. and Hoffmann, R. *Papers in Palaeontology* 2020; https://doi.org/10.1002/spp2.1311; Mehl, J. *Jahresberichte der Wetterauischen Gesellschaft für Naturkunde* 1978; 85–9; Schweigert, G. *Berliner Paläobiologische Abhandlungen* 2009; 10:321–30; Vallon, L. *New Mexico Museum of Natural History and Science Bulletin* 2012; 57:131–5.

26  Baumiller, T. K. *Annual Review of Earth and Planetary Sciences* 2008; 36:221–49; Macurda, D. B. and Meyer, D. L. *Nature* 1974; 247:394–6; Matzke, A. T. and Maisch M. W. *Neues Jahrbuch für Geologie und Palaontologie-Abhandlungen* 2019; 291:89–107.

27  Thiel, M and Gutow, L. 'The Ecology of Rafting in the Marine Environment I: The Floating Substrata'. In: Gibson, R. N. and others, eds. *Oceanography and Marine Biology: An Annual Review*. Volume 42. London: CRC Press; 2004. P. 432.

28 Hunter, A. W. & others. *Royal Society Open Science* 2020; 7:200142; McGaw, I. J. and Twitchit, T. A. *Comparative Biochemistry and Physiology A* 2012; 161:287–95; Robin, N. & others. *Palaeontology* 2018; 61:905–18; Seilacher, A. and Hauff, R. B. *PALAIOS* 2004; 19:3–16.

29 Camerini, J. R. *Isis* 1993; 84:700-727; Hunter, A. W. & others. *Paleontological Research* 2011; 15:12–22; Philippe, M. & others. *Review of Palaeobotany and Palynology* 2006; 142:15–32.

## 9. Contingency – Triassic

1 Levis, C. & others. *Science* 2017; 355:925; Lloyd, G. T. & others. *Biology Letters* 2016; 12:20160609; Moisan, P. & others. *Review of Palaeobotany and Palynology* 2012; 187:29–37; Shcherbakov, D. E. *Alavesia* 2008; 2:113–24; Voigt, S. & others. *Terrestrial Conservation Lagerstätten* 2017; 65–104.

2 Li, H. T. & others. *Nature Plants* 2019; 5:461–70; Pole, M. & others. *Palaeo3* 2016; 464:97–109.

3 Biffin, E. & others. *Proc. R. Soc. B* 2012; 279:341–8; Dobruskina, I. A. *Bulletin of the New Mexico Museum of Natural History and Science* 1995; 5:1–49.

4 Dobruskina, I. A. *Bulletin of the New Mexico Museum of Natural History and Science* 1995; 5:1–49; Fedorenko, O. A. and Miletenko, N. V. *Atlas of Lithology-Paleogeographical, Structural, Palinspastic, and Geoenvironmental Maps of Central Eurasia*. Almaty: YUGGEO; 2002; Marler, T. E. *Plant Signaling and Behavior* 2012; 7:1484–7; Moisan, P. and Voigt S. *Review of Palaeobotany and Palynology* 2013; 192:42–64; Shcherbakov, D. E. *Alavesia* 2008; 2:113–24; Shcherbakov, D. E. *Alavesia* 2008; 2:125–31; Voigt, S. & others. *Terrestrial Conservation Lagerstätten* 2017; 65–104.

5 Burtman, V. S. *Russian J. Earth Sciences* 2008; 10:ES1006; Dobruskina, I. A. *Bulletin of the New Mexico Museum of Natural History and Science* 1995; 5:1–49; Konopelko, D. & others. *Lithos* 2018; 302:405–20; Moisan, P. & others. *Review of Palaeobotany and Palynology* 2012; 187:29–37; Nevolko P. A. & others. *Ore Geology Reviews* 2019; 105:551–71; Shcherbakov, D. E. *Alavesia* 2008; 2:113–124.

6 Dyke, G. J. & others. *J. Evol. Biol.* 2006; 19:1040–43; Ericsson, L. E. *J. Aircraft* 1999; 36:349–56; Gans, C. & others. *Paleobiology* 1987; 13:415–26; Sharov, A. G. *Akad. Nauk. SSSR. Trudy Paleont. Inst.* 1971; 130:104–13.

7 Dzik, J. and Sulej, T. *Acta Palaeontologica Polonica* 2016; 61:805–23.

8 Butler, R. J. & others. *Biology Letters* 2009; 5:557–60; Chatterjee, S. and Templin, R. J. *PNAS* 2007; 104:1576–80; Fraser, N. C. & others. *J. Vert. Paleo.* 2007; 27:261–5; Simmons, N. B. & others. *Nature* 2008; 451:818–U6; Xu, X. & others.

*Nature* 2015; 521:70–U131; Zhou, Z. H. and Zhang, F. C. *PNAS* 2005; 102:18998–19002.

9  Bi, S. D. & others. *Nature* 2014; 514:579; King, B. and Beck, R. M. D. *Proc. R. Soc. B* 2020; 287:20200943; Lucas, S. G. and Luo, Z. J. *Vert. Paleo.* 1993; 13:309–34; Luo, Z. X. *Nature* 2007; 450:1011–19; Ruta, M. & others. *Proc. R. Soc. B* 2013; 280:20131865.

10  Bajdek, P. & others. *Lethaia* 2016; 49:455–77; Bown, T. M. and Kraus, M. J. 'Origin of the tribosphenic molar and metatherian and eutherian dental formulae'. In: Lillegraven, J. A. and others, eds. *Mesozoic Mammals: The First Two-Thirds of Mammalian History.* Berkeley: University of California Press; 1979. Pp. 172–81; Chudinov, P. K. 'The skin covering of therapsids'. In: Flerov, K. K., ed. *Data on the Evolution of Terrestrial Vertebrates.* Moscow: Nauka; 1970. Pp. 45–50; Maier, W. & others. *J. Zoological Systematics and Evolutionary Research* 1996; 34:9–19; Oftedal, O. T. *Journal of Mammary Gland Biology and Neoplasia* 2002; 7:225–52; Oftedal, O. T. *Journal of Mammary Gland Biology and Neoplasia* 2002; 7:253–66; Tatarinov, L. P. *Paleontological Journal* 2005; 39:192–8.

11  De Ricqles, A. & others. *Annales de Paléontologie* 2008; 94:57–76; Foth, C. & others. *BMC Evol. Biol.* 2016; 16.

12  Pritchard, A. C. and Sues, H. D. *J. Syst. Palaeo.* 2019; 17:1525–45; Renesto, S. & others. *Rivista Italiana Di Paleontologia E Stratigrafia* 2018; 124:23–33; Spiekman, S. N. F. & others. *Curr. Biol.* 2020; 30:3889–95; Wild, R. *Schweizerische Paläontologische Abhandlungen* 1973; 95:1–162.

13  Alifanov, V. R. and Kurochkin, E. N. *Paleontological Journal* 2011; 45:639–47; Gonçalves, G. S. and Sidor, C. A. *PaleoBios* 2019; 36:1–10.

14  Buatois, L. A. & others. *The Mesozoic Lacustrine Revolution. Trace-Fossil Record of Major Evolutionary Events, Vol. 2: Mesozoic and Cenozoic* 2016; 40:179–263; Dobruskina, I. A. *Bulletin of the New Mexico Museum of Natural History and Science* 1995; 5:1–49; Voigt, S. and Hoppe, D. *Ichnos* 2010; 17:1–11.

15  Dobruskina, I. A. *Bulletin of the New Mexico Museum of Natural History and Science* 1995; 5:1–49; Moisan P. & others. *Review of Palaeobotany and Palynology* 2012; 187:29–37; Schoch, R. R. & others. *PNAS* 2020; 117:11584–8; Shcherbakov, D. E. *Alavesia* 2008; 2:113–124; Wagner, P. & others. *Paleontological Research* 2018; 22:57–63.

16  Gawin, N. & others. *BMC Evol. Biol.* 2017; 17; Hengherr, S. and Schill, R. O. *J. Insect Physiology* 2011; 57:595–601; Shcherbakov, D. E. *Alavesia* 2008; 2:113–24.

17  Moser, M. and Schoch, R. R. *Palaeontology* 2007; 50:1245–66; Schoch, R. R. & others. *Zoo. J. Linn. Soc.* 2010; 160:515–30; Tatarinov, L. P. *Seymouriamorphen aus der Fauna der UdSSR.* In: Kuhn, O., ed. *Encyclopedia of Paleoherpetology*, Part 5B: Batrachosauria (Anthracosauria) Gephyrostegida-Chroniosuchida. Stuttgart: Gustav Fischer; 1972. P. 80; Voigt, S. & others. *Terrestrial Conservation Lagerstätten* 2017; 65–104; Lemanis, R. & others. *PeerJ Preprints* 2019; 7:e27476v1.

18  Buchwitz, M. and Voigt, S. *J. Vert. Paleo.* 2010; 30:1697–708; Buchwitz, M. & others. *Acta Zoologica* 2012; 93:260–80; Schoch, R. R. & others. *Zoo. J. Linn. Soc.* 2010; 160:515–30.

19  Fischer J. & others. *Paläontologie, Stratigraphie, Fazies* 2007; 15:41–6; Nakaya K. & others. *Sci. Reports* 2020; 10:12280; Vorobyeva, E. I. *Paleontological Journal* 1967; 4:102–1.

20  Fischer, J. & others. *J. Vert. Paleo.* 2011; 31:937–53; Rees, J. and Underwood, C. J. *Palaeontology* 2008; 51:117–47.

21  Kukalovapeck, J. *Can. J. Zool.* 1983; 61:1618–69; Pringle, J. W. S. *Phil. Trans. R. Soc. B* 1948; 233:347; Shcherbakov, D. E. & others. *International J. Dipterological Research* 1995; 6:76–115; Sherman, A. and Dickinson, M. H. *J. Exp. Biol.* 2003; 206:295–302.

22  Béthoux, O. *Arthropod Systematics and Phylogeny* 2007; 65:135–56; Frost, S. W. *Insect Life and Natural History.* New York, USA: Dover Publications; 1959; Gorochov, A. V. *Paleontological Journal* 2003; 37:400–406; Grimaldi, D. and Engel, M. S. *Evolution of the Insects.* Cambridge, UK: Cambridge University Press; 2005; Huang, D. Y. & others. *J. Syst. Palaeo.* 2020; 18:1217–22; Vishnyakova, V. N. *Paleontological Journal* 1998:69–76; Voigt, S. & others. *Terrestrial Conservation Lagerstätten* 2017; 65–104.

23  Buchwitz, M. and Voigt, S. *Palaeontologische Zeitschrift* 2012; 86:313-331; Unwin, D. M. & others. 'Enigmatic small reptiles from the Middle-Late Triassic of Kirgizstan'. In: Benton, M. J. and others, eds. *The Age of Dinosaurs in Russia and Mongolia.* Cambridge, UK: Cambridge University Press; 2000. Pp. 177–86.

24  Alroy, J. *PNAS* 2008; 105:11536–42; Erwin, D. H. *Annual Review of Ecology and Systematics* 1990; 21:69–91; Foth, C. & others. *BMC Evol. Biol.* 2016; 16; Monnet, C. & others. 'Evolutionary trends of Triassic ammonoids'. In: Klug, C. and others, eds. *Ammonoid Paleobiology: From macroevolution to paleogeography.* Dordrecht: Springer. Pp. 25–50.

25  Button, D. J. & others. *Nature Communications* 2017; 8; Halliday, T. J. D. & others. 'Leaving Gondwana: the changing position of the Indian Subcontinent in the global faunal network'. In: Prasad, G. V. and Patnaik, R., eds. *Biological Consequences of Plate Tectonics: New Perspectives on Post-Gondwanan Break-up – A Tribute to Ashok Sahni, Vertebrate Paleobiology and Paleoanthropology.* Switzerland: Springer; 2020. Pp. 227–49.

26  Behrensmeyer, A. K. & others. *Paleobiology* 2000; 26:103–47; Burtman, V. S. *Russian J. Earth Sciences* 2008; 10:ES1006; Padian, K. and Clemens, W. A. 'Terrestrial vertebrate diversity: episodes and insights'. In: Valentine, J., ed. *Phanerozoic Diversity Patterns: Profiles in Macroevolution.* Guildford: Princeton University Press; 1985. Pp. 41–86; Shcherbakov, D. E. *Alavesia* 2008; 2:113–24.

## 10. *Seasons – Permian*

1 Kato, K. M. & others. *Phil. Trans. R. Soc. B* 2020; 375:20190144; Tabor, N. J. & others. *Palaeo3* 2011; 299:200–213; Tsuji, L. A. & others. *J. Vert. Paleo.* 2013; 33:747–63.

2 Bendel, E. M. & others. *PLoS One* 2018; 13:e0207367; Kermack, K. A. *Phil. Trans. R. Soc. B* 1956; 240:95–133; Smiley, T. M. & others. *J. Vert. Paleo.* 2008; 28:543–7; Whitney, M. R. & others. *Jama Oncology* 2017; 3:998–1000.

3 Araujo, R. & others. *PeerJ* 2017; 5; Smith, R. M. H. & others. *Palaeo3* 2015; 440:128–41; Tabor, N. J. & others. *Palaeo3* 2011; 299:200–213.

4 Bernardi, M. & others. *Earth-Science Reviews* 2017; 175:18–43; Blakey, R. C. *Carboniferous-Permian paleogeography of the assembly of Pangaea.* 2003; Utrecht, Netherlands. Pp. 443–56; Scotese, C. R. & others. *J. Geology* 1979; 87:217–77; Tabor, N. J. & others. *J. Vert. Paleo.* 2017; 37:240–53; Vai, G. B. *Palaeo3* 2003; 196:125–55; Wu, G. X. & others. *Annales Geophysicae* 2009; 27:3631–44.

5 Chandler, M. A. & others. *Bull. Geol. Soc. Am.* 1992; 104:543–59; Kutzbach, J. E and Gallimore, R. G. *J. Geophysical Research* 1989; 94:3341–57; Shields, C. A. and Kiehl, J. T. *Palaeo3* 2018; 491:123–36.

6 Smith, R. M. H. & others. *Palaeo3* 2015; 440:128–41.

7 Looy, C. V. & others. *Palaeo3* 2016; 451:210–26.

8 Blob, R. W. *Paleobiology* 2001; 27:14–38; Brink, A. S. and Kitching, J. W. *Palaeontologica Africana* 1953; 1:1–28; Eloff, F. C. *Koedoe* 1973; 16:149–54; Kammerer, C. F. *PeerJ* 2016; 4; Kluever, B. M. & others. *Curr. Zool.* 2017; 63:121–9; Kümmell, S. B. and Frey, E. *PLoS One* 2014; 9:e113911; Smith, R. M. H. & others. *Palaeo3* 2015; 440:128–41.

9 Boitsova, E. A. & others. *Biol. J. Linn. Soc.* 2019; 128:289–310; Tabor, N. J. & others. *Palaeo3* 2011; 299:200–213; Tsuji, L. A. & others. *J. Vert. Paleo.* 2013; 33:747–63; Turner, M. L. & others. *J. Vert. Paleo.* 2015; 35:e994746; Valentini M. & others. *Neues Jahrbuch für Geologie und Palaontologie-Abhandlungen* 2009; 251:71–94.

10 Biewener, A. A. *Science* 1989; 245:45–8; Ford, D. P. and Benson, R. B. *J. Nature Ecology & Evolution* 2020; 4:57; Fuller, P. O. & others. *Zoology* 2011; 114:104–12; Langman, V. A. & others. *J. Exp. Biol.* 1995; 198:629–32; VanBuren, C. S. and Bonnan, M. *PLoS One* 2013; 8:e74842.

11 Cecil, C. B. *International J. Coal Geology* 2013; 119:21–31; Ferner, K. and Mess, A. *Respiratory Physiology & Neurobiology* 2011; 178:39–50; Gervasi, S. S. and Foufopoulos, J. *Functional Ecology* 2008; 22:100–108; Wolkers, W. F. & others. *Comparative Biochemistry and Physiology A* 2002; 131:535–43.

12 Laurin, M. and de Buffrenil, V. *Comptes Rendus Palevol* 2016; 15:115–27; Pyron, R. A. *Syst. Biol.* 2011; 60:466–81.

13 Damiani, R. & others. *J. Vert. Paleo.* 2006; 26:559–72; Liu, N. J. & others. *Zoomorphology* 2016; 135:115–20; Marjanović, D. and Laurin, M. *PeerJ* 2019; 6; Sidor,

C. A. *Comptes Rendus Palevol* 2013; 12:463–72; Sidor, C. A. & others. *Nature* 2005; 434:886–9; Stewart, J. R. 'Morphology and evolution of the egg of oviparous amniotes'. In: Sumida, S. and Martin, K., eds. *Amniote Origins – Completing the Transition to Land*. London: Academic Press; 1997. Pp. 291–326; Steyer, J. S. & others. *J. Vert. Paleo.* 2006; 26:18–28.

14  Brocklehurst, N. *PeerJ* 2017; 5; Hugot, J. P. & others. *Parasites & Vectors* 2014; 7; Modesto, S. P. & others. *J. Vert. Paleo.* 2019; 38:e1531877; O'Keefe, F. R. & others. *J. Vert. Paleo.* 2005; 25:309–19; Reisz, R. R. and Sues, H. D. 'Herbivory in Late Paleozoic and Triassic Terrestrial Vertebrates'. In: Sues, H. D., ed. *Evolution of Herbivory in Terrestrial Vertebrates*. Cambridge, UK: Cambridge University Press; 2000. Pp. 9–41; Watanabe, H. and Tokuda, G. *Cellular and Molecular Life Sciences* 2001; 58:1167–78.

15  LeBlanc, A. R. H. & others. *Sci. Reports* 2018; 8:3328; Smith, R. M. H. & others. *Palaeo3* 2015; 440:128–41.

16  Looy, C. V. & others. *Palaeo3* 2016; 451:210–26; Smith, R. M. H. & others. *Palaeo3* 2015; 440:128–41.

17  Dixon, S. J. and Sear, D. A. *Water Resources* Research 2014; 50:9194–210; Kelley, D. B. & others. *Southeastern Archaeology* 1996; 15:81–102; Watson, J. *East Texas Historical Journal* 1967; 5:104–11.

18  Fröbisch, J. *Early Evolutionary History of the Synapsida* 2014:305–19; Fröbisch, J. and Reisz, R. R. *Proc. R. Soc. B* 2009; 276:3611–18; Sennikov, A. G. and Golubev, V. K. *Paleontological Journal* 2017; 51:600–611.

19  Chandra, S. and Singh, K. J. *Review of Palaeobotany and Palynology* 1992; 75:183–218; Prevec, R. & others. *Review of Palaeobotany and Palynology* 2009; 156:454–93; Tsuji, L. A. & others. *J. Vert. Paleo.* 2013; 33:747–63.

20  Feder, A. & others. *J. Maps* 2018; 14:630–43; Looy, C. V. & others. *Palaeo3* 2016; 451:210–26; Tfwala, C. M. & others. *Agricultural and Forest Meteorology* 2019; 275:296–304.

21  Grasby, S. E. & others. *Nature Geoscience* 2011; 4:104–7.

## 11. *Fuel – Carboniferous*

1  Berner, R. A. & others. *Science* 2007; 316:557–8; Clements, T. & others. *J. Geol. Soc.* 2019; 176:1–11; Phillips, T. L. & others. *International J. Coal Geology* 1985; 5:43; Potter, P. E. and Pryor, W. A. *Geol. Soc. Am.* 1961; 72:1195–249.

2  Andrews, H. N. and Murdy, W. H. *American J. Botany* 1958; 45:552–60; DiMichele, W. A. and DeMaris, P. J. *PALAIOS* 1987; 2:146–57; Evers, R. A. *American J. Botany* 1951; 38:7317; Thomas, B. A. *New Phytologist* 1966; 65:296–303.

3  Baird, G. C. & others. *PALAIOS* 1986; 1:271-285; DiMichele, W. A. and DeMaris, P. J. *PALAIOS* 1987; 2:146–57; Thomas B. A. & others. *Geobios* 2019; 56:31–48.

4  Brown, R. *J. Geol. Soc.* 1848; 4:46–50; Eggert, D. A. and Kanemoto, N. Y. *Botanical Gazette* 1977; 138:102–11; Hetherington, A. J. & others. *PNAS* 2016; 113:6695–700.

5  Banfield, J. F. & others. *PNAS* 1999; 96:3404–11; Davies, N. S. and Gibling, M. R. *Nature Geoscience* 2011; 4:629–33; Gibling, M. R. and Davies, N. S. *Nature Geoscience* 2012; 5:99–105; Gibling, M. R. & others. *Proceedings of the Geologists Association* 2014; 125:524–33; Le Hir, G. & others. *Earth and Planetary Science Letters* 2011; 310:203–12; Pierret, A. & others. *Vadose Zone Journal* 2007; 6:269–81; Quirk, J. & others. *Biology Letters* 2012; 8:1006–11; Song, Z. L. & others. *Botanical Review* 2011; 77:208–13; Ulrich, B. 'Soil acidity and its relations to acid deposition'. In: Ulrich, B., and Pankrath, J., eds. 1982; Göttingen: Springer. Pp. 127–46.

6  Baird, G.C. & others. *PALAIOS* 1986; 1:271–85; Kuecher, G. J. & others. *Sedimentary Geology* 1990; 68:211–21; Phillips, T. L. & others. *International J. Coal Geology* 1985; 5:43; Potter, P. E. and Pryor, W. A. *Geol. Soc. Am.* 1961; 72:1195–249.

7  Armstrong, J. and Armstrong, W. *New Phytologist* 2009; 184:202–15; DiMichele, W. A. and DeMaris, P. J. *PALAIOS* 1987; 2:146–57; DiMichele, W. A. and Phillips, T. L. *Palaeo3* 1994; 106:39–90; Falcon-Lang, H. J. *J. Geol. Soc.* 1999; 156:137–48; Potter, P. E. and Pryor, W. A. *Geol. Soc. Am.* 1961; 72:1195–249.

8  Berner, R. A. & others. *Science* 2007; 316:557–8; Came, R. E. & others. *Nature* 2007; 449:198–U3; Glasspool, I. J. & others. *Frontiers in Plant Science* 2015; 6; He, T. H. and Lamont, B. B. National *Science Review* 2018; 5:237–54; Viegas, D. X. and Simeoni, A. *Fire Technology* 2011; 47:303–20.

9  Fonda, R. W. *Forest Science* 2001; 47:390–96; Keeley, J. E. & others. *Trends in Plant Science* 2011; 16:406–11; Thanos, C. A. and Rundel, P. W. *J. Ecology* 1995; 83:207–16.

10  Béthoux, O. *J. Paleontology* 2009; 83:931–7; Brockmann, H. J. & others. *Animal Behaviour* 2018; 143:177–91; Fisher, D. C. *Mazon Creek Fossils* 1979; 379–447; Mundel, P. *Mazon Creek Fossils* 1979:361–78; Tenchov, Y. G. *Geologia Croatica* 2012; 65:361–6.

11  Aslan, A. and Behrensmeyer, A. K. *PALAIOS* 1996; 11:411–21; Behrensmeyer, A. K. & others. *Paleobiology* 2000; 26:103–47; Clements, T. & others. *J. Geol. Soc.* 2019; 176:1–11; Coombs, W. P. and Deméré, T. A. *J. Paleontology* 1996; 70:311–26; Foster, M. W. *Mazon Creek Fossils* 1979; 191–267; Jablonski, N. G. & others. *Hist. Biol.* 2012; 24:527–36; Kjellesvig-Waering, E. N. *State of Illinois Scientific Papers* 1948; 3:1–48; Mann, A. and Gee, B. M. *J. Vert. Paleo.* 2020:39;e1727490; Pfefferkorn, H. W. *Mazon Creek Fossils* 1979; 129–42; Shabica, C. *Mazon Creek Fossils* 1979; 13–40.

12 Boyce, C. K. and DiMichele, W. A. *Review of Palaeobotany and Palynology* 2016; 227:97–110; DiMichele, W. A. and DeMaris, P. J. *PALAIOS* 1987; 2:146–57; Poorter, L. & others. *J. Ecology* 2005; 93:268–78.

13 Beattie, A. *The Danube: A Cultural History.* Oxford, UK: Oxford University Press; 2010; Castendyk, D. N. & others. *Global and Planetary Change* 2016; 144:213–27; Fagan, W. E. & others. *American Naturalist* 1999; 153:165–82; Harris, L. D. *Conservation Biology* 1988; 2:330–32; McLaughlin, F. A. & others. *J. Geophysical Research* 1996; 101:1183–97; Partch, E. N. and Smith, J. D. *Estuarine and Coastal Marine Science* 1978; 6:3–19.

14 Wedel, M. *J. Morphology* 2007; 268:1147.

15 Clements, T. & others. *Nature* 2016; 532:500; Foster, M. W. *Mazon Creek Fossils* 1979:269–301; Johnson, R. G. and Richardson, E. S. *J. Geology* 1966; 74:626–31; Johnson, R. G. and Richardson, E. S. *Fieldiana Geol* 1969; 12:119–49; Rauhut, O. W. M. & others. *PeerJ* 2018; 6; McCoy, V. E. & others. *Nature* 2016; 532:496.

16 Coad, B. *Encyclopedia of Canadian Fishes.* Waterdown, Ontario: Canadian Museum of Nature: Canadian Sportfishing Productions; 1995; Delamotte, I. and Burkhardt, D. *Naturwissenschaften* 1983; 70:451–61; Herring, P. J. *J. of the Marine Biological Association of the United Kingdom* 2007; 87:829–42; Moser, H. G. 'Morphological and functional aspects of marine fish larvae'. In: Lasker, R., ed. *Marine Fish Larvae: Morphology, Ecology, and Relation to Fisheries.* Washington: Sea Grant Program; 1981. Pp. 90–131; Sallan, L. & others. *Palaeontology* 2017; 60:149–57.

17 Clements, T. & others. *J. Geol. Soc.* 2019; 176:1–11.

18 Cascales-Miñana, B. and Cleal, C. J. *Terra Nova* 2014; 26:195–200; Dunne, E. M. & others. *Proc. R. Soc. B* 2018; 285:20172730; Feulner, G. *PNAS* 2017; 114:11333–7; Nelsen, M. P. & others. *PNAS* 2016; 113:2442–7; Robinson, J. M. *Geology* 1990; 18:607–10; Weng, J. K. and Chapple, C. *New Phytologist* 2010; 187:273–85.

## 12. Collaboration – Devonian

1 Gabrielsen, R. H. & others. *J. Geol. Soc.* 2015; 172:777–91; Hall, A. M. *Trans. R. Soc. Edinburgh – Earth Sciences* 1991; 82:1–26; Miller, S. R. & others. *Earth and Planetary Science Letters* 2013; 369:1–12; Rast N. & others. *Geological Society Special Publications* 1988; 38:111–22.

2 Burg, J. P. and Podladchikov, Y. *International J. Earth Sciences* 1999; 88:190–200; Dewey, J. F. 'The geology of the southern termination of the Caledonides'. In: Nairn, A. E. M. and Stehli, F. G., eds. *The Ocean Basins and Margins: vol 2 The North Atlantic.* Boston, MA, USA: Springer; 1974. Pp. 205–31; Dewey, J. F.

and Kidd, W. S. F. *Geology* 1974; 2:543–6; Fossen, H. & others. *Geology* 2014; 42:791–4; Gee, D. G. & others. *Episodes* 2008; 31:44–51; Hacker, B. R. & others. *Annual Review of Earth and Planetary Sciences* 2015; 43:167–205; Johnson, J. G. & others. *Bull. Geol. Soc. Am.* 1985; 96:567–87; Lehtovaara, J. *Bull. Geol. Soc. Finland* 1989; 61:189–95; Mueller, P. A. & others. *Gondwana Research* 2014; 26:365–73; Nance R. D. & others. *Gondwana Research* 2014; 25:4–29; Pickering, K. T. & others. *Trans. R. Soc. Edinburgh – Earth Sciences* 1988; 79:361–82; Redfern, R. *Origins: The Evolution of Continents, Oceans, and Life*. University of Oklahoma Press; 2001; Stone, P. *Journal of the Open University Geological Society* 2012; 33:29–36; Ziegler, P. A. *CSPG Special Publications*; 1988. Pp. 15–48.

3  Charlesworth, J. K. *Proc. R. Irish Acad. B* 1921; 36:174–314; Chew, D. M. and Strachan, R. A. *New Perspectives on the Caledonides of Scandinavia and Related Areas* 2014; 390:45–91; Lehtovaara, J. J. *Fennia* 1985; 163:365–8; Lehtovaara, J. *Bull. Geol. Soc. Finland* 1989; 61:189–95.

4  Dahl, T. W. & others. *PNAS* 2010; 107:17911–15; Hastie, A. R. & others. *Geology* 2016; 44:855–8.

5  Edwards, D. & others. *Phil. Trans. R. Soc. B* 2018; 373:20160489; Mark, D. F. & others. *Geochimica Et Cosmochimica Acta* 2011; 75:555–69; Trewin, N. H. and Rice, C. M. *Scottish J. Geology* 1992; 28:37–47.

6  Rice, C. M. & others. *J. Geol. Soc.* 2002; 159:203–14; Strullu-Derrien, C. & others. *Curr. Biol.* 2019; 29:461; Wellman, C. H. & others. *Palz* 2019; 93:387–93.

7  Burt, R. M. *The geology of Ben Nevis, south-west Highlands, Scotland*: University of St Andrews; 1994; Moore, I. and Kokelaar, P. *J. Geol. Soc.* 1997; 154:765–8; Rice, C. M. & others. *J. Geol. Soc.* 1995; 152:229–50; Trewin, N. H. *Earth and Environmental Science Transactions of the Royal Society of Edinburgh* 1993; 84:433–42; Trewin, N. H. *Evolution of Hydrothermal Ecosystems on Earth (and Mars?)* 1996; 202:131–49; Trewin, N. H. & others. *Can. J. Earth Sci.* 2003; 40:1697–712.

8  Channing, A. *Phil. Trans. R. Soc. B* 2018; 373:20160490; Wellman, C. H. *Phil. Trans. R. Soc. B* 2018; 373:20160491.

9  Cox, A. & others. *Chemical Geology* 2011; 280:344–51; Gorlenko, V. & others. *Int. J. Syst. Evol. Microbiol.* 2004; 54:739-743; Nugent, P. W. & others. *Applied Optics* 2015; 54:B128–B139; Saiki, T. & others. *Agricultural and Biological Chemistry* 1972; 36:2357–66.

10  Krings, M. and Sergeev, V. N. *Review of Palaeobotany and Palynology* 2019; 268:65–71; Sompong, U. & others. *Fems Microbiology Ecology* 2005; 52:365–76; Sugiura, M. & others. *Microbes and Environments* 2001; 16:255–61.

11  Channing, A. and Edwards, D. *Plant Ecology & Diversity* 2009; 2:111–43; Edgecombe, G. D. & others. *PNAS* 2020; 117:8966–72; Powell, C. L. & others. *Geological Society of London Special Publications* 2000; 180:439–57; Trewin, N. H. *Evolution of Hydrothermal Ecosystems on Earth (and Mars?)* 1996; 202:131–49.

12 Channning, A. and Edwards, D. *Trans. R. Soc. Edinburgh – Earth Sciences* 2004; 94:503–21.

13 Berbee, M. L. and Taylor, J. W. *Mol. Biol. Evol.* 1992; 9:278–84; Harrington, T. C. & others. *Mycologia* 2001; 93:111–36; Honegger, R. & others. *Phil. Trans. R. Soc. B* 2018; 373:20170146; Hueber, F. M. *Review of Palaeobotany and Palynology* 2001; 116:123–58; O'Donnell, K. & others. *Mycologia* 1997; 89:48–65; Retallack, G. J. and Landing E. *Mycologia* 2014; 106:1143–58; Taylor, J. W. & others. *Syst. Biol.* 1993; 42:440–57.

14 Nash, T. H. *Lichen Biology*. Cambridge, UK: Cambridge University Press; 1996.

15 Boyce, C. K. & others. *Geology* 2007; 35:399–402; Hueber, F. M. *Review of Palaeobotany and Palynology* 2001; 116:123–58; Labandeira, C. *Insect Science* 2007; 14:259–75; Retallack, G. J. and Landing E. *Mycologia* 2014; 106:1143–58.

16 Ahmadjian, V. *The Lichen Symbiosis*. New York: John Wiley and Sons; 1993; Friedl, T. *Lichenologist* 1987; 19:183–91; Jones, G. P. *J. Experimental Marine Biology and Ecology* 1992; 159:217–35; Karatygin, I. V. & others. *Paleontological Journal* 2009; 43:107–14; Offenberg, J. *Behavioral Ecology and Sociobiology* 2001; 49:304–10; Rytter, W. and Shik, J. Z. *Animal Behaviour* 2016; 117:179–86; Taylor T. N. & others. *American J. Botany* 1997; 84:992–1004; Schneider, S. A. *The meat-farming ants: predatory mutualism between* Melissotarsus *ants (Hymenoptera: Formicidae) and armored scale insects (Hemiptera: Diaspididae)*. Amherst: UM Amherst; 2016.

17 Edwards, D. S. *Bot. J. Linn. Soc.* 1986; 93:173–204; Remy, W. & others. *PNAS* 1994; 91:11841–3; Schüßler, A. & others. *Mycological Research* 2001; 105:1413–21.

18 Haig, D. *Botanical Review* 2008; 74:395–418.

19 Brown, R. C. and Lemmon, B. E. *New Phytologist* 2011; 190:875–81.

20 Gambardella, R. *Planta* 1987; 172:431–8; Mascarenhas, J. P. *Plant Cell* 1989; 1:657–64; Rosenstiel, T. N. & others. *Nature* 2012; 489:431–3.

21 Remy, W. and Hass, H. *Review of Palaeobotany and Palynology* 1996; 90:175–93.

22 Babikova, Z. & others. *Ecology Letters* 2013; 16:835–43; Daviero-Gomez, V. & others. *International J. Plant Sciences* 2005; 166:319–26.

23 Hetherington, A. J. and Dolan L. *Current Opinion in Plant Biology* 2019; 47:119–26; Kerp, H. & others. *International J. Plant Sciences* 2013; 174:293–308; Roth-Nebelsick, A. & others. *Paleobiology* 2000; 26:405–18; Wilson, J. P. and Fischer, W. W. *Geobiology* 2011; 9:121–30.

24 Ahlberg, P. E. *Zoo. J. Linn. Soc.* 1998; 122:99–141; Smithson T. R. & others. *PNAS* 2012; 109:4532–7; Taylor T. N. & others. *Mycologia* 2004; 96:1403–19.

25 Dunlop, J. A. and Garwood, R. J. *Phil. Trans. R. Soc. B* 2018; 373:20160493; Jezkova, T. and Wiens, J. J. *American Naturalist* 2017; 189:201–12; Wendruff, A. J. & others. *Sci. Reports* 2020; 10:20441; Zhao F. C. & others. *Science China* 2010; 53:1784–99.

26  Davies, W. M. *Quarterly J. Microscopical Science* 1927; 71:15–30; Freitas, L. & others. *J. Evol. Biol.* 2018; 31:1623–31; Whalley, P. and Jarzembowski, E. A. *Nature* 1981; 291:317–17.

27  Kim, H. Y. & others. *Physical Review Fluids* 2017; 2:100505.

28  Claridge, M. F. and Lyon, A. G. *Nature* 1961; 191:1190–91; Dunlop, J. A. and Garwood, R. J. *Phil. Trans. R. Soc. B* 2018; 373:20160493; Dunlop, J. A. & others. *Zoomorphology* 2009; 128:305–13.

29  Fayers, S. R. and Trewin, N. H. *Trans. R. Soc. Edinburgh* 2003; 93:355–82; Scourfield, D. J. *Phil. Trans. R. Soc. B* 1926; 214:153–87; Womack, T. & others. *Palaeo3* 2012; 344:39–48.

30  Kelman, R. & others. *Trans. R. Soc. Edinburgh* 2004; 94:445–55; Strullu-Derrien, C. & others. *PLoS One* 2016; 11:e0167301; Taylor, T. N. & others. *Mycologia* 1992; 84:901–10.

31  Karling, J. S. *American J. Botany* 1928; 15:485–U7; Taylor, T. N. & others. *Nature* 1992; 357:493–4.

32  Kerp, H. & others. 'New data on *Nothia aphylla* Lyon 1964 ex El-Saadawy et Lacey 1979, a poorly known plant from the Lower Devonian Rhynie chert'. In: Gensel, P. G. and Edwards, D., eds. *Plants Invade the Land – Evolutionary and Environmental Perspectives.* New York, NY, USA: Columbia University Press; 2001. Pp. 52–82; Krings, M. & others. *New Phytologist* 2007; 174:648–57; Poinar, G. & others. *Nematology* 2008; 10:9–14; Krings, M. & others. *Plant Signaling and Behaviour* 2007:125–6.

## 13. *Depths – Silurian*

1  Graening, G. O. and Brown, A. V. *J. the American Water Resources Association* 2003; 39:1497–507; Noltie, D. B. and Wicks, C. M. *Environmental Biology of Fishes* 2001; 62:171–194; Ramsey, E. E. *J. Comparative Neurology* 1901; 11:40–47.

2  Broek, H. W. *J. Physical Oceanography* 2005; 35:388–94; del Giorgio, P. A. and Duarte, C. M. *Nature* 2002; 420:379–84; Lee, Z. & others. *J. Geophysical Research-Oceans* 2007; 112:C03009; Lorenzen, C. J. *ICES J. Marine Science* 1972; 34:262–7; Morita, T. *Annals of the New York Academy of Sciences* 2010; 1189:91–4; Saunders, P. M. *J. Physical Oceanography* 1981; 11:573–4.

3  Clough, L. M. & others. *Deep-Sea Research Part II-Topical Studies in Oceanography* 1997; 44:1683–704; Lonsdale, P. *Deep-Sea Research* 1977; 24:857; Scheckenbach, F. & others. *PNAS* 2010; 107:115–20.

4  Bazhenov, M. L. & others. *Gondwana Research* 2012; 22:974–91; Brewer, P. G. and Hester, K. *Oceanography* 2009; 22:86–93; Dziak, R. P. & others.

*Oceanography* 2017; 30:186–97; Filippova, I. B. & others. *Russian J. Earth Sciences* 2001; 3:405–26; Maslennikov, V. V. & others. *The trace element zonation in vent chimneys from the Silurian Yaman-Kasy VHMS deposit in the Southern Ural, Russia: insights from laser ablation inductively coupled plasma mass-spectrometry (LA-ICP-MS).* Eliopoulous, D. G., ed. Netherlands: Millpress; 2003. Pp. 151–4. Ryazantsev, A. V. & others. *Geotectonics* 2016; 50:553-578; Seltmann, R. & others. *J. Asian Earth Sciences* 2014; 79:810–41; Simonov, V. A. & others. *Geology of Ore Deposits* 2006; 48:369–83.

5 Beatty, J. T. & others. *PNAS* 2005; 102:9306–10; Van Dover, C. L. & others. *Geophysical Research Letters* 1996; 23:2049–52.

6 Burle, S. 04/06. Flood Map (www.floodmap.net). Accessed 2020 04/06; Charette, M. A. and Smith, W. H. F. *Oceanography* 2010; 23:112–14; Haq, B. U. and Schutter, S. R. *Science* 2008; 322:64–8.

7 Maslennikov, V. V. & others. *The trace element zonation in vent chimneys from the Silurian Yaman-Kasy VHMS deposit in the Southern Ural, Russia: insights from laser ablation inductively coupled plasma mass-spectrometry (LA-ICP-MS).* Eliopoulous, D. G., ed. Netherlands: Millpress; 2003. Pp. 151–4. 151–4; Zaikov V. V. & others. *Geology of Ore Deposits* 1995; 37:446–63.

8 Georgieva, M. N. & others. *J. Systematic Palaeontology* 2019; 17:287–329; Little, C. T. S. & others. *Palaeontology* 1999; 42:1043–78; Ravaux, J. & others. *Cahiers de Biologie Marine* 1998; 39:325–6; Schulze, A. *Zoologica Scripta* 2003; 32:321–42.

9 Allen, J. F. F. & others. *Trends in Plant Science* 2011; 16:645–55; McFadden, G. I. *Plant Physiology* 2001; 125:50–53; Pfannschmidt, T. *Trends in Plant Science* 2003; 8:33–41; Raven, J. A. and Allen, J. F. *Genome Biology* 2003; 4:209.

10 Breusing, C. & others. *PLoS One* 2020; 15:e0227053; Bright, M. and Sorgo, A. *Invertebrate Biology* 2003; 122:347–68; Cowart, D. A. & others. *PLoS One* 2017; 12:e0172543; Forget, N. L. & others. *Marine Ecology* 2015; 36:35–44; Georgieva, M. N. & others. *Proc. R. Soc. B* 2018; 285:20182004; Miyamoto, N. & others. *PLoS One* 2013; 8:e55151; Zal, F. & others. *Cahiers de Biologie Marine* 2000; 41:413–23.

11 Maslennikov, V. V. & others. *The trace element zonation in vent chimneys from the Silurian Yaman-Kasy VHMS deposit in the Southern Ural, Russia: insights from laser ablation inductively coupled plasma mass-spectrometry (LA-ICP-MS).* Eliopoulous, D. G., ed. Rotterdam, Netherlands: Millpress; 2003. Pp. 151–4. Nakamura, R. & others. *Angewandte Chemie* 2010; 49:7692–4; Novoselov, K. A. & others. *Mineralogy and Petrology* 2006; 87:327–349.

12 Belka, Z .and Berkowski, B. *Acta Geologica Polonica* 2005; 55:1–7; Little, C. T. S. and Vrijenhoek R. C. *Trends in Ecology & Evolution* 2003; 18:582–8.

13 Adams, D. K. & others. *Oceanography* 2012; 25:256–68; Levins, R. *Bull. Entomol. Soc. Am.* 1969; 15:237–40; Sylvan, J. B. & others. *mBio* 2012; 3:e00279-11; Vrijenhoek, R. C. *Molecular Ecology* 2010; 19:4391–411.

14 Finnegan, S. & others. *Proc. R. Soc. B* 2016; 283:20160007; Finnegan, S. & others. *Biology Letters* 2017; 13:20170400; Little, C. T. S. & others. *Palaeontology* 1999; 42:1043–78; Rong, J. Y. and Shen, S. Z. *Palaeo3* 2002; 188:25–38; Sheehan, P. M. and Coorough, P. J. *Palaeozoic Palaeogeography and Biogeography* 1990; 12:181–7; Sutton, M. D. & others. *Nature* 2005; 436:1013–15.

15 Jollivet, D. *Biodiversity and Conservation* 1996; 5:1619–53; Little, C. T. S. & others. *Nature* 1997; 385:146–8; Vrijenhoek, R. C. *Deep-Sea Research* Part II-*Topical Studies in Oceanography* 2013; 92:189–200.

16 Ashford, O. S. & others. *Proc. R. Soc. B* 2018; 285:20180923; Stratmann, T. & others. *Limnology and Oceanography* 2018; 63:2140–53; Tsurumi, M. *Global Ecology and Biogeography* 2003; 12:181–90; Van Dover, C. L. *Biological Bulletin* 1994; 186:134–5.

17 McNichol, J. & others. *PNAS* 2018; 115:6756–61; Nagano, Y. and Nagahama, T. *Fungal Ecology* 2012; 5:463–71; Orcutt, B. N. & others. *Frontiers in Microbiology* 2015; 6.

18 Bonnett, A. *Off the Map: Lost Space, Invisible Cities, Forgotten Islands, Feral Places, and What They Tell Us about the World.* London: Aurum Press; 2014; Jutzeler, M. & others. *Nature Communications* 2014; 5; Maschmeyer, C. H. & others. *Geosciences* 2019; 9:245; Maslennikov, V. V. & others. *The trace element zonation in vent chimneys from the Silurian Yaman-Kasy VHMS deposit in the Southern Ural, Russia: insights from laser ablation inductively coupled plasma mass-spectrometry (LA-ICP-MS).* Eliopoulous, D. G., ed. Millpress, Rotterdam, Netherlands: Millpress; 2003. Pp. 151–4.

19 Lindberg, D. R. *Evolution: Education and Outreach* 2009; 2:191–203; Little, C. T. S. & others. *Palaeontology* 1999; 42:1043–78.

20 Gubanov, A. P. and Peel, J. S. *American Malacological Bulletin* 2000; 15:139–45; Hilgers, L. & others. *Mol. Biol. Evol.* 2018; 35:1638–52.

21 Fara, E. *Geological Journal* 2001; 36:291–303; Lemche, H. *Nature* 1957; 179:413–16; Lindberg, D. R. *Evolution: Education and Outreach* 2009; 2:191–203; Lü, J. & others. *Nature Communications* 2017; 8; Smith J. L. B. *Trans. R. Soc. S. Afr.* 1939; 27:47–50; Zhu, M. and Yu, X. B. *Biology Letters* 2009; 5:372–5.

22 Van Roy, P. & others. *J. Geol. Soc.* 2015; 172:541–9.

23 Faure, G. *Origin of Igneous Rocks: The Isotopic Evidence.* Berlin: Springer; 2001; Folinsbee, R. E. & others. *Geochimica Et Cosmochimica Acta* 1956; 10:60–68; Lancelot, J. & others. *Earth and Planetary Science Letters* 1976; 29:357–66; Larsen, E. S. & others. *Bull. Geol. Soc. Am.* 1952; 63:1045–52.

24 Tomczak, M. and Godfrey, J. S. *Regional Oceanography: an Introduction.* Pergamon; 1994; Webb, P. *Introduction to Oceanography.* Roger Williams University; 2019.

25 Jedlovszky, P. and Vallauri, R. *J. Chemical Physics* 2001; 115:3750–62; Moore, G. T. & others. *Geology* 1993; 21:17–20; Sanchez-Vidal, A. & others. *PLoS One* 2012; 7:e30395.

26  Duval, S. & others. *Interface Focus* 2019; 9:20190063; Lane, N. *Bioessays* 2017; 39:1600217; Lane, N. & others. *BioEssays* 2010; 32:271–80; Martin, W. and Russell, M. J. *Phil. Trans. R. Soc. B* 2007; 362:1887–925.

27  Lipmann, F. *Advances in Enzymology and Related Subjects of Biochemistry* 1941; 1:99–162.

## 14. *Transformation – Ordovician*

1  Blignault, H. J. and Theron, J. N. *S. Afr. J. Geol.* 2010; 113:335–60; Bromwich, D. H. *Bull. Am. Meteorological Soc.* 1989; 70:738–49; Gabbott, S. E. & others. *Geology* 2010; 38:1103–6; Naumann, A. K. & others. *Cryosphere* 2012; 6:729–41; Sansiviero, M. & others. *J. Marine Systems* 2017; 166:4–25.

2  Fountain, A. G. & others. *International J. Climatology* 2010; 30:633–42; Gabbott, S. E. & others. *Geology* 2010; 38:1103–6; Leroux, C. and Fily, M. *J. Geophysical Research – Planets* 1998; 103:25779–88; Smalley, I. J. *J. Sedimentary Research* 1966; 36:669–76.

3  Bindoff, N. L. & others. *Papers and Proceedings of the Royal Society of Tasmania* 2000; 133:51–6; Cordes, E. E. & others. *Oceanography* 2016; 29:30–31; Lappegard, G. & others. *J. Glaciology* 2006; 52:137–48; Parsons, D. R. & others. *Geology* 2010; 38:1063–6; Urbanski, J. A. & others. *Sci. Reports* 2017; 7:43999; Vrbka, L. and Jungwirth, P. *J. Molecular Liquids* 2007; 134:64–70.

4  Blignault, H. J. and Theron, J. N. *S. Afr. J. Geol.* 2010; 113:335–60; Deane, G. B. & others. *Acoustics Today* 2019; 15:12–19; Müller, C. & others. *Science* 2005; 310:1299; Pettit, E. C. & others. *Geophysical Research Letters* 2015; 42:2309–16; Scholander, P. F. and Nutt, D. C. *J. Glaciology* 1960; 3:671–8; Severinghaus, J. P. and Brook, E. J. *Science* 1999; 286:930–34.

5  Leu, E. & others. *Progress in Oceanography* 2015; 139:151–70; Lovejoy, C. & others. *Aquatic Microbial Ecology* 2002; 29:267–78; Moore, G. W. K. & others. *J. Physical Oceanography* 2002; 32:1685–98; Price, P. B. *Science* 1995; 267:1802–4.

6  Bassett, M. G. & others. *J. Paleontology* 2009; 83:614–23; Gabbott, S. E. *Palaeontology* 1998; 41:631–67; Gabbott, S. E. & others. *Geology* 2010; 38:1103–6; Moore, G. W. K. & others. *J. Physical Oceanography* 2002; 32:1685-1698; Smith, R. E. H. & others. *Microbial Ecology* 1989; 17:63–76; von Quillfeldt, C. H. *J. Marine Systems* 1997; 10:211–40.

7  Clarke, A. and North, A. W. 'Is the growth of polar fish limited by temperature?' In: di Prisco, G. and others, eds. *Biology of Antarctic Fish*. Berlin: Springer; 1991. Pp. 54–69; Kim, S. & others. *Integrative and Comparative Biology* 2010; 50:1031–40.

8  Blignault, H. J. and Theron, J. N. *S. Afr. J. Geol.* 2010; 113:335–60; Gabbott, S. E. & others. *J. Geol. Soc.* 2017; 174:1–9; Harper, D. A. T. *Palaeo3* 2006; 232:148–66; Le Heron, D. P. & others. 'The Early Palaeozoic Glacial Deposits of Gondwana: Overview, Chronology, and Controversies'. *Past Glacial Environments*, 2nd edition 2018; 47–73; Pohl, A. & others. *Paleoceanography* 2016; 31:800–821; Rohrssen, M. & others. *Geology* 2013; 41:127–30; Servais, T. & others. *Palaeo3* 2010; 294:99–119; Sheehan, P. M. *Annual Review of Earth and Planetary Sciences* 2001; 29:331–64; Summerhayes, C. P. 'Measuring and Modelling $CO_2$ Back Through Time: $CO_2$, temperature, solar luminosity, and the Ordovician Glaciation'. *Paleoclimatology: From Snowball Earth to the Anthropocene*: John Wiley and Sons Ltd; 2020. Pp. 204–15.

9  Finlay, A. J. & others. *Earth and Planetary Science Letters* 2010; 293:339–48; Ling, M. X. & others. *Solid Earth Sciences* 2019; 4:190–98; Patzkowsky, M. E. & others. *Geology* 1997; 25:911–14; Servais, T. & others. *Palaeo3* 2019; 534; Sheehan, P. M. *Annual Review of Earth and Planetary Sciences* 2001; 29:331–64; Shen, J. H. & others. *Nature Geoscience* 2018; 11:510.

10  Chiarenza, A. A. & others. *PNAS* 2020:1–10; Lindsey, H. A. & others. *Nature* 2013; 494:463–7; Reichow, M. K. & others. *Earth and Planetary Science Letters* 2009; 277:9–20; Zou, C. N. & others. *Geology* 2018; 46:535–8.

11  Gabbott, S. E. *Palaeontology* 1999; 42:123–48; Gabbott, S. E. & others. *Proceedings of the Yorkshire Geological Society* 2001; 53:237–44; Gough, A. J. & others. *J. Glaciology* 2012; 58:38–50; Price, P. B. *Science* 1995; 267:1802–4.

12  Cocks, L. R. M. and Fortey, R. A. *Geological Magazine* 1986; 123:437–44; Gabbott, S. E. *Palaeontology* 1999; 42:123–48; Goudemand, N. & others. *PNAS* 2011; 108:8720–24; Lovejoy, C. & others. *Aquatic Microbial Ecology* 2002; 29:267–78; Price, P. B. *Science* 1995; 267:1802–4; Rohrssen, M. & others. *Geology* 2013; 41:127–30; Whittle, R. J. & others. *Palaeontology* 2009; 52:561–7; Williams, A. & others. *Phil. Trans. R. Soc. B* 1992; 337:83–104.

13  Klug, C. & others. *Lethaia* 2015; 48:267–88; LoDuca, S. T. & others. *Geobiology* 2017; 15:588–616; Rohrssen, M. & others. *Geology* 2013; 41:127–30; Seilacher, A. *Palaeo3* 1968; 4:279.

14  Braddy, S. J. & others. *Palaeontology* 1995; 38:563–81; Braddy, S. J. & others. *Biology Letters* 2008; 4:106–9; Lamsdell, J. C. & others. J. *Systematic Palaeontology* 2010; 8:49–61.

15  Braddy, S. J. & others. *Palaeontology* 1995; 38:563–81; Budd, G. E. *Nature* 2002; 417:271–5; Hughes, C. L. & others. *Evolution & Development* 2002; 4:459–99.

16  Aldridge, R. J. & others. 'The Soom Shale'. In: Briggs, D. E. G. and Crowther, P. R., eds. *Palaeobiology II*: Blackwell Science Ltd; 2001. Pp. 340–42.

17  Aldridge, R. J. & others. *Phil. Trans. R. Soc. B* 1993; 340:405–21; Bergström, S. M. and Ferretti, A. *Lethaia* 2017; 50:424–39; Chernykh, V. V. & others.

*J. Paleontology* 1997; 71:162–4; Ellison, S. P. *AAPG Bulletin* 1946; 30:93–110; Yin, H. F. & others. *Episodes* 2001; 24:102–14.

18  Aldridge, R. J. & others. *Phil. Trans. R. Soc. B* 1993; 340:405–21; George, J. C. and Stevens, E. D. *Environmental Biology of Fishes* 1978; 3:185–91; Nishida, J. and Nishida, T. *British Poultry Science* 1985; 26:105–15; Pridmore, P. A. & others. *Lethaia* 1996; 29:317–28; Suman, S. P. and Joseph, P. *Annual Review of Food Science and Technology*, Vol. 4 2013; 4:79–99.

19  Gabbott, S. E. & others. *Geology* 2010; 38:1103–6.

20  Blignault, H. J. and Theron, J. N. *S. Afr. J. Geol.* 2010; 113:335–60; Clark, J. A. *Geology* 1976; 4:310–12.

21  Allegre, C. J. & others. *Nature* 1984; 307:17–22; Barth, G. A. and Mutter J. C. *J. Geophysical Research-Solid Earth* 1996; 101:17951–75; Chambat, F. and Valette, B. *Physics of the Earth and Planetary Interiors* 2001; 124:237–53; Shennan, I. & others. *J. Quaternary Science* 2006; 21:585–99.

22  Bradley, S. L. & others. *J. Quaternary Science* 2011; 26:541–52; de Geer, G. *Geologiska Föreningen i Stockholm Förhandlingar* 1924; 46:316–24.

23  Ross, J. R. and Ross, C. A. 'Ordovician sea-level fluctuations'. In: Webby, B. D. and Laurie, J. R., eds. *Global Perspectives on Ordovician Geology*. Rotterdam: A. Balkema; 1992. Pp. 327–35; Saupe, E. E. & others. *Nature Geoscience* 2020; 13:65.

24  Saupe, E. E. & others. *Nature Geoscience* 2020; 13:65; Scotese, C. R. & others. *J. African Earth Sciences* 1999; 28:99–114; Smith, R. E. H. & others. *Microbial Ecology* 1989; 17:63–76; Wiens, J. J. & others. *Ecology Letters* 2010; 13:1310–24.

25  Bennett, M. M. and Glasser, N. F. *Glacial Geology: Ice Sheets and Landforms*. 2nd ed. Oxford, UK: John Wiley and Sons; 2009. P. 385; Blignault, H. J. and Theron, J. N. *S. Afr. J. Geol.* 2010; 113:335-360; Blignault, H. J. and Theron, J. N. *S. Afr. J. Geology* 2017; 120:209–22; Goldstein, R. M. & others. *Science* 1993; 262:1525–30; Ragan, D. M. *The J. Geology* 1969; 77:647–67.

## 15. *Consumers – Cambrian*

1  Berner, R. A. and Kothavala, Z. *American J. Science* 2001; 301:182–204; Han, J. & others. *Gondwana Research* 2008; 14:269-276; Hearing, T. W. & others. *Science Advances* 2018; 4:eaar5690; Hou, X. and Bergström, J. *Paleontological Research* 2003; 7:55–70; Labandeira, C. C. *Trends in Ecology & Evolution* 2005; 20:253–62; National Research Council of the United States – Committee on Toxicology. *Carbon Dioxide. Emergency and continuous exposure guidance levels for selected submarine contaminants*. Volume 1. Washington, DC, USA: The National Academies Press; 2007. Pp. 46–66.

2  Daczko, N. R. & others. *Sci. Reports* 2018; 8:8371; Dott, R. H. *Geology* 1974; 2:243–46; Haq, B. U. and Schutter, S. R. *Science* 2008; 322:64–8; Hou, X. and Bergström, J. *Paleontological Research* 2003; 7:55–70.

3  Hou, X. and Bergström, J. *Paleontological Research* 2003; 7:55–70; MacKenzie, L. A. & others. *Palaeo3* 2015; 420:96–115; Peters, S. E. and Loss, D. P. *Geology* 2012; 40:511–14.

4  Bergström, J. & others. *GFF* 2008; 130:189–201; Briggs, D. E. G. *Phil. Trans. R. Soc. B* 1981; 291:541–84; Hou, X. G. & others. *Zoologica Scripta* 1991; 20:395–411; Zhang, X. L. & others. *Alcheringa* 2002; 26:1–8.

5  Chen, A. L. & others. *Palaeoworld* 2015; 24:46–54; Hou, X. G. & others. *Zoologica Scripta* 1991; 20:395–411; Hu, S. X. & others. *Acta Geologica Sinica* 2008; 82:244–8; Huang, D. Y. & others. *Palaeo3* 2014; 398:154–64; Ou, Q. & others. *PNAS* 2017; 114:8835–40; Vannier, J. and Martin, E. L. O. *Palaeo3* 2017; 468:373–87; Zhang, X. G. & others. *Geological Magazine* 2006; 143:743–8; Zhang, Z. F. & others. *Acta Geologica Sinica* 2003; 77:288–93; Zhang, Z. F. & others. *Proc. R. Soc. B* 2010; 277:175–81.

6  Budd, G. E. and Jackson, I. S. C. *Phil. Trans. R. Soc. B* 2016; 371:20150287; Conci, N. & others. *Genome Biology and Evolution* 2019; 11:3068–81; Landing, E. & others. *Geology* 2010; 38:547–50; Ortega-Hernández, J. *Biological Reviews* 2016; 91:255–73; Paterson, J. R. & others. *PNAS* 2019; 116:4394–9; Satoh, N. & others. *Evolution & Development* 2012; 14:56–75.

7  Akam, M. *Cell* 1989; 57:347–9; Akam, M. *Phil. Trans. R. Soc. B* 1995; 349:313–19; Jezkova, T. and Wiens, J. J. *American Naturalist* 2017; 189:201–12.

8  Hughes, N. C. *Integrative and Comparative Biology* 2003; 43:185–206; Parat, A. *Les Grottes de la Cure côte d'Arcy XXI. Bull. Soc. Sci. Hist. & Nat. de l'Yonne* 1903, 1–53; Shu, D. G. & others. *Nature* 1999; 402:42–6; *The Illustrated London News*, September 22nd, 1949, pages 190, 201, 204; Vannier, J. & others. *Sci. Reports* 2019; 9:14941.

9  Dai, T. and Zhang, X. L. *Alcheringa* 2008; 32:465–8; Hou, X. G. & others; *Earth and Environmental Science* Transactions of the Royal Society of Edinburgh 2009; 99:213–23.

10  Bromham, L. and Penny, D. *Nature Reviews Genetics* 2003; 4:216–24.

11  Dos Reis, M. & others. *Curr. Biol.* 2015; 25:2939–50.

12  Káldy, J. & others. *Genes* 2020; 11:753.

13  Erwin, D. H. *Palaeontology* 2007; 50:57–73.

14  Budd, G. E. and Jackson, I. S. C. *Phil. Trans. R. Soc. B* 2016; 371:20150287.

15  Dunne, J. A. & others. *PLoS Biology* 2008; 6:693–708; Penny, A. M. & others. *Science* 2014; 344:1504–6.

16  Läderach, P. & others. *Climatic Change* 2013; 119:841–54; Lagad, R. A. & others. *Analytical Methods* 2013; 5:1604–11; Potrel, A. & others. *J. Geol. Soc.* 1996; 153:507–10; Wooldridge, S. W. and Smetham, D. J. *The Geographical Journal* 1931; 78:243–65;

Wright, J. B. & others. *Geology and Mineral Resources of West Africa.* Netherlands: Springer; 1985; Zhao, F. C. & others. *Geological Magazine* 2015; 152:378–82.

17   Bryson, B. *A Short History of Nearly Everything.* London: Black Swan; 2004; Koren, I. & others. *Environmental Research Letters* 2006; 1:014005.

18   Chen, J. Y. and Zhou, G. Q. *Collection and Research* 1997; 10:11–105; Dunne, J. A. & others. *PLoS Biology* 2008; 6:693–708; Han, J. & others. *Alcheringa* 2006; 30:1–10; Han, J. A. & others. *PALAIOS* 2007; 22:691-694; Hou, X. G. & others. *GFF* 1995; 117:163–83.

19   Baer, A. and Mayer, G. *J. Morphology* 2012; 273:1079–88; Barnes, A. and Daniels, S. R. *Zoologica Scripta* 2019; 48:243–62; Dunne, J. A. & others. *PLoS Biology* 2008; 6:693–708; Morris, S. C. *Palaeontology* 1977; 20:623–40; Hou, X. G. & others. *Zoologica Scripta* 1991; 20:395–411; Ramsköld, L. *Lethaia* 1992; 25:221–4; Smith, M. R. and Ortega-Hernández, J. *Nature* 2014; 514:363.

20   Liu, J. N. & others. *Gondwana Research* 2008; 14:277–83; Smith, M. R. and Caron, J. B. *Nature* 2015; 523:75; Vannier, J. & others. *Nature Communications* 2014; 5.

21   Fenchel, T. *Microbiology UK* 1994; 140:3109–16; Galvão, V. C. and Fankhauser, C. *Current Opinion in Neurobiology* 2015; 34:46–53; Jury, S. H. & others. *J. Experimental Marine Biology and Ecology* 1994; 180:23–37; Magnuson, J. J. & others. *American Zoologist* 1979; 19:331–43; Mollo, E. & others. *Natural Product Reports* 2017; 34:496–513; Murayama, T. & others. *Curr. Biol.* 2013; 23:1007–12; Nordzieke, D. E. & others. *New Phytologist* 2019; 224:1600–1612; Rozhok, A. *Orientation and Navigation in Vertebrates.* Berlin: Springer; 2008.

22   Galvão, V. C. and Fankhauser, C. *Current Opinion in Neurobiology* 2015; 34:46–53; Ma, X. Y. & others. *Arthropod Structure & Development* 2012; 41:495–504.

23   Clarkson, E. N. K. and Levi-Setti, R. *Nature* 1975; 254:663–7; Clarkson, E. & others. *Arthropod Structure & Development* 2006; 35:247–59; Gál, J. & others. *Hist. Biol.* 2000; 14:193–204; Ma, X. Y. & others. *Nature* 2012; 490:258; Richdale, K. & others. *Optometry and Vision Science* 2012; 89:1507–11.

24   Hou, X. G. & others. *Geological Journal* 2006; 41:259–69; Ortega-Hernández, J. *Biological Reviews* 2016; 91:255–73; University of Bristol Press Release. 2016. https://www.bristol.ac.uk/news/2016/september/penisworm.html; Vinther, J. & others. *Palaeontology* 2016; 59:841–9.

25   Chen, J. Y. and Zhou, G. Q. *Collection and Research* 1997; 10:11–105; Chen, J. Y. & others. *Lethaia* 2004; 37:3–20; Tanaka, G. & others. *Nature* 2013; 502:364.

26   Duan, Y. H. & others. *Gondwana Research* 2014; 25:983–90; Fu, D. J. & others. *BMC Evol. Biol.* 2018; 18; Shu, D. G. & others. *Lethaia* 1999; 32:279–98.

27   Promislow, D. E. L. and Harvey, P. H. *J. Zool.* 1990; 220:417–37.

28   Gabbott, S. E. & others. *Geology* 2004; 32:901–4; Zhu, M. Y. & others. *Acta Palaeontologica Sinica* 2001; 40:80–105.

29   Cuthill, J. F. H. and Han, J. *Palaeontology* 2018; 61:813–23.

## 16. Emergence – Ediacaran

1 Fujioka, T. & others. *Geology* 2009; 37:51–4; Giles, D. & others. *Tectonophysics* 2004; 380:27–41; Haines, P. W. and Flottmann T. *Australian J. Earth Sciences* 1998; 45:559–70; MacKellar, D. *My Country. The Witch-Maid and Other Verses.* London: J. M. Dent and Sons; 1914. P. 29; Williams, P. J. *Economic Geology and the Bulletin of the Society of Economic Geologists* 1998; 93:1120–31.

2 Glansdorff, N. & others. *Biology Direct* 2008; 3; Goin, F. J. & others. *Revista de la Asociacion Geologica Argentina* 2007; 62:597–603; Hamm, G. & others. *Nature* 2016; 539:280; Hiscock, P. & others. *Australian Archaeology* 2016; 82:2–11; Palci, A. & others. *Royal Society Open Science* 2018; 5:172012; Wells, R. T. and Camens, A. B. *PLoS One* 2018; 13:e0208020.

3 Jenkins, R. J. F. & others. *J. Geol. Soc.* Australia 1983; 30:101–19.

4 Ielpi, A. & others. *Sedimentary Geology* 2018; 372:140–72; Kamber, B. S. and Webb, G. E. *Geochimica Et Cosmochimica Acta* 2001; 65:2509–25; Santosh, M. & others. *Geoscience Frontiers* 2017; 8:309–27.

5 Abuter, R. & others. *Astronomy & Astrophysics* 2019; 625; Bond, H. E. & others. *Astrophysical Journal* 2017; 840:70; Che, X. & others. *Astrophysical Journal* 2011; 732:68; Dolan, M. M. & others. *Astrophysical Journal* 2016; 819:7; García-Sánchez, J. & others. *Astronomy & Astrophysics* 2001; 379:634–59; Hummel, C. A. & others. *Astronomy & Astrophysics* 2013; 554; Innanen, K. A. & others. *Astrophysics and Space Science* 1978; 57:511–15; Nagataki, S. & others. *Astrophysical Journal* 1998; 492:L45–L48; Przybilla, N. & others. *Astronomy & Astrophysics* 2006; 445:1099–126; Quillen, A. C. and Minchev, I. *Astronomical Journal* 2005; 130:576–85; Rhee, J. H. & others. *Astrophysical Journal* 2007; 660:1556–71; Tetzlaff, N. & others. *Monthly Notices of the Royal Astronomical Society* 2011; 410:190–200; Voss, R. & others. *Astronomy & Astrophysics* 2010; 520; Wielen, R. & others. *Astronomy & Astrophysics* 2000; 360:399–410; Zasche, P. & others. *Astronomical Journal* 2009; 138:664–79; Zorec, J. & others. *Astronomy & Astrophysics* 2005; 441:235–U120.

6 Stevenson, D. J. and Halliday, A. N. *Phil. Trans. R. Soc. A* 2014; 372:20140289; Williams, G. E. *Reviews of Geophysics* 2000; 38:37–59.

7 Cloud, P. *Economic Geology* 1973; 68:1135–43; Godderis, Y. & others. *Geological Record of Neoproterozoic Glaciations* 2011; 36:151–61; Hoffman, P. F. and Schrag, D. P. *Terra Nova* 2002; 14:129–55; Johnson, B. W. & others. *Nature Communications* 2017; 8; Luo, G. M. & others. *Science Advances* 2016; 2:e1600134; Tashiro, T. & others. *Nature* 2017; 549:516.

8 Brocks, J. J. & others. *Nature* 2017; 548:578; Lechte M. A. & others. *PNAS* 2019; 116:25478–83; Herron, M. D. & others. *Sci. Reports* 2019; 9:2328; Sahoo, S. K. &

others. *Geobiology* 2016; 14:457–68; Wood, R. & others. *Nature Ecology & Evolution* 2019; 3:528–38.

9 Gibson, T. M. & others. *Geology* 2018; 46:135–8; Tang, Q. & others. *Nature Ecology and Evolution* 2020; 4:543–9.

10 Ispolatov, I. & others. *Proc. R. Soc. B* 2012; 279:1768–76; Maliet, O. & others. *Biology Letters* 2015; 11:20150157.

11 Cocks, L. R. M. and Fortey, R. A. *Geological Society of London Special Publications* 2009; 325:141–55; of Monmouth, G. *The History of the Kings of Britain*. 1136 (Penguin edition, 1966).

12 Clapham, M. E. & others. *Paleobiology* 2003; 29:527–44; Shen, B. & others. *Science* 2008; 319:81–4.

13 Jenkins, R. J. F. & others. *J. Geol. Soc. Australia* 1983; 30:101–19; Zhu, M. Y. & others. *Geology* 2008; 36:867–70.

14 Gehling, J. G. *PALAIOS* 1999; 14:40–57; Jenkins, R. J. F. & others. *J. Geol. Soc. Australia* 1983; 30:101–19; Lemon, N. M. *Precambrian Research* 2000; 100:109–20; Noffke, N. & others. *Geology* 2006; 34:253–6; Schneider, D. & others. *PLoS One* 2013; 8:e66662.

15 Droser, M. L. & others. *Australian J. Earth Sciences* 2018; 67:915–21; Feng, T. & others. *Acta Geologica Sinica* 2008; 82:27-34; Gehling, J. G. and Droser, M. L. *Episodes* 2012; 35:236–46; Tang, F. & others. *Evolution & Development* 2011; 13:408–14; Wang, Y. & others. *Paleontological Research* 2020; 24:1–13; Welch, V. L. & others. *Curr. Biol.* 2005; 15:R985–R986; Zhao, Y. & others. *Curr. Biol.* 2019; 29:1112.

16 Dunn, F. S. & others. *Biological Reviews* 2018; 93:914–32; Dunn, F. S. & others. Papers in *Palaeontology* 2019; 5:157–76; Fedonkin, M. A. 'Vendian body fossils and trace fossils'. In: Bengtson, S., editor. *Early Life on Earth*. New York: Columbia University Press; 1994. Pp. 370–88; Ford, T. D. *Yorkshire Geological Society Proceedings* 1958; 31:211–17; Liu, A. G. and Dunn, F. S. *Curr. Biol.* 2020; 30:1322–8; Mason, R. 'The discovery of *Charnia masoni*'. In: *Leicester's fossil celebrity: Charnia and the evolution of early life*. Leicester Literary and Philosophical Society Section C Symposium, 10th March 2007; Narbonne, G. M. and Gehling, J. G. *Geology* 2003; 31:27–30; Nedin, C. and Jenkins, R. J. F. *Alcheringa* 1998; 22:315–16; Sprigg, R. C. *Trans. Roy. Soc. S. Aust.* 1947; 72:212–24.

17 Droser, M. L. and Gehling, J. G. *Science* 2008; 319:1660–62; Gibson, T. M. & others. *Geology* 2018; 46:135–8; Hartfield, M. *J. Evol. Biol.* 2016; 29:5–22; Normark, B. B. & others. *Biol. J. Linn. Soc.* 2003; 79:69–84; Pence, C. H. and Ramsey, G. *Philosophy of Science* 2015; 82:1081–91; Smith, J. M. *J. Theor. Biol.* 1971; 30:319.

18 Bobrovskiy, I. & others. *Science* 2018; 361:1246; Dunn, F. S. & others. *Biological Reviews* 2018; 93:914–32; Evans, S. D. & others. *PLoS One* 2017; 12:e0176874; Gehling, J. G. & others. *Evolving Form and Function: Fossils and Development*

2005:43–66; Sperling, E. A. and Vinther, J. *Evolution & Development* 2010; 12:201–9.

19  Chen, Z. & others. *Science Advances* 2018; 4:eaao6691; Evans, S. D. & others. *PNAS* 2020; 117:7845-7850; Gehling, J. G. and Droser, M. L. *Emerging Topics in Life Science* 2018; 2:213–22.

20  Clites, E. C. & others. *Geology* 2012; 40:307–10; Coutts, F. J. & others. *Alcheringa* 2016; 40:407–21; Droser, M. L. and Gehling, J. G. *PNAS* 2015; 112:4865–70; Gehling, J. G.and Droser, M. L. *Episodes* 2012; 35:236-246; Joel, L. V. & others. *J. Paleontology* 2014; 88:253–62; Mitchell, E. G. & others. *Ecology Letters* 2019; 22:2028–38.

21  Wade, M. *Lethaia* 1968; 1:238–67; Zhu, M. Y, & others. *Geology* 2008; 36:867–70.

22  Ivantsov, A. Y. *Paleontological Journal* 2009; 43:601–11.

23  Fedonkin, M. A. & others. *Geological Society of London Special Publications* 2007; 286:157–79.

24  Budd, G. E. and Jensen, S. *Biological Reviews* 2017; 92:446–73; Erwin, D. H. and Tweedt, S. *Evolutionary Ecology* 2012; 26:417–33; Shu, D. G. & others. *Science* 2006; 312:731–4.

25  Dunn, F. & others. *5th International Paleontological Congress*. Paris 2018. P. 289.

26  Medina, M. & others. *Int. J. Astrobiol.* 2003; 2:203–11; Xiao, S. H. & others. *American J. Botany* 2004; 91:214–27.

27  Burns, B. P. & others. *Env. Microbiol.* 2004; 6:1096–101; Lowe, D. R. *Nature* 1980; 284:441–3; Puchkova, N. N. & others. *Int. J. Syst. Evol. Microbiol.* 2000; 50:1441–7.

## Epilogue – A Town Called Hope

1  Mills, W. J. *Hope Bay. Exploring Polar Frontiers: A Historical Encyclopedia*. Santa Barbara, California, USA: ABC Clio; 2003. Pp. 308–9.

2  Birkenmajer, K. *Polish Polar Research* 1992; 13:215–40; de Souza Carvalho, I. & others. *Ichnos* 2005; 12:191–200; Erwin, D. H. *Annual Review of Ecology and Systematics* 1990; 21:69–91.

3  De Souza Carvalho, I. & others. *Ichnos* 2005; 12:191–200; Hays, L. E. & others. *Palaeoworld* 2007; 16:39–50; Penn, J. L. & others. 2018; 362:1130; Xiang, L. & others. *Palaeo3* 2020; 544; Zhang, G. J. & others. *PNAS* 2017; 114:1806–10.

4  Keeling, R. F. and Garcia, H. E. *PNAS* 2002; 99:7847–53; Ren, A. S. & others. *Sci. Reports* 2018; 8:7290; Schmidtko, S. & others. *Nature* 2017; 542:335.

5  Breitburg, D. & others. *Science* 2018; 359:46.

6  Jurikova, H. & others. *Nature Geoscience* 2020; 13:745–50.

7  Feely, R. A. & others. *Sci. Brief* April 2006:1–3; Hoegh-Guldberg, O. & others. *Frontiers in Marine Science* 2017; 4; Kleypas J. A. & others. *Impacts of Ocean*

*Acidification on Coral Reefs and Other Marine Calcifiers: A Guide for Future Research* 2006. National Science Foundation Report; van Woesik, R. & others. *PeerJ* 2013; 1:e208.

8  Fillinger, L. & others. *Curr. Biol.* 2013; 23:1330–34; Leys, S. P. & others. *Marine Ecology Progress Series* 2004; 283:133–49; Maldonado, M. & others. 'Sponge grounds as key marine habitats: a synthetic review of types, structure, functional roles, and conservation concerns'. In: *Marine Animal Forests: The Ecology of Benthic Biodiversity Hotspots* (Rossi, S. and others, eds.) Berlin: Springer, 2017. Pp. 145–83; Saito, T. & others. *J. Mar. Biol. Ass.* UK 2001; 81:789–97.

9  Clem, K. R. & others. *Nature Climate Change* 2020; 10:762–70; Zhang, L. & others. *Earth-Science Reviews* 2019; 189:147–58.

10  Kim, B. M. & others. *Nature Communications* 2014; 5:4646; Overland, J. E. and Wang, M. *International Journal of Climatology* 2019; 39:5815–21; Robinson, S. A. & others. *Global Change Biology* 2020; 26:3178–80.

11  Meehl, G. A. & others. *Science* 2005; 307:1769–72; Meehl, G. A. & others. 'Global Climate Projections'. In: *Climate Change 2007: The Physical Science Basis. Contribution of Working Group I to the Fourth Assessment Report of the Intergovernmental Panel on Climate Change* (Solomon, S. and others, eds.). Cambridge UK: Cambridge University Press; 2007; O'Brien, C. L. & others. *PNAS* 2020; 117:25302–9.

12  Pugh, T. A. M. & others. *PNAS* 2019; 116:4382–7; Scott, V. & others. *Nature Climate Change* 2015; 5:419–23; Terrer, C. & others. *Nature Climate Change* 2019; 9:684–9.

13  Couture, N. J. & others. *Journal of Geophysical Research Biogeosciences* 2018; 123:406–22; Friedlingstein, P. & others. *Earth Syst Sci Data* 2019; 11:1783–838; Nichols, J. E. and Peteet, D. M. *Nature Geoscience* 2019; 12:917–21.

14  Fujii, K. & others. *Arctic Antarctic and Alpine Research* 2020; 52:47–59; Olid, C. & others. *Global Change Biology* 2020; 26:5886–98.

15  Bolch, T. & others. 'Status and change of the cryosphere in the extended Hindu Kush Himalaya region'. In: Wester, P. and others, eds., *The Hindu Kush Himalaya Assessment.* Cham: Springer; 2019 Church, J. A. & others. *Journal of Climate* 1991; 4:438 P.56; Kulp, S. A. and Strauss, B. H. *Nature Communications* 2019; 10; Loo, Y. Y. & others. *Geoscience Frontiers* 2015; 6:817 P.23; Nepal, S and Shrestha, A. B. *International Journal of Water Resources Development* 2015; 31:201 P.18; Yi, S. & others. *The Cryosphere Discussions* 2019. https://doi.org/10.5194/tc-2019–211.

16  Muntean, M. & others. *Fossil $CO_2$ emissions of all world countries.* Publications Office of the European Union 2018. DOI: 10.2760/30158.

17  Friends of the Earth. *Overconsumption? Our use of the world's natural resources.* 2009. 1–36.

18  Avery-Gomm, S. & others. *Marine Pollution Bulletin* 2013; 72:257–9; Beaumont, N. J. & others. *Marine Pollution Bulletin* 2019; 142:189–95.

19  Russell, J. R. & others. *Applied and Environmental Microbiology* 2011; 77:6076–84; Tanasupawat, S. & others. *Int. J. Syst. Evol. Microbiol.* 2016; 66:2813–18; Taniguchi, I. & others. *Acs Catalysis* 2019; 9:4089–105.

20  Polito, M. J. & others. *American Geophysical Union Fall Meeting 2018.* Abstract #PP13C–1340.

21  Habel, J. C. & others. 'Review refugial areas and postglacial colonizations in the Western Palearctic'. In: Habel, J. C. & others, eds. *Relict Species.* 2010. Springer, Berlin, Heidelberg: Pp. 189–97; Roberts, C. P. & others. *Nature Climate Change* 2019; 9:562.

22  Cardoso, G. C. and Atwell, J. W. *Animal Behaviour* 2011; 82:831–6; Martín, J. and López, P. *Functional Ecology* 2013; 27:1332–40; Owen, M. A. & others. *J. Zool.* 2015; 295:36–43.

23  Bar-On, Y. M. & others. *PNAS* 2018; 115:6506–11; Bennett, C. E. & others. *Royal Society Open Science* 2018; 5:180325; Elhacham, E. & others. *Nature* 2020; doi. org/10.1038/s41586-020-3010-5; Giuliano, W. M. & others. *Urban Ecosystems* 2004; 7:361–70; WWF. *Living Planet Report – 2018: Aiming Higher.* Grooten, M. and Almond, R. E. A., eds. Gland, Switzerland: WWF; 2018.

24  Kleinen, T. & others. *Holocene* 2011; 21:723–34; Summerhayes, G. R. *IPPA Bulletin* 2009; 29:109–23.

25  Abate, R. S. and Kronk, E. A. *Climate change and Indigenous peoples: The search for legal remedies.* 2013. Cheltenham UK: Edward Elgar; 2013; Ahmed, N. *Entangled Earth.* Third Text 2013; 27:44–53.

26  Associated Press in St Petersburg, Florida. 'Hurricane Iota is 13th hurricane of record-breaking Atlantic season'. *Guardian,* 15 November 2020; González-Alemán, J. J. & others. *Geophysical Research Letters* 2019; 46:1754–64; Knutson, T. R. & others. *Nature Geoscience* 2010; 3:157–63.

27  Hodbod, J. & others. *Ambio* 2019; 48:1099–115; Michaelson, R. ' "It'll cause a water war": divisions run deep as filling of Nile dam nears'. *Guardian,* 23 April 2020; Spohr, K. 'The race to conquer the Arctic – the world's final frontier'. *New Statesman,* 12 March 2018.

28  UK Environment Agency. *TE2100 5 Year Review Non-technical Summary.* 2016:1–7; Secretariat of the Multilateral Fund for the Implementation of the Montreal Protocol on Substances that Deplete the Ozone Layer. *Creating a real change for the environment.* 2007:1–24.

29  Henley, J. 'Iceland holds funeral for first glacier lost to climate change'. *Guardian,* 22 July 2019.

# Acknowledgements

Like this book, perhaps I should thank those involved in the time-line that led to this book in reverse chronological order. Without the input from my editors Laura Stickney and Rowan Cope at Penguin Press, Hilary Redmon at Random House, and Nick Garrison at Penguin Canada, this would surely have been a drearier, more saltatory, and more technical read. Working with Beth Zaiken as she created the incredible chapterhead images of the species in this book has been a dream, from the first drafts that, had I not been told were works in progress, I would have been happy to accept as finalised. I find the final results to be utterly breathtaking.

I thank Marion Boyars, Dr Alice Tarbuck, the Society of Biblical Literature, Miguelángel Meza, Tracy K. Lewis, John Curl, Canongate Books, Columbia University Press, Laurel Rasplica Rodd, Shambhala, the University of Western Australia Press, Rachael Mead, Lascaux Publishers, Robert Ziller, the family of Natalia Molchanova, Viktor Hilkevich, UNESCO, Diego Arguedas Ortiz, BBC Futures, and Taylor and Francis for permission to use extracts from works to which they created or hold copyright for. The extract from *Life and Fate* by Vasily Grossman is published by Vintage. Copyright © Editions L'Age d'Homme, 1980. English translation copyright © Collins Harvill, 1985. Reproduced by permission of The Random House Group Ltd. The extract of *The Epic of Gilgamesh* is copyright © 1985 by the Board of Trustees of the Leland Stanford Jr. University. All rights reserved. Used by permission of the publisher, Stanford University Press, sup. org. The extract from *The Aeneid* by Virgil published by The Penguin Group. Translation and Introduction copyright © David West 1990, 2003. Reproduced by permission of Penguin Books Ltd. ©. The extract of *Miss Peregrine's Home for Peculiar Children* (2013, Quirk Books, Ransom Riggs) is courtesy of Ransom Riggs and Quirk Books. I thank Emma Brown for her work in sourcing these permissions. Dr John Halliday kindly translated the fragment of Nietzsche's *Dionysus Dithyrambs* from German. I thank members of the African-Caribbean Research Collective for their advice concerning the appropriate referencing of Matthew A. Henson's 1912 autobiography *A Negro Explorer at the North Pole*, and Dr Sam Giles for putting me in contact with them. Every effort has been made to trace all copyright holders and to obtain their permission for the use of copyright material.

My agent, Catherine Clarke, along with the rest of the team at Felicity Bryan Associates, has done wonderful work getting this book from the proposal stage to the right publishers and helping me to make what has been a wonderful decision. Thanks, too, to the many subagents who helped this book become a reality outwith the confines of the UK – Zoë Pagnamenta of ZP Agency, and Barbara Barbieri, Juliana Galvis, Sabine Pfannenstiel, Marei Pittner, Rachael Sharples, Ludmilla Sushkova, Susan Xia, and Jackie Yang of Andrew Nurnberg Associates.

From a practical perspective, I must thank those who made it possible to write this book thanks to provision of space and time. Much of this was written while lodging with Chris Bryan or Jenny Ainsworth in the working week away from my family. The British Library and Enfield Library were both fantastic spaces not just for writing, but for access to useful material. Since much of this was written during lockdown, I must also thank the Biodiversity Heritage Library; without their commitment to open access to older scientific material, I would not have been able to adequately research this book.

Thanks, too, to the many friends and family who read drafts of these chapters and provided me with very helpful feedback – Dr Catherine Ainsworth, Eugenie Aitchison, Dr Gemma Benevento, Dr Andrew Button, Ivan Brett, Prof. Hugh Bowden, Andrew Dickson, Martin Dowling, Charlotte Halliday, Marianne Johnson, Johnny Mindlin, Dr Travis Park, Tammela Platt, Roxanne Scott, and Steve Wright.

One cannot know everything about the history of life, so my

sincere thanks to my palaeobiologist colleagues who gave their time to provide additional expertise on the text and on the illustrated reconstructions, helping me to avoid egregious errors in localities, times, species, and subjects in which their expertise outweighs mine: Dr Chris Basu, Dr Gemma Benevento, Dr Neil Brocklehurst, Dr Thomas Clements, Dr Mario Coiro, Dr Darin Croft, Dr Emma Dunne, Dr Daniel Field, Prof. Sarah Gabbott, Dr Maggie Georgieva, Dr Sandy Hetherington, Dr Lars van den Hoek Ostende, Dr Dan Ksepka, Dr Liz Martin-Silverstone, Dr Emily Mitchell, Dr Elsa Panciroli, Dr Stephanie Smith, and Dr Zhang Hanwen (Steven Zhang). Any errors that remain in the text are of course my own. My thanks to Dr Douglas Boubert for his advice on astronomical matters, and to Dr Will Tattersdill for introducing me to the texts of the first Victorian popularisers of geology, which has been of immense help for me to understand the earlier strata of this writing form.

I would like to extend additional thanks to Dr Elsa Panciroli, without whom I would not have entered the Hugh Miller Writing Competition, and also to the other judges of that competition, particularly Larissa Reid. My experience in that competition is the direct and proximal cause of the first halting sentences of this book. Ivan Brett pointed me in the right direction for how to write a book proposal, setting me properly on this path, which also deserves thanks.

I am forever indebted to the support, mentoring, and guidance of my primary doctoral supervisor Prof. Anjali Goswami, not only during my time as a PhD student but also as a postdoctoral researcher and a member of her field team in India and Argentina. For their time in my development as a scientist, I must also thank Professors Paul Upchurch and Ziheng Yang, who mentored me during my doctoral and first postdoctoral positions, as well as those who supervised or mentored me during other research projects – Professors Richard Butler, Mike Benton, and Andrew Balmford. Throughout, my friends and colleagues, too many to comprehensively name, have made palaeontology one of the most rewarding communities to be a part of.

An interest in the natural world must be maintained by sterling educators, and so I cannot undersell the influence of the lectures of Dr Rob Asher, Prof. Nick Davies, and the late, great, Professor Jenny Clack, nor the biology lessons of Geoff Morgan and Fiona Graham, whether in the classroom or the field. Speaking of the field, I must also mention Neo Kim Seng, who first taught me to dive and introduced me to the spaces below the waves, and Dr Federico Agnolin, Dr Andrew Cuff, Dr Ryan Felice, Prof. Anjali Goswami, Javier Ochoa, Prof. Guntupalli Prasad, Dr Agustín Scanferla, MS Thanglemmoi, and Dr Aki Watanabe for their companionship and advice while hunting fossils.

Deep in my own Proterozoic, I give thanks to my primary school teachers who taught me how to be wrong and endured my nine-year-old self giving a presentation on Linnaean classification. My deepest thanks of all are to my parents, who in answer to a child's question, even if, looking back, I think they knew the answer, would always tell me to look it up in a book. They and my grandparents got me birdwatching, or identifying trees, hunting for mushrooms, or measuring rainfall every morning. Whether picking up football-sized lumps of milky quartzite from the mountain that weighed the earth, watching the baby ospreys from an attic telescope with one pair of grandparents, or helping the others feed their doos and imitate the calls of their garden, my family certainly made it easy to watch and listen to the natural world.

Finally, I must acknowledge those from whose land these fossils have been excavated, and thank the scientists themselves. Without the countless hours of research conducted by the thousands of women and men who have deciphered the rock record such that it can be read at all, this book would not even be possible. The direct references in this book alone refer to the work of over 4,000 individual scientists. Of those, the work of the following has been cited a substantial number of times: Josep Alcover, Mike Benton, René Bobe, Darin Croft, Michael Engel, John Flynn, Andrzej Gaździcki, Javier Gelfo, Phil Gingerich, Anjali Goswami, Dale Guthrie, Kirk Johnson, Conrad Labandeira, Meave Leakey, Sally Leys, Lü Junchang, Fredrik Manthi, Sergio Marenssi, Jean-Michel Mazin,

*Acknowledgements*

Marcelo Reguero, Ren Dong, Sergio Santillana, Gustav Schweigert, Claudia Tambussi, Carol Ward, Lars Werdelin, Greg Wilson, Andy Wyss, James Zachos, and Zhang Haichun. To all those who have ever quarried, dissolved, scanned, and sieved out wonders, included in these pages or otherwise, I give thanks.

Leaping back to the present and into the future again, I must thank my wife Charlotte for her support during the writing of this book, and her ability to ask exactly the right questions to spark each next idea. After all this, I dedicate this book to my sons – when you are old enough to read this, the world will already have changed. Let us hope it is for the better.

# Permissions

Page xi. Oodgeroo Noonuccal, 'The Past', *The Dawn is at Hand* (Marion Boyars Publishers, 1990).

Page xi. Carson McCullers, 'Look Homeward, Americans' (*Vogue*, December 1940).

Page 1. Vasily Grossman, *Life and Fate*, trans. Robert Chandler (Vintage, 2017).

Page 1. H. A. Hoffner, *Hittite Myths* (Society of Biblical Literature, 1990).

Page 21. J. K. Kassagam, *What is this Bird Saying?* (Binary Computer Services, 1997).

Page 21. Miguelángel Meza, 'Ko'ê', trans. Tracy K. Lewis (*Words Without Borders*, July 2020).

Page 57. *Epic of Gilgamesh*, tr. Maureen Kovacs (Stanford University Press, 1985).

Page 77. Virgil, *Aeneid*, trans. David West (Penguin Random House, 2003).

Page 95. Ransom Riggs, *Miss Peregrine's Home for Peculiar Children* (Quirk Books, 2013).

Page 115. Nezahualcoyōtl, *Ancient American Poets*, trans. John Curl (Bilingual Press, 2005).

Page 135. Rachel Carson, *The Sea Around Us* (Oxford University Press, 1951).

Page 135. Ichiyō Higuchi, 'Koigokoro', trans. L. Rasplica Rodd, *The Modern Murasaki*, eds. Rebecca L. Copeland and Melek Ortabasi (Colombia University Press, 2006).

Page 155. Han Shan, *Cold Mountain Poems*, trans. J. P. Seaton (Shambhala, 2009).

Page 171. Rachael Mead, 'Kati Thanda/Lake Eyre', *The Flaw in the Pattern* (University of Western Australia Press, 2018).

Page 185. Jean-Joseph Rabearivelo, *Traduit de la Nuit*, trans. Robert Ziller (Lascaux Editions, 2007).

Page 219. Natalia Molchanova, 'И осознала я небытие', trans. Victor Hilkevich (http://molchanova.ru/ru).

Page 235. Abu al-Rayhan al-Biruni, *Chronology of Ancient Nations*, trans. Bobojon Ghafurov (*UNESCO Courier*, June 1974).

Page 285. Diego Arguedas Ortiz, 'Is it wrong to be hopeful about climate change?' (*BBC Future*, 10th January 2020).

# Index

animals – *cont'd.*

competitive exclusion concept, 35*, 53, 63, 100, 281

crepuscular, 91

diversification of amphibian and reptile, 214

earliest chordate, 256–7

earliest definite relative of vertebrates, 255–7

in Early Cretaceous, 115–19, 121–4, 125–30

in Early Devonian Rhynie, 204, 208, 214–17, 264

in Eocene Antarctica, 77, 78, 81–3, 84–93

farming-like activities by, 210

first tetrapod vertebrates on land, 214

first to brood eggs until they hatch, 265

first tree-climbing vertebrate, 182

grazing, 8, 10, 25, 29–30, 61, 72–3

growth marks in, 3

human-adapted in present day, 297

initial divergences during Ediacaran, 282–3

initial diversification of animal life, 214

island dwarfism, 48–50, 52, 92

of Jurassic Europe, 135–9, 136, 141–6, 147–51, 289

loss of vertebrates (1978–2018), 297

Mediterranean as barrier, 42–3

in Miocene Mediterranean, 43–7, 48–50, 52, 92

muscle use in, 245

mythological appeal of earliest Paleocene, 109–10

notion of a single individual, 212

of Oligocene Tinguiririca, 60–4, 65, 69

opposable thumbs, 182

in Paleocene Americas, 101–6, 108–11

in Permian Moradi, 171–3, 172, 175–6, 178–9, 180–1, 182, 183

in Pleistocene epoch, 1–5, 6, 7–11, 12–17, 18–19, 26, 182, 271

in Pliocene Africa, 24–8, 29–32, 34–5, 36

predator–prey relationships emerge in Cambrian, 259–60, 267

in pre-Phanerozoic oceans, 273–4

pretending to be something dangerous, 6–7

processing of grasses for food, 71, 72–3

random colonization over water, 44

range of disparate forms in Triassic, 160–7

sensational depictions of extinct creature, xxi

sexual reproduction, 211, 265–6, 277–8

size of on Miocene islands, 48–50, 52, 53–4, 92

small per-centage of as wild in present day, 297

survival of Chicxulub asteroid strike, xvi, 97, 98, 101–12

taste and smell, 262

and Thermal Maximum, 80, 86

traces of historical associations, 6–7

of Triassic Central Asia, 156, 158–67

trilobite body plan, 256

*see also* marine/aquatic animals and entries for groups/species

Anning, Mary, 138

anomalocaridids, 230–1, 261, 264

*Anomalocaris* (great appendage arthropod), 264

Antarctica, 15*

# About the Author

THOMAS HALLIDAY is a paleontologist and evolutionary biologist. He holds a Leverhulme Early Career Fellowship at the University of Birmingham and is a scientific associate of the Natural History Museum in London. His research combines theoretical and real data to investigate long-term patterns in the fossil record, particularly in mammals. Halliday was the winner of the Linnean Society's John C. Marsden Medal in 2016 and the Hugh Miller Writing Competition in 2018.

thomashalliday.co.uk
Twitter: @TJDHalliday

# About the Type

This book was set in Dante, a typeface designed by Giovanni Mardersteig (1892–1977). Conceived as a private type for the Officina Bodoni in Verona, Italy, Dante was originally cut only for hand composition by Charles Malin, the famous Parisian punch cutter, between 1946 and 1952. Its first use was in an edition of Boccaccio's *Trattatello in laude di Dante* that appeared in 1954. The Monotype Corporation's version of Dante followed in 1957. Though modeled on the Aldine type used for Pietro Cardinal Bembo's treatise *De Aetna* in 1495, Dante is a thoroughly modern interpretation of that venerable face.